AF676293

Advanced Innovative Solutions for Final Design in Terms of Energy Sustainability of Nearly/Net Zero Energy Buildings (nZEB)

Advanced Innovative Solutions for Final Design in Terms of Energy Sustainability of Nearly/Net Zero Energy Buildings (nZEB)

Guest Editors

Domenico Mazzeo
Giuseppe Oliveti

MDPI

Basel • Beijing • Wuhan • Barcelona • Belgrade • Novi Sad • Cluj • Manchester

Guest Editors
Domenico Mazzeo
Department of Energy
Politecnico di Milano
Milano
Italy

Giuseppe Oliveti
Department of Mechanical,
Energy and Management
Engineering (DIMEG)
University of Calabria
Rende
Italy

Editorial Office
MDPI AG
Grosspeteranlage 5
4052 Basel, Switzerland

This is a reprint of the Special Issue, published open access by the journal *Sustainability* (ISSN 2071-1050), freely accessible at: https://www.mdpi.com/journal/sustainability/special_issues/Advanced_Innovative_Solutions.

For citation purposes, cite each article independently as indicated on the article page online and as indicated below:

Lastname, A.A.; Lastname, B.B. Article Title. *Journal Name* **Year**, *Volume Number*, Page Range.

ISBN 978-3-7258-5875-0 (Hbk)
ISBN 978-3-7258-5876-7 (PDF)
https://doi.org/10.3390/books978-3-7258-5876-7

© 2025 by the authors. Articles in this book are Open Access and distributed under the Creative Commons Attribution (CC BY) license. The book as a whole is distributed by MDPI under the terms and conditions of the Creative Commons Attribution-NonCommercial-NoDerivs (CC BY-NC-ND) license (https://creativecommons.org/licenses/by-nc-nd/4.0/).

Contents

About the Editors

Domenico Mazzeo

Domenico Mazzeo is an Assistant Professor in the Department of Energy of the Politecnico di Milano, Italy, where he teaches Thermodynamics, Heat Transfer, and Basic Energy Engineering. He holds a PhD in Civil and Industrial Engineering from the University of Calabria, where his doctoral thesis on the dynamic thermal behavior of building walls with Phase-change materials received national recognition. His research focuses on heat transfer, thermal energy storage, photovoltaic and building integration, smart energy systems, and the application of artificial intelligence to renewable energy. He has authored more than one hundred scientific publications in high-impact journals and conferences and has supervised numerous master's and doctoral theses. According to Scopus, his record includes about 90 indexed documents, more than 2800 citations, and an h-index of 33. His work has been acknowledged with some prizes, including the ATI Award from the Italian Thermotechnical Association, the 2024 *Energies* Young Investigator Award, and the 2022 MDPI *Sustainability* Travel Award. He has also been listed among the World's Top 2% Scientists in the last five years according to the Elsevier/Stanford ranking. He has carried out research collaborations with several international universities and research centers, including a visiting period at the University of Lleida in Spain and teaching activities at the University of Waterloo in Canada. Domenico Mazzeo is a member of multiple professional associations, including Associazione della Fisica Tecnica Italiana, IBPSA-Italy, IEEE, and he previously belonged to AiCARR, ISES, and IABP.

Giuseppe Oliveti

Giuseppe Oliveti is Honorary Professor at the University of Calabria, where he served as Full Professor in Environmental Technical Physics in the Department of Mechanical, Energy and Management Engineering. He graduated in Mechanical Engineering from the Politecnico di Torino, and he has dedicated his career to the study of heat transfer, energy efficiency in buildings, and renewable energy technologies. His research spans solar energy utilization, thermal storage, atmospheric radiation measurements, dynamic behavior of building envelopes, and phase-change material applications, as well as the development of innovative trigeneration systems for buildings. He has coordinated and participated in numerous national and international research projects funded by the European Commission, ENEA, CNR, and the Italian Ministry of Education, including several PRIN programs on solar energy and building energy retrofitting. He is the author of more than 160 scientific publications, including articles in leading international journals, conference proceedings, and textbooks such as Ingegneria Solare and Tecnologie Solari negli Edifici Esistenti. He has supervised doctoral and master's theses and contributed to the foundation of degree programs in Energy Engineering and Environmental Engineering at the University of Calabria. Throughout his career, he has held leadership roles as a Director of the Department of Mechanical Engineering, a member of the Academic Senate, and President of the Calabrian Section of the Italian Thermotechnical Association (ATI).

Preface

The Reprint "Advanced Innovative Solutions for Final Design in Terms of Energy Sustainability of Nearly/Net Zero Energy Buildings" (nZEB) gathers a selection of high-quality contributions that advance the knowledge and practice of sustainable building design. Its subject is the study and application of innovative materials, technologies, and design strategies aimed at reducing energy consumption, improving thermal comfort, and supporting the achievement of nearly and net-zero-energy buildings.

The scope of this Reprint is broad, encompassing passive and active solutions for the building envelope, advanced systems for heating and cooling, the integration of renewable energy sources, and the evaluation of comfort and performance standards. By bringing together experimental investigations, simulation-based approaches, and policy-oriented studies, the Reprint provides a multidisciplinary perspective on how innovation can enhance energy sustainability in the built environment.

The aim and purpose of this collection are to highlight effective pathways for bridging the gap between research and practice, and to demonstrate how scientific advances can be applied to different contexts and climates. The Reprint seeks to inspire further research, guide practitioners in the adoption of advanced solutions, and support policymakers in the definition of frameworks that encourage sustainable design.

The motivation for preparing this Reprint stems from the pressing global need to reduce greenhouse gas emissions, limit energy demand, and improve resilience in the building sector. Buildings account for a significant share of energy use worldwide, and the technologies and strategies presented here represent concrete steps toward a more sustainable future.

This Reprint is addressed to a wide audience: researchers working on energy-efficient materials and systems, engineers and architects engaged in sustainable design, and policymakers and stakeholders who shape regulations and strategies for the built environment.

We are grateful to the authors for their contributions, to the reviewers for their constructive feedback, and to the editorial team for their support in the realization of this Reprint.

Domenico Mazzeo and Giuseppe Oliveti
Guest Editors

sustainability

MDPI

Editorial

Advanced Innovative Solutions for Final Design in Terms of Energy Sustainability of Nearly/Net Zero Energy Buildings (nZEB)

Domenico Mazzeo * and Giuseppe Oliveti

Department of Mechanical, Energy and Management Engineering (DIMEG), University of Calabria, 87036 Rende, Italy; giuseppe.oliveti@unical.it

* Correspondence: domenico.mazzeo@unical.it; Tel.: +39-0984-494841

Received: 6 December 2020; Accepted: 8 December 2020; Published: 12 December 2020

1. Overview

EU Directives have reinforced both studies and research for the development of innovative technological solutions to improve building energy performance and to achieve a reduction in total energy consumption, with benefits in terms of reducing greenhouse gas emissions, as well as in economic terms [1–3].

To realize nearly/net Zero Energy Buildings (nZEB), sustainability has become a significant aspect and has been integrated into the design, construction and operation of buildings. For this reason, innovative sustainable technological solutions have been proposed to improve the thermal performance of the building envelope and to increase the use of energy from renewable sources. For example, they concern: (i) the stratigraphy of walls and the windows [4–7], through the use of phase change materials [8–12], new silica aerogel-based insulating [13,14], green vertical system [15,16] and green [17–19] and cool roofs [20,21], etc.; (ii) the integration of systems that employ renewable heat sources, namely solar radiation, wind, geothermal, etc. [22,23].

The aim of sustainable buildings is the ability to integrate and adapt buildings to environmental factors and climate conditions and convert them for space and comfort. The selection of proper sustainable building materials or technologies is an important issue in the building design. A correct selection should lead to high durability, energy efficiency, recyclability, maintainability, economic saving, and use of local materials to reduce the environmental impact of construction. In this context, there is a lack of common understanding in the field, and this Special Issue aims to collect contributions regarding the evaluation of the sustainability of nZEB. Overall, 30 original articles were submitted to this Special Issue, of which 15 high-quality papers successfully overcame the single-blind peer-review process and are summarized in this Editorial.

2. A Short Review of the Contributions to This Special Issue

Various topics were numerically and experimentally developed in this Special Issue. The study of the energy performance of some innovative window and opaque walls in new, existing, or laboratory prototypes was addressed worldwide in the context of achievement of Nearly Zero Energy Buildings (nZEB) target. For the numerical analyses, new and standard procedures were implemented in new simulation tools or existing ones. The main indices considered are the building heating and the cooling energy demand, and indoor comfort. The studies involved different climatic conditions, such as Italy, China, United Kingdom, South Korea, Norway, Poland, and Chile.

As regards the opaque envelope, six articles studied the impact of implementing passive solutions, such as phase change materials (PCM), green roofs, dynamic insulations, high-performance insulation materials and sunshades, flexible canopies and Prefabricated Straw Bale Construction (PSBL), on building energy performance and saving [24–29].

Lee et al. [24] used Esp-r simulation software to evaluate the heating and cooling load reduction and Predicted Mean Vote (PMV) for human comfort adopting multi variable PCM types as the building wall composite in a modernized Korean Hanok. They concluded that PCM helps reduce energy consumption also of 53.1%. Piro et al. [25] analyzed the hydraulics and energy performances of a real scale model of an extensive green roof at the University of Calabria (Italy) in Mediterranean area This study confirms that green roofs contribute to reducing thermal fluctuation on the external and internal surfaces, thereby lowering the energy consumption for the maintenance of indoor comfort conditions. From the point of view of stormwater management, a green roof system can significantly reduce stormwater runoff generation in terms of runoff volume reduction and peak attenuation. Gruner and Matusiak [26] tested dynamic insulation which has the task to potentially reduce heat loss through highly insulating windows in the Nordic climate. The new system was both numerically simulated using THERM and tested in full-scale in a Hot-box apparatus in a steady-state condition at the laboratory of SINTEF Building and Infrastructure in Trondheim, Norway. The results are promising and showed improvement of the thermal performance of a poorly insulated window. J.-H. Kim et al. [27] attempted to quantify the consumers' willingness to pay for the governmental pilot net-zero energy apartments (NZEA), which use high-performance insulation materials and sunshades to minimize the use of household cooling and heating energy. Kwon et al. [28] analyzed the environmental role of a flexible canopy as a microclimate modifier in balancing indoor energy demands and outdoor thermal comfort using several environmental software packages (Ecotect, Rayman, WinAir, DaySim, and EDSL TAS). The results showed that the flexible canopy played a crucial role in reducing indoor energy demands (heating and electricity for lighting) and increasing outdoor thermal comfort under the canopy area. To analyze the performance of canopies, simulations using different transparencies and orientations were performed for flexible and fixed canopies in London, Manchester, and Glasgow, having different climates. The flexible canopy with higher transparency helped increase outdoor thermal comfort in Glasgow, while one with lower transparency showed better performance during summer in London. Yin et al. [29] presented an analysis of Prefabricated Straw Bale Construction (PSBC) in residential buildings to reduce energy consumption in this building type in severe cold regions and cold regions in northern China. The results show that the buildings with PSBC reduce both heating and cooling energy uses, as well as heating intensities, across all climatic regions in northern China, compared to the domestic buildings with conventional constructions.

Instead, two articles aimed to study the thermal performance of enhanced window systems [30,31]. Oh et al. [30] compared the impact of the recently developed electrochromic glazing technology on load reduction by comparing it with the double-glazing and shading devices that are available commercially for high-rise residential buildings in Korea. Using EnergyPlus software, electrochromic glazing was found to be more effective for reducing the cooling loads compared with the heating loads. Koo et al. [31] searched for a solution to the high-cost problem of the existing double-skin façade system by proposing a double-skin façade window with a smaller ventilated air cavity. Windows were installed in chambers of an artificial solar lab and real building test cells. The artificial solar lab test results showed that the solar heat gain coefficient can be lowered through ventilation by 28% to 52.9%. In addition, the cumulative power consumption was reduced by around 3% through ventilation in the air cavity.

The other four articles applied and compared different common international standards related to energy and comfort performance indices [32–35]. Y. Kim et al. [32] calculated energy requirements that satisfied the insulation standards, for each of the Korean regional climate conditions, of the Building Energy Conservation Code (BECC), a standard of mandatory energy-related minimum requirements to specify mandatory and energy performance indices of energy-saving design. For this issue, they used the methodology of ISO 13790. Finally, this study derived that a return of investment can be achieved in 10 years, which was determined through comprehensive economic feasibility analysis due to strengthening insulation performances, proving the rationalization of the legal strengthening. The same authors presented a methodology and process to establish a mandatory policy of zero-energy buildings (ZEBs) in Korea [33]. A self-sufficiency rate was also proposed to strengthen the passive standard of the buildings as well as to encourage new and renewable energy production. The overall scope is to define ZEBs following the technical and economic circumstances in Korea and to describe the study methodology for establishing the mandatory policies and criteria to this end. It also aims to investigate the characteristics of certified buildings through case analyses on zero-energy-certified buildings. Piasecki et al. [34] contributed to the actual state of knowledge as a "small and specific scale" validation of the existing thermal comfort model. They compared the thermal comfort measurement method based on the ISO-Fanger model with the actual comfort results obtained by the panelists in the experimental building of the Małopolska Laboratory of Energy Efficient Building in Cracow, Poland. Pérez-Fargallo et al. [35] analyzed the standards recommended by Chile's Construction with Sustainability Criteria for the building of housing, in a social housing prototype located in the region of the Bio-Bio and analyzed under two international standards—the ASHRAE 55-2017 and the EN 15251:2007 adaptive comfort models—and one national standard, the Chilean ECCS. Results in numerical terms can be different depending on the standard, but all of them will point in the same direction.

Other topics treated in this Special Issue are the visual comfort while investigating energy consumption [36], innovative air-conditioning systems [37] and monitoring system [38] in nZEBs. Zhang et al. [36] presented an optimization method and industrial internet of things (IIoT) technology that is proposed for nearly zero energy buildings (nZEB), providing a comfortable visual environment by only utilizing natural light while improving its induced indoor air conditioner energy consumption. Five modes were defined and investigated for different applications, and the best visual comfort and energy consumption conditions were identified. Chuah et al. [37] experimentally studied an integrated dedicated outdoor air system (IDOAS), which integrates the enthalpy exchange and outdoor air cooling into a unitary system, to investigate a mean of saving air-conditioning energy while maintaining good indoor air quality. The test results show that about 44% of the energy needed to condition the outdoor air can be saved and the cooling coefficient of performance (COP) was improved by about 46%. Huh et al. [38] describe a monitoring solution, a power line communication (PLC) and Android Application, for the management of decrepit buildings or apartments. This platform technology for smart metering is proposed as a PLC-based monitoring system in zero energy building management. This technology allows a system operator or an operation center to monitor, calculate, or manage power use remotely through the existing power line so that individual suppliers will be able to grasp the present condition of power use and respond to any unexpected incidents. The authors expect that the proposed platform technology will be one of the promising technologies for sustainable building, since its uninterrupted power usage monitoring and remote power control/management functions will be quite useful for economizing power with convenience.

This Special Issue may help bridge the gap between sustainability and improvement of the building energy efficiency. Other Special Issues on the topic should be organized periodically to update the state of the art.

Author Contributions: Conceptualization, D.M.; writing—original draft preparation, D.M.; writing—review and editing, D.M.; visualization, D.M and G.O. Both authors have read and agreed to the published version of the manuscript.

Funding: This research received no external funding.

Acknowledgments: The Editors would like to thank all the researchers for their high-quality papers, the reviewers for the suggestions provided in the scheduled deadlines, the editorial team of Sustainability and MDPI Branch Office.

Conflicts of Interest: The authors declare no conflict of interest.

References

1. Directive 2010/31/EU. European Parliament and of the Council of 19 May 2010 on the energy performance of buildings (recast). *Off. J. Eur. Union* **2010**, *13–35*, L153.

2. Directive 2012/27/EU. European Parliament and of the Council of 25 October 2012 on energy efficiency, amending Directives 2009/125/EC and 2010/30/EU and repealing directives 2004/8/EC and 2006/32/EC. *Off. J. Eur. Union* **2012**, *1–56*, L315/1.
3. Directive 2018/844/EU. European Parliament and of the Council of 30 May 2018 Amending Directive 2010/31/EU on the Energy Performance of Buildings and Directive 2012/27/EU on Energy Efficiency. 2018. Available online: https://eur-lex.europa.eu/legal-content/EN/TXT/?uri=uriserv%3AOJ.L_.2018.156.01.0075.01.ENG (accessed on 11 December 2020).
4. Mazzeo, D.; Kontoleon, K.J. The role of inclination and orientation of different building roof typologies on indoor and outdoor environment thermal comfort in Italy and Greece. *Sustain. Cities Soc.* **2020**, *60*, 102111. [CrossRef]
5. Mazzeo, D.; Oliveti, G.; Arcuri, N. Mapping of the seasonal dynamic properties of building walls in actual periodic conditions and effects produced by solar radiation incident on the outer and inner surfaces of the wall. *Appl. Therm. Eng.* **2016**, *102*, 1157–1174. [CrossRef]
6. Congedo, P.M.; Baglivo, C.; Centonze, G. Walls comparative evaluation for the thermal performance improvement of low-rise residential buildings in warm Mediterranean climate. *J. Build. Eng.* **2020**, *28*, 101059. [CrossRef]
7. Zhao, J.; Du, Y. Multi-objective optimization design for windows and shading configuration considering energy consumption and thermal comfort: A case study for office building in different climatic regions of China. *Sol. Energy* **2020**, *206*, 997–1017. [CrossRef]
8. Li, D.; Wu, Y.; Wang, B.; Liu, C.; Arıcı, M. Optical and thermal performance of glazing units containing PCM in buildings: A review. *Constr. Build. Mater.* **2020**, *233*, 117327. [CrossRef]
9. Elarga, H.; Fantucci, S.; Serra, V.; Zecchin, R.; Benini, E. Experimental and numerical analyses on thermal performance of different typologies of PCMs integrated in the roof space. *Energy Build.* **2017**, *150*, 546–557. [CrossRef]
10. Mazzeo, D.; Matera, N.; Cornaro, C.; Oliveti, G.; Romagnoni, P.; De Santoli, L. EnergyPlus, IDA ICE and TRNSYS predictive simulation accuracy for building thermal behaviour evaluation by using an experimental campaign in solar test boxes with and without a PCM module. *Energy Build.* **2020**, *212*, 109812. [CrossRef]
11. Mazzeo, D.; Oliveti, G.; de Gracia, A.; Coma, J.; Solé, A.; Cabeza, L.F. Experimental validation of the exact analytical solution to the steady periodic heat transfer problem in a PCM layer. *Energy* **2017**, *140–141*, 1131–1147. [CrossRef]
12. Mazzeo, D.; Oliveti, G. Thermal field and heat storage in a cyclic phase change process caused by several moving melting and solidification interfaces in the layer. *Int. J. Therm. Sci.* **2018**, *129*, 462–488. [CrossRef]
13. Zeng, Q.; Mao, T.; Li, H.; Peng, Y. Thermally insulating lightweight cement-based composites incorporating glass beads and nano-silica aerogels for sustainably energy-saving buildings. *Energy Build.* **2018**, *174*, 97–110. [CrossRef]
14. Ibrahim, M.; Biwole, P.H.; Wurtz, E.; Achard, P. A study on the thermal performance of exterior walls covered with a recently patented silica-aerogel-based insulating coating. *Build. Environ.* **2014**, *81*, 112–122. [CrossRef]
15. Pérez, G.; Coma, J.; Martorell, I.; Cabeza, L.F. Vertical Greenery Systems (VGS) for energy saving in buildings: A review. *Renew. Sustain. Energy Rev.* **2014**, *39*, 139–165. [CrossRef]
16. Coma, J.; Pérez, G.; de Gracia, A.; Burés, S.; Urrestarazu, M.; Cabeza, L.F. Vertical greenery systems for energy savings in buildings: A comparative study between green walls and green facades. *Build. Environ.* **2017**, *111*, 228–237. [CrossRef]
17. Palermo, S.A.; Turco, M.; Principato, F.; Piro, P. Hydrological effectiveness of an extensive green roof in Mediterranean climate. *Water* **2019**, *11*, 1378. [CrossRef]
18. Mazzeo, D.; Bevilacqua, P.; De Simone, M.; Arcuri, N. A new simulation tool for the evaluation of energy performances of green roofs. *Build. Simul. Appl.* **2015**, 313–321. Available online: http://www.ibpsa.org/proceedings/BSA2015/9788860460745_39.pdf (accessed on 11 December 2020).
19. Pirouz, B.; Palermo, S.A.; Maiolo, M.; Arcuri, N.; Piro, P. Decreasing Water Footprint of Electricity and Heat by Extensive Green Roofs: Case of Southern Italy. *Sustainability* **2020**, *12*, 10178. [CrossRef]
20. Saafi, K.; Daouas, N. A life-cycle cost analysis for an optimum combination of cool coating and thermal insulation of residential building roofs in Tunisia. *Energy* **2018**, *152*, 925–938. [CrossRef]

21. Fang, H.; Zhao, D.; Yuan, J.; Aili, A.; Yin, X.; Yang, R.; Tan, G. Performance evaluation of a metamaterial-based new cool roof using improved Roof Thermal Transfer Value model. *Appl. Energy* **2019**, *248*, 589–599. [CrossRef]
22. Mazzeo, D.; Matera, N.; De Luca, P.; Baglivo, C.; Congedo, P.M.; Oliveti, G. Worldwide geographical mapping and optimization of stand-alone and grid-connected hybrid renewable system techno-economic performance across Köppen-Geiger climates. *Appl. Energy* **2020**, *276*, 115507. [CrossRef]
23. Li, K.; Liu, C.; Jiang, S.; Chen, Y. Review on hybrid geothermal and solar power systems. *J. Clean. Prod.* **2020**, *250*, 119481. [CrossRef]
24. Lee, J.; Park, J. Phase Change Material (PCM) Application in a Modernized Korean Traditional House (Hanok). *Sustainability* **2018**, *10*, 948. [CrossRef]
25. Piro, P.; Carbone, M.; De Simone, M.; Maiolo, M.; Bevilacqua, P.; Arcuri, N. Energy and Hydraulic Performance of a Vegetated Roof in Sub-Mediterranean Climate. *Sustainability* **2018**, *10*, 3473. [CrossRef]
26. Gruner, M.; Matusiak, B. A Novel Dynamic Insulation System for Windows. *Sustainability* **2018**, *10*, 2907. [CrossRef]
27. Kim, J.; Kim, H.; Yoo, S. Consumers' Willingness to Pay for Net-Zero Energy Apartment in South Korea. *Sustainability* **2018**, *10*, 1564. [CrossRef]
28. Kwon, C.; Lee, K.; Cho, S. Numerical Study of Balancing between Indoor Building Energy and Outdoor Thermal Comfort with a Flexible Building Element. *Sustainability* **2019**, *11*, 6654. [CrossRef]
29. Yin, X.; Dong, Q.; Zhou, S.; Yu, J.; Huang, L.; Sun, C. Energy-Saving Potential of Applying Prefabricated Straw Bale Construction (PSBC) in Domestic Buildings in Northern China. *Sustainability* **2020**, *12*, 3464. [CrossRef]
30. Oh, M.; Tae, S.; Hwang, S. Analysis of Heating and Cooling Loads of Electrochromic Glazing in High-Rise Residential Buildings in South Korea. *Sustainability* **2018**, *10*, 1121. [CrossRef]
31. Koo, B.; Lee, K.; An, Y.; Lee, K. Solar Heat Gain Reduction of Ventilated Double Skin Windows without a Shading Device. *Sustainability* **2018**, *10*, 64. [CrossRef]
32. Kim, Y.; Yu, K. Study on the Certification Policy of Zero-Energy Buildings in Korea. *Sustainability* **2020**, *12*, 5172. [CrossRef]
33. Kim, Y.; Yu, K. Study on Policy Marking of Passive Level Insulation Standards for Non-Residential Buildings in South Korea. *Sustainability* **2018**, *10*, 2554. [CrossRef]
34. Piasecki, M.; Fedorczak-Cisak, M.; Furtak, M.; Biskupski, J. Experimental Confirmation of the Reliability of Fanger's Thermal Comfort Model—Case Study of a Near-Zero Energy Building (NZEB) Office Building. *Sustainability* **2019**, *11*, 2461. [CrossRef]
35. Pérez-Fargallo, A.; Rubio-Bellido, C.; Pulido-Arcas, J.; Gallego-Maya, I.; Guevara-García, F. Influence of Adaptive Comfort Models on Energy Improvement for Housing in Cold Areas. *Sustainability* **2018**, *10*, 859. [CrossRef]
36. Zhang, Y.; Zhong, Y.; Gong, Y.; Zheng, L. The Optimization of Visual Comfort and Energy Consumption Induced by Natural Light Based on PSO. *Sustainability* **2019**, *11*, 49. [CrossRef]
37. Chuah, Y.; Yang, J. A Integrated Dedicated Outdoor Air System to Optimize Energy Saving. *Sustainability* **2020**, *12*, 1051. [CrossRef]
38. Huh, J.; Park, J. Decrepit Building Monitoring Solution for Zero Energy Building Management Using PLC and Android Application. *Sustainability* **2020**, *12*, 1993. [CrossRef]

Publisher's Note: MDPI stays neutral with regard to jurisdictional claims in published maps and institutional affiliations.

© 2020 by the authors. Licensee MDPI, Basel, Switzerland. This article is an open access article distributed under the terms and conditions of the Creative Commons Attribution (CC BY) license (http://creativecommons.org/licenses/by/4.0/).

sustainability

MDPI

Article

Study on the Certification Policy of Zero-Energy Buildings in Korea

Yeweon Kim and Ki-Hyung Yu *

Department of Living and Built Environment Research, Korea Institute of Civil Engineering and Building Technology, Ilsan 411-712, Korea; yeweon.kim@kict.re.kr

* Correspondence: raytrace@kict.re.kr; Tel.: +82-2-910-0347

Received: 20 April 2020; Accepted: 17 June 2020; Published: 24 June 2020

Abstract: This study presents a methodology and process to establish a mandatory policy of zero-energy buildings (ZEBs) in Korea. To determine the mandatory level to acquire the rating of a ZEB in Korea, this study was conducted under the assumption that the criteria of ZEB was a top 5% building considering the building's energy-efficiency rating, which was certified through a quantitative building energy analysis. A self-sufficiency rate was also proposed to strengthen the passive standard of the buildings as well as to encourage new and renewable energy production. Accordingly, zero-energy buildings (ZEBs) in Korea are defined as having 60 kWh/(m^2·yr) of non-renewable primary energy (NRPE) consumption in residential buildings and 80 kWh/(m^2·yr) in non-residential buildings, and the self-reliance rate should be more than 20% of the renewable energy consumption as compared to the total energy consumption of the buildings. In addition, the mandatory installation of building energy management systems (BEMS) was promoted to investigate the energy behavior in buildings to be certified as zero-energy in the future. This study also investigated the number of ZEB certificates during the demonstration period from 2017 to 2019 to analyze the energy demand, non-renewable primary energy, renewable primary energy, and self-sufficiency rate as compared to those under the previous standards. For ZEB Grade 1 as compared to the existing building energy-efficiency rating, the sum of the NRPE decreased more than 50%, and renewable energy consumption increased more than four times.

Keywords: zero-energy building (ZEB), building energy policy; self-sufficiency rate; certification criteria of ZEBs

1. Introduction

1.1. Background of the Study

As greenhouse gas emissions have emerged as an important issue in the building sector, nations around the world have implanted various policies and systems to promote energy efficiency. In particular, a mandatory minimum level required for energy efficiency in new buildings has been enforced and a mandatory zero-energy building (ZEB) level, which is defined according to each nation, will be targeted by 2020. Accordingly, Korea has led the way and started mandatory ZEB construction from 2017 [1].

1.2. Current Status of Energy Building Policies Overseas

European Union (EU) nations established and executed a target goal and published detailed criteria in accordance with their circumstances based on the energy performance of buildings. In 2010, the adoption of the Energy Performance of Buildings Directive—EBPD, Directive 2010/31/EU—presented the building industry in the member states (MSs) with new challenges. One of the most prominent among them, as far as new buildings are concerned, is the progress towards nearly zero-energy buildings (NZEBs) by 2021 (or, in the case of public buildings, by 2019) [2]. Denmark established the Energy Policy Vision 2025, by which the proportion of new and renewable energy (NRE) accounts for up to 30% of all energy consumption. Consumers who use more than 100 MWh/yr. must install a smart meter. Although no mandatory regulations for residential customers have been made, Denmark has provided guidelines for remote metering [3]. Germany has applied building energy-saving design criteria since 2002 after acquiring approval of the Energy Saving Act under the Energy Saving Ordinance (EnEV) issued by the federal government. The Act was revised in 2016. As part of the stepwise measures toward nearly zero-energy buildings (NZEBs), the measures state that the primary energy of building permits submitted after January 1, 2016, should not exceed 75% of the reference building's required amount, and the NRE allocation criteria for new buildings under the Renewable Energy Heat Act were added in 2009. Germany has also set the target of 100% smart meter use by 2020 in pursuit of EU guidelines [4]. France has set a mandatory Bâtiments à Energie POSitive (BEPOS) standard rather than benchmarking ZEBs as the final target and the enforcement for this will begin for new buildings from 2020. This standard state the energy used for a building should not exceed the required energy amount of a reference building (BEPOSmax) and satisfy a certain amount of energy produced using NRE [5]. Japan has launched the "Research Institution for ZEB Achievement and Distribution" in May 2009 to investigate a detailed measurement in the pursuit of ZEBs and to build zero-energy houses (ZEHs), thereby proposing an overall ZEB vision. The Japanese quantitative definition of ZEB is as follows. First, buildings must not exceed more than 50% of the primary energy consumption based on a reference building, except for a building in which renewable energy was saved (zero-energy building ready level). Second, buildings must save more than 100% of the primary energy consumption through the use of NREs or achieve a 50% reduction of the primary energy consumption by adding NREs [6]. In the USA, the Obama administration aimed for residential ZEBs by 2020, and non-residential ZEBs by 2025, focusing on the increase in clean energy sources through the "All of the Above Energy Strategy" policy and a stable energy supply through demand management, and this policy has been adapted and promulgated by individual states as verified in the NBI Code and Policy. For example, the state of California targeted ZEBs for new residential buildings by 2020 and non-residential buildings by 2030. [7].

1.3. Objectives of the Study

This study aims to define ZEBs in accordance with the technical and economic circumstances in Korea and to describe the study methodology for establishing the mandatory policies and criteria to this end. It also aims to investigate the characteristics of certified buildings through case analyses on zero-energy-certified buildings.

2. Main Body

2.1. Mandatory ZEB Road Map

There are two methodologies to achieve ZEBs: a passive methodology to reduce the energy used in the building as much as possible and an active methodology that increases the use of NRE, thereby replacing fossil fuel sources with NRE sources, at least to the specified minimal energy used in buildings. The Ministry of Land, Infrastructure, and Transport in the Korean government supervises the direction to reduce the required energy amount used in buildings, and the Ministry of Trade, Industry and Energy plays a leading role in reducing the use of fossil fuels through the promotion of NRE projects. Figure 1 shows the roles of the government ministries for ZEBs.

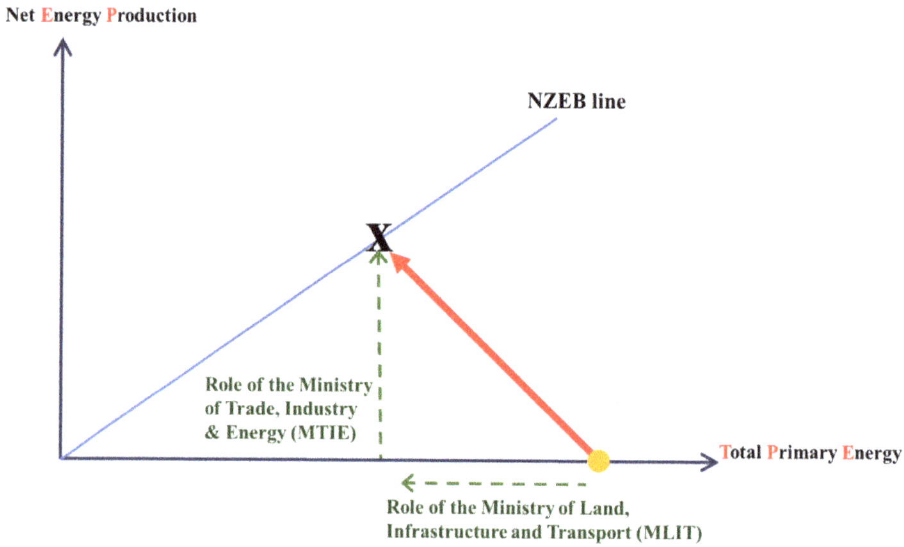

Figure 1. Roles of the government ministries for zero-energy buildings (ZEBs).

To play these roles, the Ministries have announced a mandatory road map for ZEBs through a mandatory certification promotion basis by amending the Green Building Creation Support Act in April 2019 [8]. Since ZEB standards in Korea are positioned under the Green Building Construction Support Act, a building that satisfies the energy requirements in addition to the requirements of the environmental elements, which belong to the part of the general green building definition, are also defined as a green building. Figure 2 shows the mandatory road map for ZEBs from 2017 to 2030. From 2017 to 2019, market acceptability and problems associated with ZEBs were identified through pilot projects, through the step-by-step mandatory ZEBs implementation according to the area and building purpose from 2020. Korea ultimately targets mandatory ZEBs for all new buildings and additional building extensions whose area is 500 m^2 or greater by 2030.

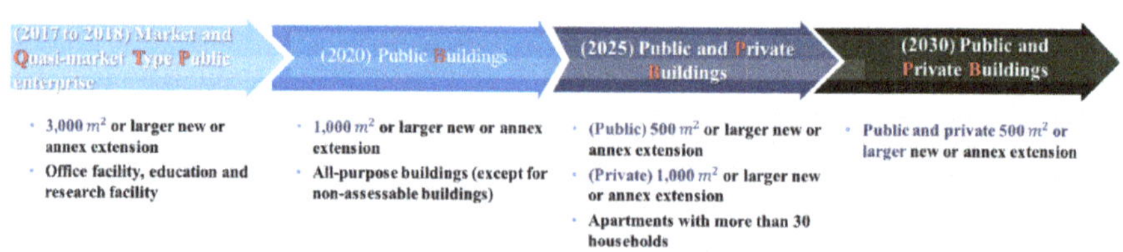

Figure 2. Road map of mandatory zero-energy buildings [8].

2.2. *Certification System of Building Energy in Korea*

2.2.1. Building Energy Efficiency Rating

The certification system of the building energy-efficiency ratings in Korea was established in 2001 and this assessed the energy-saving rate of buildings and the required energy-saving amounts to assign and certify the energy-efficiency ratings to the building. Since then, the certificate ratings have been re-evaluated based on the total energy saving rate compared to that of a standard house built in 2010. However, with the enactment of the Green Building Creation Support Act in 2013, the building energy-efficiency rating certificate applications filed after September 2013 were evaluated based on the non-renewable primary energy per unit area using the ECO2 software, which is a quantitative energy simulation tool [9].

Similar to other advanced building energy nations, such as Europe, the USA, and Japan, Korea has implemented a stepwise market direction toward ZEBs as the national direction rather than establishing an immediate regulation requiring net-zero-energy buildings in the current market.

To assess whether buildings comply with the direction, this study aims to minimize the confusion of the system by designing an assessment system that can be used for existing building energy assessment and a ZEB existing building energy-efficiency rating system.

2.2.2. ECO2, a Building Energy Assessment Tool

The ECO2 program provides a building energy assessment technique based on ISO 13790 and DIN V18599 [10] and quantitatively evaluates five items: heating, cooling, lighting, hot water, and ventilation systems, based on monthly mean weather data, and which is used in the building's energy-efficiency rating system [11]. The assessment results of the five items in ECO2 are calculated to determine the energy need, the energy use, non-renewable primary energy (NRPE), renewable primary energy (RPE), and the total primary energy. The primary energy conversion coefficient to calculate the primary energy is 1.1 for fuel, 2.75 for electricity, 0.728 for district heating, and 0.937 for district cooling, in which weights according to the usage profile and the heating method are applied to calculate the NRPE for the final rating. The default values are provided according to building use time, operating time, minimum inflow of outdoor air, required hot water amount, lighting time setting, heat-producing sources from persons and work auxiliary equipment, cooling and heating indoor air temperature, and the number of use days in a month. For weather data, standard profiles are retrieved from the ECO2 server to set the outdoor air temperature and the horizontal and vertical global radiations for any of the 13 regions in Korea. For the weather data of ECO2, monthly mean values are calculated based on weather data according to the typical meteorological year (TMY) methodology [12]. Four NRE inputs (geothermal, solar photovoltaic, solar thermal, and combined heat and power) can be entered, and when calculating the NRPE, the RPE amount is deducted from the cooling, heating, hot water, lighting, and ventilation items from the primary energy.

The calculated results have 10 labels, presented in Table 1, for the annual non-renewable primary energy per unit area for the rating, which are divided into residential and non-residential buildings [13].

Table 1. Building energy-efficiency rating table.

Label	Annual Non-Renewable Primary Energy (kWh/m^2)	
	Residential Buildings	**Non-Residential Buildings**
1+++	Under 60	Under 80
1++	More than 60 and under 90	More than 80 and under 140
1+	More than 90 and under 120	More than 140 and under 200
1	More than 120 and under 150	More than 200 and under 260
2	More than 150 and under 190	More than 260 and under 320
3	More than 190 and under 273	More than 320 and under 380
4	More than 230 and under 270	More than 380 and under 450
5	More than 270 and under 320	More than 450 and under 520
6	More than 320 and under 370	More than 520 and under 610
7	More than 370 and under 420	More than 610 and under 700

2.3. *Study on Definition of ZEBs in Korea*

2.3.1. Definition of ZEB Level

The definition of a ZEB in Korea is a green building that minimizes energy loads and minimizes the required energy amount by utilizing NREs according to the Green Building Creation Support Act, which was amended in 2017 [14].

To define the mandatory rating level of zero energy according to the 2020 road map, this study aimed to define the 2017 NZEB (nearly zero-energy building) levels using the database of building energy-efficiency ratings from the Korea Energy Agency. Figure 3 shows the analyzed data of the energy levels in 300 public buildings that had the building energy efficiency rating certificates issued from 2010 to 2015. In the left figure of Figure 3, the NRPE was distributed between 80 kWh/m^2, which was Rating 1+++, to 320 kWh/m^2, which was Rating 2. All the certified energy ratings for the public buildings were a Rating 2 or higher.

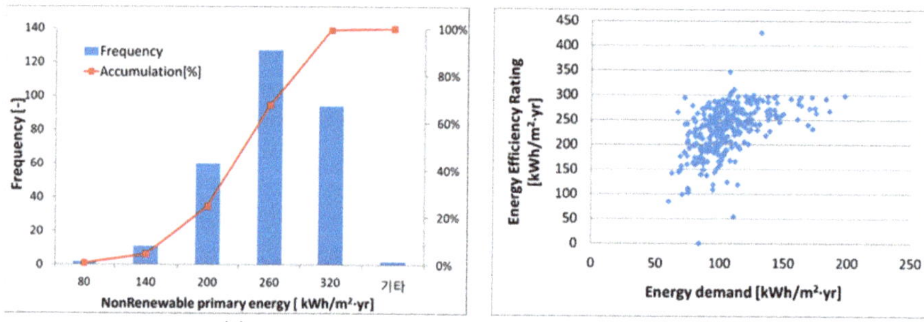

(a) (b)

Figure 3. (**a**) Energy level of the efficiency rating of buildings, and the frequency of certification by rating (**b**) distribution of the non-renewable primary energy (NRPE).

The range of energy performance that can be achieved depends on the minimum amount of effort according to expert opinions to reach the same standards as those buildings in the top 5%, and are based on the NRPE per unit area in the elements: heating, cooling, hot water supply, lighting, and ventilation. The mandatory level of ZEBs were restricted to 138 kWh/m^2. This value corresponds to a building energy-efficiency rating of 1++. To define the level of ZEBs, the energy amount produced by the new and renewable sources of energy is also highly meaningful in addition to the NRPE. Thus, a mandatory level in relation to a certain amount of NRE production at a nearly zero-energy level was important. Thus, the production of NRE in buildings whose rating was 1++ was found to be below 20% of the total consumption.

2.3.2. Level of ZEB Certification

The 1++ level of the energy-efficiency rating, which was the top 5% of buildings considering the energy-efficiency ratings, was based on the five energy elements: heating, cooling, hot water, lighting, and ventilation, and defined as the minimum certificate level for the meeting the definitions of a ZEB in Korea and had to satisfy an energy self-sufficiency rate, which represents the NRE production compared to fossil-fuel energy use. This was added to achieve the link with NRE at sites where buildings are being constructed.

Figure 4 shows the diagram of the ZEB rating levels. This figure depicts the compatibility between the current building energy efficiency ratings and a zero-energy rating, in which ZEB 5 was defined as the minimum rating of ZEBs that corresponds to the current building energy-efficiency rating of 1++, as well as satisfying the 20% or more NRE production requirement. As the NRE production increased, the zero-energy rating was raised. This is a methodology that can improve the certification level from NZEB (ZEB 5) to NZEB (ZEB 1).

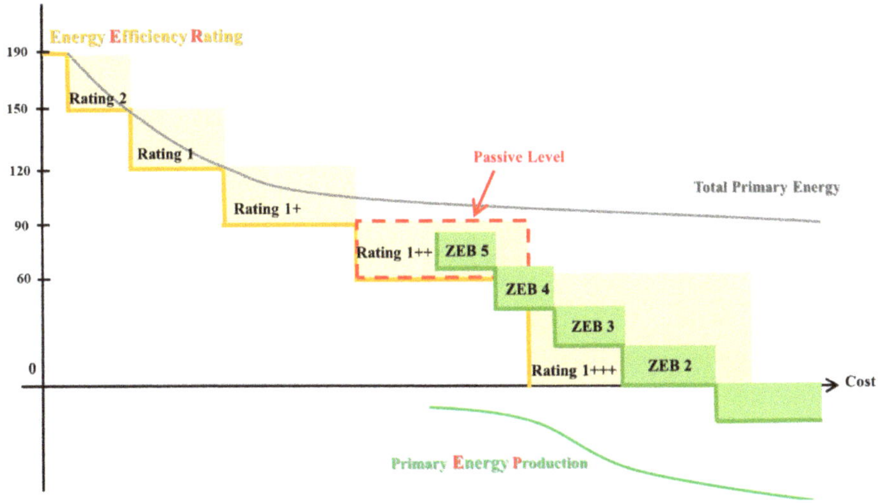

Figure 4. Rating level of the zero-energy buildings.

2.3.3. Total Primary Energy, NRPE, RPE, and Energy Self-Sufficiency Rate

When the level (target of the building energy-efficiency rating) of the reduction of the total primary energy is as great as possible in the building elements, the building is defined as having achieved the passive level as shown in Figure 4, thus satisfying the stepwise ZEB rating through the promotion of NREs and would, therefore, allow the building to acquire a ZEB certification. The concept of the energy self-sufficiency rate is defined as RPE (renewable primary energy) compared to the total primary energy, which is an indicator of the NRE performance and is evaluated using a stepwise rating system. Here, the energy self-sufficiency rate was calculated by determining the ratio of the RPE as compared to the total primary energy per unit area, and the total primary energy was a sum of the required fossil fuel energy (NRPE: non-renewable primary energy) and the NRE production (RPE: renewable primary energy) in the building. This is equivalent to the concept of the required building energy in the existing building energy efficiency rating system and is calculated by subtracting the NRE production from the energy amount (energy consumption) used in the entire building.

RPE refers to the energy production produced through NRE systems, excluding the fossil-fuel energy used to drive the system. For example, in the case of a geothermal heat pump shown in Figure 5, the value of the heat production produced by the system, and excluding the energy consumption for the pump working power and the driving power of the heat pump, is accepted as the energy production. This can be expressed by the following equation.

$$\text{Energy self}-\text{sufficiency rate}(\%) = \frac{\text{RPE per unit area}}{\text{Total primary energy per unit area}} \times 100. \quad (1)$$

Note (1): RPE (renewable primary energy per unit area (kWh/m^2)

= ∑((NRE production - energy amount required to produce NRE) × corresponding primary energy conversion coefficient)/evaluated area.

Note (2): Total primary energy per unit area (kWh/m^2)

= Non-renewable primary energy (NRPE) per unit area + renewable primary energy (RPE) per unit area.

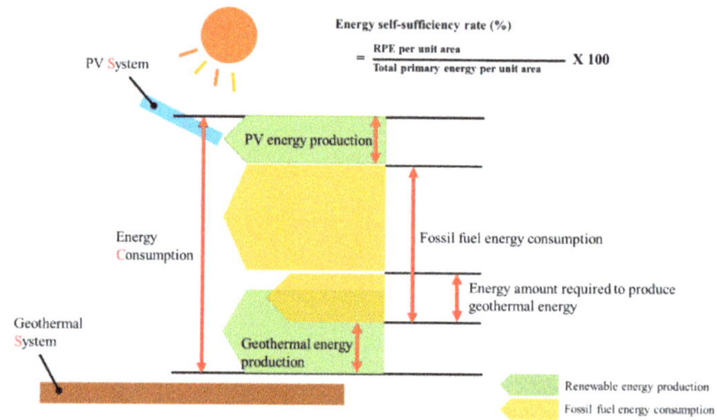

Figure 5. Flow diagram of the components for the self-sufficiency rating by renewable energy source.

2.3.4. Other Certification Criteria of ZEBs

The current mandatory system of ZEBs is based on the limited building performance under the standard conditions. Thus, this study added a criterion of the mandatory installation of BEMS by which energy usage can be read or electronic meters can allow constant remote reading of a meter for each energy source to investigate the operating conditions of the buildings or the operating patterns of the residents in the future. Thus, the final certification criteria of the ZEBs are shown in Figure 6.

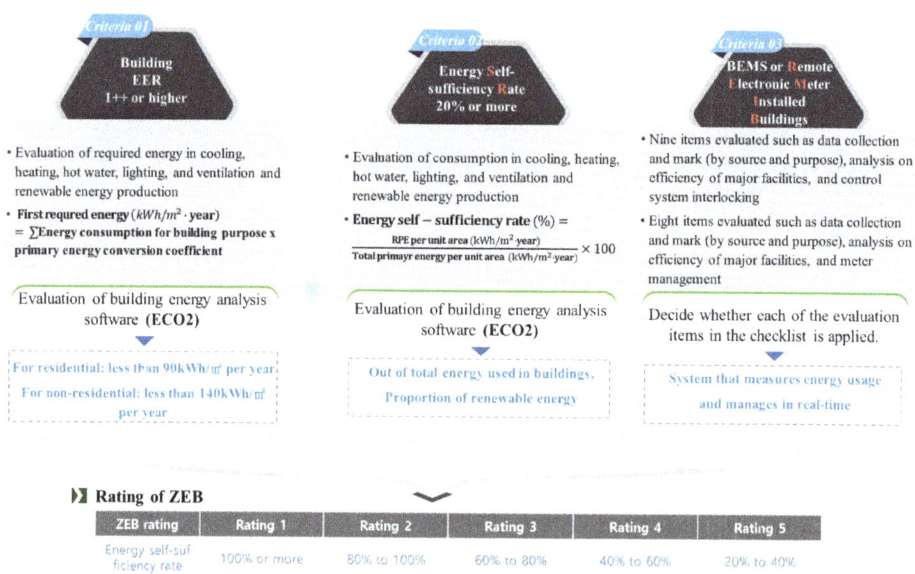

Figure 6. ZEB certification criteria.

2.4. *Current Status of ZEB Certification*

To analyze the ZEB certification trends through pilot projects prior to making the ZEB certification mandatory in 2020, the current status of the buildings certified as a ZEB in 2017 to 2019 was analyzed [15]. The number of certificates for three years, as presented in Table 2, was 83, in which nine cases were for residential buildings and 75 cases were for non-residential buildings, showing a much greater number of certificates issued for non-residential buildings. This was because the mandatory target was concentrated on market-type public companies and residential buildings were affected by only the pilot complex project. The proportion of Rating 5, which is the minimum criterion of the mandatory level, was 74% (64 cases). The higher ratings had smaller numbers of certified cases.

Table 2. Current status of ZEB certificates.

Rating	Rating 5	Rating 4	Rating 3	Rating 2	Rating 1
No. of certificates (–)	64	11	8	1	3
Proportion (%)	74	13	9	1	3

The comparison results of the NRPE and RPE, which were two of the five energy elements, are considered according to each rating as shown in Figure 7, and indicated that the primary heating energy was reduced by 59.2% from 69.7 kWh/m^2 to 28.4 kWh/m^2 as the rating was raised from Rating 1 to ZEB 5; the primary cooling energy was reduced by 69.38% from 58.6 kWh/m^2 to 17.7 kWh/m^2; and the primary lighting energy was reduced by 54.2% from 46.7 kWh/m^2 to 21.4 kWh/m^2. The primary ventilation energy was reduced by 64% from 43.8 kWh/m^2 to 15.8 kWh/m^2, and the RPE increased by 403% from 6.5 kWh/m^2 to 32.6 kWh/m^2. These results indicated that a total sum of the primary energy was reduced by more than 50% whereas the NRE production more than quadrupled as the rating was raised.

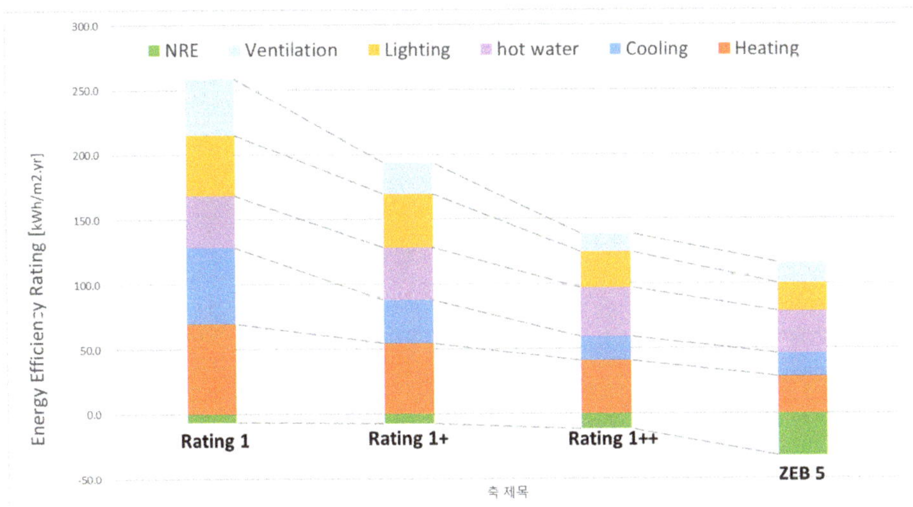

Figure 7. Five energy elements according to the certification criteria, NRPE, and RPE.

This can be seen as an increase in production as the installed NRE capacity increased according to the rating level as shown in Figure 8.

To install the NRE to increase the RPE, sites must be provided. Currently, NRE is accepted only when it is produced on-site in Korea. Thus, those energy sources are accepted when considering the primary energy coefficient. Thus, for all 83 ZEB certification cases, photovoltaic systems were installed as the NRE source. This is due to the fact that the primary energy conversion coefficient of the electricity (2.75) was multiplied, adding to the convenience of installation and price competitiveness. However, as mentioned in the above, since only on-site installation was accepted, the self-sufficiency rating level (level of certification rating) tended to be inversely proportional to the total area, as shown in Figure 9.

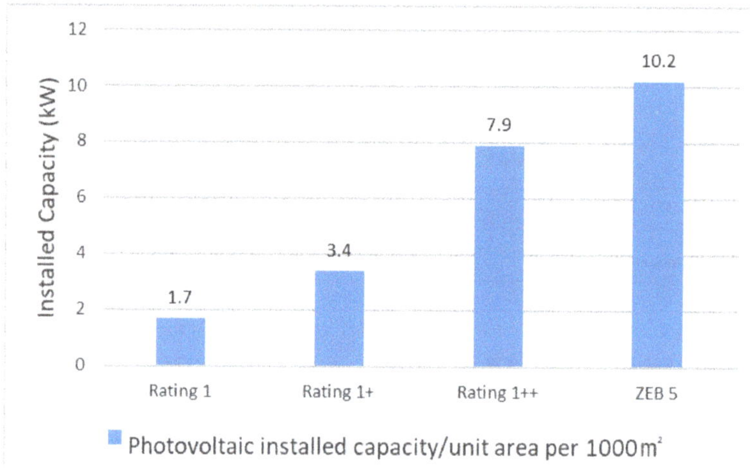

Figure 8. Photovoltaic installed capacity according to the ZEB rating.

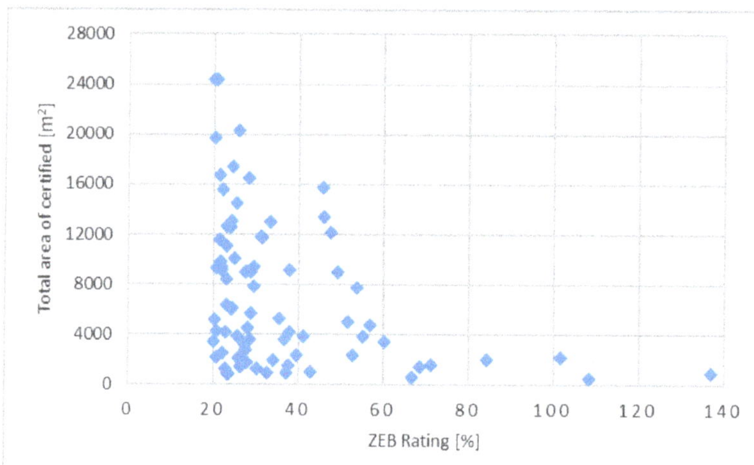

Figure 9. ZEB rating according to the total area of certified buildings.

3. Conclusions

This study proposed the definition of a ZEB and certification levels that complied with the definition of the International Zero-Energy Building and is relevant in Korea. This study also analyzed the current status of the buildings that acquired ZEB certification. The study results are summarized as follows:

(1) The ZEB certification in Korea proposes a stepwise certification level that can be certified from the nearly ZEB (NZEB) level to a plus level. The rating level of the ZEB was defined as those meeting a 1++ building level, which were the top 5% of buildings based on the existing building energy efficiency rating database.

(2) The rating was defined as satisfying the passive level of the building and differentiating the NRE production. The rating was assigned for every 20% section based on the NRE self-sufficiency rate, which is determined through the total primary energy, non-renewable primary energy (NRPE), renewable primary energy (RPE), and the self-sufficiency rate.

(3) The current certification status of the ZEBs in Korea exhibited that as the rating increased, the total sum of the required first energy amount was reduced by 50% in the case of Rating ZEB 1 compared to that of Rating 1++ of building energy-efficiency rating, and the NRE production more than quadrupled.
(4) The NRE self-sufficiency rate level (level of certification rating) was inversely proportional to the total area, which showed a limitation according to the building environments since the ZEBs in Korea are evaluated based on the NRE produced only on-site.

For future studies, constant methodology developments of off-site evaluations in addition to on-site evaluations are needed, and a study on measures to extend the currently limited certification assessment in heating, cooling, hot water supply, lighting, and ventilation into other areas, such as household appliances and transportation, is also needed.

Author Contributions: Y.K. analyzed the study outcomes; K.-H.Y. conceived the concept. All authors have read and agreed to the published version of the manuscript.

Funding: This research was funded by [Infrastructure and Transport of the Korean government] grant number [19AUDP-B079104-06].

Acknowledgments: This research was supported by a grant (19AUDP-B079104-06) from Architecture & Urban Development Research Program funded by Ministry of Land, Infrastructure and Transport of the Korean government.

Conflicts of Interest: The authors declare no conflict of interest.

References

1. International Energy Agency. *Energy Policies of IEA Countries—The Republic of Korea 2012 Review*; International Energy Agency: Paris, France, 2012.
2. Concerted Action EPBD. *2016 Implementation the Energy Performance of Buildings Directive (EPBD): Featuring Country Reports*; ADENE: Lisboa, Portugal, 2015; p. 594. Available online: https://epbd-ca.eu/ca-outcomes/2011-2015 (accessed on 20 April 2020).
3. Danish Transport and Construction Agency. *Danish Building Regulations 2015, 2015.12*; Danish Transport and Construction Agency: Copenhagen, Denmark, 2015.
4. Feist, W. *Lernzielkatalog "Zertifizierter PassivhausPlaner"*; Passive Haus Institute: Darmstadt, Germany, 2015.
5. IPEEC. *Building Energy Efficiency Requirements in Building Codes, Energy Efficiency Policies for New Buildings*; International Energy Agency (IEA): Paris, France, 2008; pp. 477–488.
6. Cui, C.; Niu, J.J. Energy-saving technology strategy 2011 and Enlightenment to us. *Energy China* **2011**, *33*, 10–14.
7. American Society of Heating, Refrigerating, and Air-Conditioning Engineers. *ANSI/ASHRAE/IES Standard 90.1-2016: Energy Standard for Buildings Except Low-Rise Residential Buildings*; ASHRAE: Atlanta, GA, USA, 2016.
8. Joint Related Ministries. *2030 Modification & Complementation of the Greenhouse Gas Reduction Roadmap*; Ministry of Environment: New Delhi, India, 2018; p. 3.
9. Green Buildings Construction Support Act No. 16418. Available online: http://www.law.go.kr/lsSc.do?tabMenuId=tab18&section=&eventGubun=060101&query=%EB%85%B9%EC%83%89%EA%B1%B4%EC%B6%95#liBgcolor0 (accessed on 20 April 2020).
10. Andaloro, A.P.E.; Salomone, R.; Ioppolo, G.; Andaloro, L. Energy certification of buildings: A comparative analysis of progress towards implementation in European countries. *Energy Policy* **2010**, *28*, 5840–5866. [CrossRef]
11. Jo, J.-H. A Study on Evaluation Criteria and Evaluation Tool for Energy Performance of internal and external Building, Construction Technique/Sangyong, Special Project II—Zero Energy Building Policies and Trends of Internal and External Building. 2017, p. 23. Available online: http://www.auric.or.kr/User/Rdoc/DocRdoc.aspx?BrsDataBaseName=RDOC&PrimaryKey=350201&DBNM=RDCR#.XvGVO3duKUl (accessed on 20 April 2020).
12. Kim, Y.; Jang, H.K.; Yu, K.H. Study on Extension of Standard Meteorological Data for Cities in South Korea Using ISO 15927-4. *Atmosphere* **2017**, *8*, 220. [CrossRef]

13. Certification Standards of Building Energy Efficiency Rating and Zero Energy Building, Annexed Table 2. In *Notice No. 2018-675 of the Ministry of Land, Infrastructure and Transport*. Available online: http://www.law.go.kr/admRulSc.do?tabMenuId=tab107&query=%EA%B1%B4%EC%B6%95%EB%AC%BC%EC%97%90%EB%84%88%EC%A7%80%20%ED%9A%A8%EC%9C%A8%EB%93%B1%EA%B8%89#AJAX (accessed on 20 April 2020).
14. Ministry of Land. The Second Green Building Basic Plan, Green Architecture Division. 2020; pp. 1–12. Available online: http://www.molit.go.kr/USR/NEWS/m_71/dtl.jsp?id=95083381 (accessed on 20 April 2020).
15. Zero-Energy Building Information Site in Korea Energy Agency. Available online: https://zeb.energy.or.kr/BC/BC00/BC00_01_001.do (accessed on 20 April 2020).

© 2020 by the authors. Licensee MDPI, Basel, Switzerland. This article is an open access article distributed under the terms and conditions of the Creative Commons Attribution (CC BY) license (http://creativecommons.org/licenses/by/4.0/).

sustainability

MDPI

Article

Energy-Saving Potential of Applying Prefabricated Straw Bale Construction (PSBC) in Domestic Buildings in Northern China

Xunzhi Yin [1,2], Qi Dong [1,2], Siyuan Zhou [1,2], Jiaqi Yu [1,2], Lu Huang [1,2] and Cheng Sun [1,2,*]

1 School of Architecture, Harbin Institute of Technology, Harbin 150001, China; x.yin@hit.edu.cn (X.Y.); dongqi@hit.edu.cn (Q.D.); 18S034024@stu.hit.edu.cn (S.Z.); 19S034015@stu.hit.edu.cn (J.Y.); 19S134163@stu.hit.edu.cn (L.H.)

2 Key Laboratory of Cold Region Urban and Rural Human Settlement Environment Science and Technology, Ministry of Industry and Information Technology, Harbin 150001, China

* Correspondence: suncheng@hit.edu.cn; Tel.: +86-0451-8628-1089

Received: 22 February 2020; Accepted: 22 April 2020; Published: 24 April 2020

Abstract: The Prefabricated Straw Bale Construction (PSBC) has been proven as one of the most efficient construction methods to achieve low-energy buildings with low environmental impacts. This research presents analysis of the rationale for using straw bale constructions in northern China and a discussion of feasible constructions of PSBC to meet the local building codes following evaluations of potential energy performance of domestic buildings with PSBC in severe cold regions and cold regions in China. The results show that the buildings with PSBC reduce both heating and cooling energy uses, as well as heating intensities across the severe cold and cold regions, compared to the domestic buildings with conventional constructions. The findings of this research will contribute to reducing energy consumption in building industries in China.

Keywords: low-carbon design; bio-based building materials; straw bale; prefabricated straw bale construction (PSBC); operational energy; energy simulation

1. Introduction

1.1. Research Background

As an outcome of the boom in economic development over the last two decades, China has experienced a rapid increase in urbanization. Urbanization has increased from 46% in 2008 to a predicted value of 60% by 2020 [1]. During the urbanization process, the major energy source has been coal [1]. Due to the inevitable increase in energy demand from the urbanization process, Green House Gas emissions in China will continue to increase significantly until this urbanization process is completed [2]. Energy consumption in residential buildings makes a significant contribution to the GHG emissions generated by the urbanization process in China [3]. As a result, the building industry has been a major obstacle in reducing energy consumption and GHG emissions from China [4,5]. The current situation with building energy efficiency levels is to ensure high energy efficiency standards for new construction [6]. However, the rapid growth of energy demand in Northern China may make it more difficult to achieve these standards [7].

To deliver the commitment that the Chinese government made in the Paris Agreement, policies adopted by the building industry have largely been defined and supervised by the Ministry of Housing and Urban–Rural Development of the People's Republic of China (MOHURD). The majority of policies are enacted through statutory and industrial standards [8]. The strategies to reduce GHG emissions caused by the building industry focus on increasing the energy efficiency levels of existing buildings and reducing energy use in newly built buildings [6]. The strategies for increasing the energy efficiency levels of existing buildings involve refurbishment of existing buildings and replacing existing building service systems [8]. The reductions in energy use in newly built buildings involve higher standards of energy efficiency levels, encouraging the use of prefabricated construction in buildings and promoting the number of green buildings amongst newly built buildings [8]. These strategies have been applied as a priority in northern China due to the higher energy reduction potential in this region compared to those in other regions [9,10].

Nearly Zero-Energy Buildings (NZEBs) are defined as low-energy buildings with superior building performances than conventional buildings, and renewable energy resources are often applied to meet the energy requirements of NZEBs [11,12]. As the ultimate target of NZEBs is to achievable zero energy consumption, the designs and constructions of NZEBs require high energy efficiency levels of the NZEB and passive strategies for improving the energy efficiency levels [13]. As part of the passive strategies for improving energy efficiency levels of NZEBs, the building envelopes have notable impact on both the built environment and methods of building designs and constructions [14].

1.2. Climatic Features and Energy Consumptions of Buildings in Northern China

The climate in China encompasses a wide range of air temperatures and humidity. MOHURD published the national code for thermal design of civil buildings, which identifies a number of different climatic regions [15]. The design and construction of buildings in China is informed by five climate regions differentiated by the climatic characteristics of the regions (Figure 1). Northern China is in the climatic areas defined as "Severe Cold" and "Cold". In these two regions, the primary energy demand in buildings is for winter heating [15].

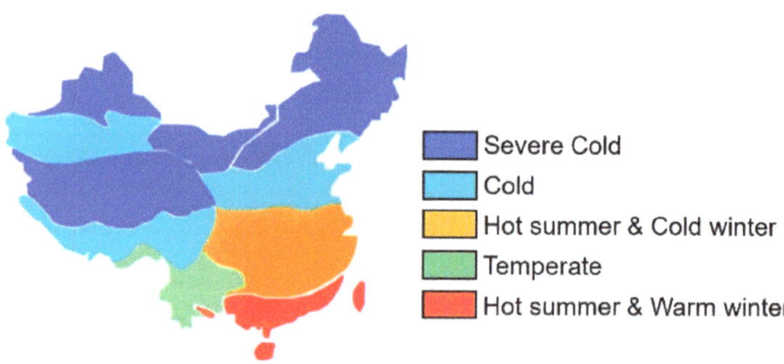

Figure 1. Climatic regionalization in the GB50178-93 (reproduced from [15]).

Both severe cold and cold regions are in the "Temperate" monsoon climate. The climatic features of the two climatic regions represent two typical climatic characteristics [15]: Firstly, both the air temperature and humidity are at high levels in summer months. The daily high temperatures are around 30 °C in summer months in severe cold regions, whereas they are over 30 °C in cold regions. Rainfalls are expected mostly in summer months, and they lead to the high air humidity. Secondly, the air temperatures are expected to be below freezing during the whole winter period in the two climatic regions, and highest monthly air humidity levels are present during the same period of time. However, the high air humidity levels do not result in a humid environment inside and outside buildings. As the low temperatures decrease absolute water vapor pressures in the air, the relative humidity levels in winter months are significantly higher than other months annually in the severe cold and cold regions.

The need for improved building energy efficiency is most critical in northern China due to the high winter heating demand caused by the regional climatic conditions. The Climate Policy Initiative established that the floor area of urban residential buildings has doubled between 2000 and 2008 [16]. As a result, residential buildings are expected to account for up to 90% of total building floor area in China by 2030 [16]. Due to this dominant position of domestic buildings, end-use energy consumption is a significant contributor to the total in-use energy of buildings in China (Figure 2). According to the predictions for 2021, in comparison with the 1996 energy consumption levels, energy use increased at its most rapid rate in China in 2008 [16]. The heating energy demand by domestic buildings more than doubled between 1996 and 2008, and has become the most crucial factor for energy use in domestic buildings in China.

Heating demand is notably high in northern China due to the requirement for thermal comfort in the winter months. As a result, in this region, the proportion of energy consumption associated with heating is much greater than the national average. The latest versions of building regulations commit to a reduction of 50% in the heating energy consumption of buildings against the 1980 levels in northern China [17,18]. The regulations specify the thermal conductivity of the building envelope, with U-values ranging from 0.25 to 0.7 W/m^{2}K, depending on the heights and locations of residential buildings in northern China [17,18].

1.3. Rationale for Straw Bale Construction in Northern China

Straw bale construction uses agricultural co-products in the construction of buildings. It was originally developed due to a shortage of building materials in Nebraska in the late 19th century [19]. Straw bale buildings were originally used as temporary buildings [20]. This building type was replaced through the development of more industrialized building materials in the Mid-West of America in the early 20th century [19]. In the 1970s, the oil crisis prompted the development of concepts for more energy-efficient buildings in the USA [19]. Straw bale buildings are characterized by a combination of low cost, quick construction process, and high thermal insulation [20]. This construction technique was re-introduced in the 1980s in west America, and the construction method has since become popular worldwide [21]. Straw bale construction is now recognized globally and has developed into a contemporary building typology and construction method in the Western world [19]. The construction technique was initially introduced to northern China by the Adventist Development and Relief Agency (ADRA) in 1998 [22]. This project was funded by the ADRA/China, Central Government, and Local Government in northern China, and more than 600 straw bale buildings were completed by 2006 [23]. Informed by the practices of the ADRA project, a different design for straw bale buildings was developed in Jilin province [24].

There are three significant advantages to straw bale construction in the context of conditions in China.

First of all, straw bales are a carbon-sink building material with significantly low embodied energy and embodied carbon [25]. The atmospheric CO_2 is sequestered in the body of the plant through the process of photosynthesis [25–28]. It can be calculated through stoichiometry that 1 kg of carbon sequestered in the straw stems requires the removal of 3.67 kg of CO_2 from the atmosphere. This amount of adsorbed carbon will not be released into the atmosphere until the straw bale buildings are demolished [29].

Secondly, due to the high thermal insulation properties of straw bale walls, straw bale houses have low heating energy demand [30]. The U-value of typical prefabricated straw bale walls can be as low as from 0.11 to 0.19 Wm^{2}K^{-1} for 450 mm thick wall panels [31]. In comparison with the thermal performance stipulated in current Chinese standards for walls, the thermal resistances of these panels are 50–300% better than those of standard wall constructions. Air pollution is also exacerbated by the use of coal to heat buildings in Northern China [32–34]. The high thermal insulation properties of straw bale buildings notably decrease heating demand; therefore, less coal is required for winter heating in northern China.

Thirdly, the use of straw in the building industry will benefit the agricultural economy of northern China. Straw is considered a waste material in the farming process for rice and wheat in northern China [35]. The total annual rice production in northeast China is approximately 203 million metric tons [36]. Due to the associated large amount of waste straw, disposal of the material has been a concern in China for decades. Currently, the most common approach for disposal of the straw is for it to be burnt in fields [37]. This issue demands more environmentally friendly disposal solutions for straw as an alternative to burning [37].

1.4. Prefabricated Straw Bale Construction (PSBC)

Prefabricated Straw Bale Construction (PSBC) is a prefabricated construction technique for utilizing straw bales in buildings [38]. This building technique combines conventional straw bales with the superior characteristics of controlled prefabricated construction processes [38]. In comparison with conventional straw bale construction, the main benefit of PSBC is that this construction technique minimizes the risks associated with wet weather on construction sites. It also offers better quality control than onsite construction of straw bale walls [39]. Additional benefits of PSBC include a reduction in onsite construction time, no site waste removal, and lower risks of fire on the construction site due to elimination of loose straw [38].

ModCell is one of the first companies to produce PSBC. This PSBC is in the form of PSBC panels. The panels typically consist of engineered timber frames, in-fill straw bales, and lime render [39]. The dimensions of the engineered frames are typically 100 x 480 mm to accommodate the dimensions of the in-fill straw bales. The sizes of the panels vary according to different building projects, but are typically 3.0 (width) × 3.2 m (height). The in-fill straw bales are stacked to form walls of densities of around 110 kg/m^3 [39], and these are pre-compressed during the process to increase stability and reduce thermal bridging arising from gaps between the bales and the frames [40]. The bales are finished with a 30 mm lime render, which is the minimum thickness of lime render possible without the use of a rain screen in the climatic environment of the UK [41].

The ModCell panels have been used in several projects. One of the more high-profile applications of this product was in the construction of the Low-Impact Living Affordable Community (LILAC) project in Leeds. This project included a 20 household community in Bramley, west Leeds on a former school site [42]. The ModCell PSBC panels were used for the walls of the domestic buildings in the LILAC project. The project featured low-energy-consumption residential buildings [42]. The average energy use in the LILAC buildings can be categorized as follows: 35.73 kWh/m^2 for space heating, 39.22 kWh/m^2 for domestic hot water, and 30.00 kWh/m^2 for lighting [43]. Comparison between these energy demands and other benchmarks shows that the energy consumption level of the LILAC project residences is greatly superior to the Chartered Institution of Building Services Engineers (CIBSE) Guide F benchmarks, and is comparable to PassivHaus standards in terms of space heating energy demand (Figure 2).

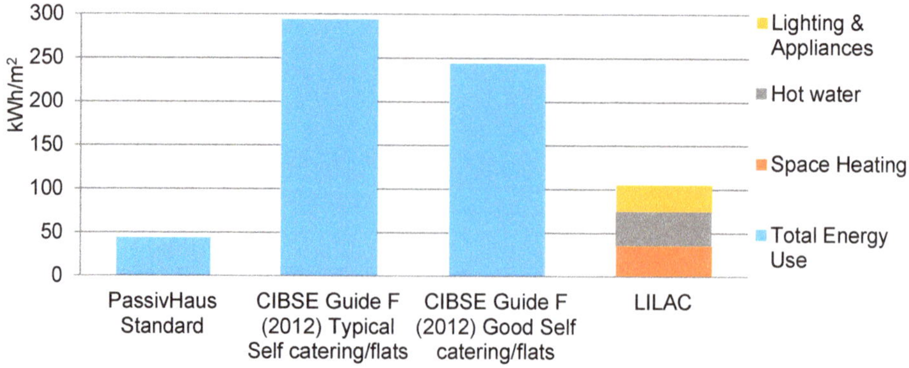

Figure 2. Comparative energy consumption of Low-Impact Living Affordable Community (LILAC) and other benchmarks (data of benchmarks from [44]).

Apart from the ModCell PSBC, another prefabrication method for straw bales was developed by Ecococon in Lithuania. The Ecococon prefabrication system consists of in-fill straw, timber frames, and a plastering and rendering layer. The assembly of the PSBC walls required only simple tools and standard screws for the building of Ecococon walls [45]. The prefabricated Ecococon panels use in-fill loose straw rather than straw bales. Straw stems are stacked in several layers on the timber frames and compressed during the stacking process [45]. According to Ecococon, this stacking method can effectively reduce the gaps between the straw and results in a high uniform density of the in-fill bale walls [45].

The design of the in-fill method of straw and the design of the timber frames results in a high thermal resistance for the Ecococon PSBC panels. Typical U-values for the ModCell panels are 0.190 W/m^{2}K [40]. The thermal conductivity of the Ecococon products is 0.056 W/mK, with a resulting overall U-value of 0.107 W/m^{2}K [45]. The panels were certified as PassiveHaus building materials, as they meet the PassiveHaus standards [45].

1.5. Research Scope and Objective

With a relatively long history of straw bale buildings, Prefabricated Straw Bale Construction (PSBC) has developed in recent years. Current research verifies the rationale for applying straw bale buildings in accordance with the policies and climatic features of northern China. However, the discussion of feasibility and the potential benefits of this relatively innovative walling construction method has been limited.

This research discusses the feasibility of PSBC in residential buildings to reduce energy consumption in this building type in northern China. In order to establish overall estimates of the energy consumption of residential buildings with PSBC, this research discusses the operational energy load (heating and cooling) of reference buildings in the climatic conditions for fifteen major cities in northern China. This is followed by a comparison of the operational heating and cooling energy demand of buildings with PSBC and with standard wall construction.

2. Research Methodology

This study commences by determining the reference buildings and the application methods for PSBC in the reference buildings. Energy modeling and the simulation processes for the reference buildings are based on the IESVE-2018, which is computational simulation software produced by the Integrated Environmental Solutions Limited (Glasgow, UK). The detailed research methods are as follows.

2.1. Typical Residential Buildings

In the urban environment of China, the majority of residential buildings are multiple-floor apartment buildings and high-rise apartment buildings [46,47]. There are two types of typical layouts for high-rise domestic building designs in northern China. The first layout consists of several apartment blocks [48]. In each block, stairs and lifts are available to every apartment on each floor [48]. The second layout is for tower-type high-rise residential buildings [48]. The floor plan for these tower-type residential buildings is more centralized than that for the block-type high-rise residential buildings, and is generally suitable for the design of small apartments. However, in practice, since the floor area of residential buildings has been increasing since the 1980s [46], the block layout for residential buildings is most common.

In order to representing the most common form of residential buildings, this research is based on apartment residential buildings of different building heights. The reference floor plan consists of two blocks with two one-bedroom apartments and one two-bedroom apartment (in blue and green), with a corridor and stairs (in red) to each apartment in the building (Figure 3). As required by GB50096-2011, residential buildings with less than six floors are identified as multiple-floor residential buildings, whereas high-rise buildings have more than seven floors [48]. Skyscraper residential buildings are defined in the standard as being over 100 m high [48]. However, due to the severe requirements for fire evacuation and the related economic implications, this type of residential building is not widely constructed in China. As a result, the reference building heights for this study are defined as 6, 12, and 18 floors.

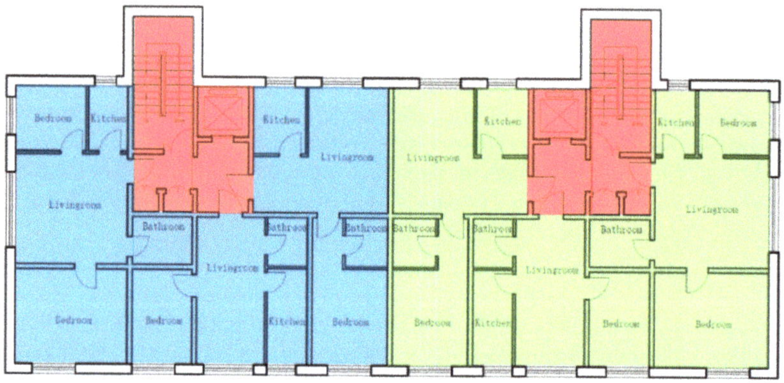

Figure 3. Typical floor plan of the reference residential building in this research (corridor space in red; flat space in blue and green).

2.2. Application of PSBC

The method of application of PSBC wall panels is referenced from the standard building details in 16J110-2/16G333. This standard building detail was published by the MOHURD in 2008. This published standard introduces the standard construction details for application methods for prefabricated concrete wall systems in China [49]. Based on the 16J110-2/16SG333, the PSBC wall panels can be used for in-fill construction between building frames. In this technique, the PSBC wall panels are directly placed between the main structural frames (Figure 4). The weight of the PSBC walling panels is directly contained by the main structure. As a result, metal connections are only required to fix the PSBC wall panels to the structural frame of the building. The structural frame requires a separate layer of thermal insulation materials in the in-fill PSBC walling façade.

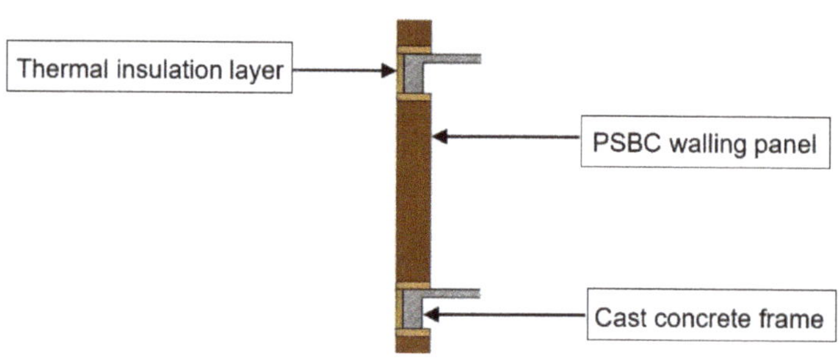

Figure 4. Construction of a Prefabricated Straw Bale Construction (PSBC) walling façade with in-fill PSBC panels.

The detail designs between the PSBC wall panels and the structural frame can be referenced from standard designs for metal fixing in 16J110-2/16G333. One detail design for metal fixing is shown in Figure 5. The metal fixings include three components, which are the part embedded in the concrete frame, the part embedded in the prefabricated wall panels, and the connection between the prefabricated wall panels and the cast concrete frame. In consideration of infiltration issues and the fire resistance requirements for the prefabricated wall panels, fire resistance materials are applied between the prefabricated wall panels.

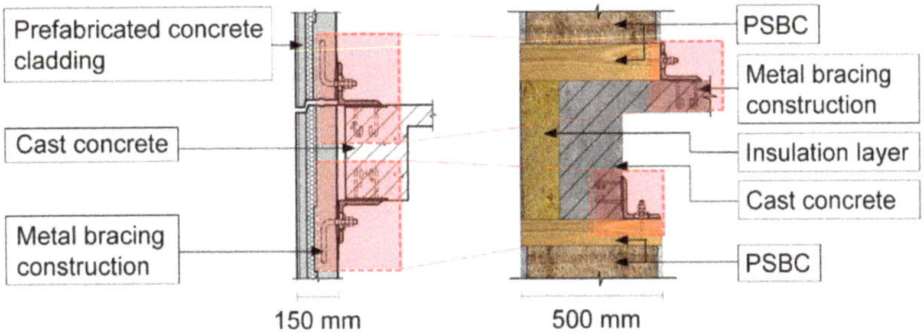

Figure 5. Detailed design of PSBC walling in reference building (right) from the 16J110-2/16G333 (left) [49].

The construction of the in-fill PSBC wall panels is different from that of the prefabricated concrete wall panels. As a result, the detailed connections between the cast concrete frame and the PSBC wall panels are modified from the standard metal fixing designs in the 08SJ110-2/08SG333. Since the PSBC panels are in-filled in the space between the concrete frames, there is no need for fire resistance sealing between these walling panels. The in-fill construction of the PSBC wall façade requires a thermal insulation layer on the external surface of the concrete frames. The U-value of the thermal insulation layer can be referenced from the design U-value of the PSBC walling panels to avoid thermal bridging through the concrete frames.

There are two locations for the metal fixing between the PSBC wall panels and the structural frames. The metal fixing should be applied both between the top plate of the structural frames for the PSBC wall panels and the cast concrete frame, and between the base plate of the structural frames for the PSBC wall panels and the cast concrete frame. The embedded parts for metal fixing of the wall panels are different to accommodate the different construction processes for prefabricated concrete wall panels and PSBC wall panels. The embedded parts for the PSBC wall panels are screws rather than the rebar used in prefabricated concrete wall panels. The screws have an M30 thread and are 80 mm long [49].

As per the design for the PSBC wall construction, there are two potential thermal bridging concerns. The first is the linear bridging between the PSBC wall panels and the thermal insulation layer. The second is the timber frame for the PSBC wall panels. The linear bridging issues can be solved through proper sealing between the insulation layers and the PSBC wall panels during installation of the wall panels. For the timber frame bridging, due to the relatively low thermal conductivity of timber, the bridging may not be significant.

2.3. IESVE Simulation Process

The simulation process is based on the IESVE-2018, which is widely used in accessing the energy efficiency levels of building projects for research purposes. There are three key elements in the modeling process in this research: The choosing of simulated cities, construction details of reference buildings, and the occupancy behaviors. Each element is discussed as follows.

The reference buildings will be examined in 15 major cities across the severe cold regions and the cold regions (Table 1). In the simulation process, each sub-region contains three examined cities to provide more accurate estimation of the energy requirements of reference buildings in the region. For the same reason, the examined cities are also selected across both longitude and latitude of the severe cold regions and the cold regions in the simulation process.

Table 1. Examined cities of the IESVE-based simulation of reference buildings.

15 Study Cities by Climate Zones					
Climatic Regions	**Sub-Regions**	**Index of Heating Degree Day (HDD) and Cooling Degree Day (CDD) (JGJ26-2010)**	**City**	**Longitude**	**Latitude**
Severe Cold Regions	Region 1A	6000 ≤ HDD18	Mohe	122.53° E	52.97°N
			Hailar	119.70° E	49.25°N
			Nenjiang	125.23° E	49.17°N
	Region 1B	5000 ≤ HDD18 < 6000	Dunhua	128.20° E	43.37°N
			Qiqihar	123.92° E	47.24°N
			Harbin	126.57° E	45.93°N
	Region 1C	3800 ≤ HDD18 < 5000	Changchun	125.68° E	44.00°N
			Shenyang	123.52° E	41.73°N
			Wulumuqi	87.62° E	43.78°N
Cold Regions	Region 2A	2000 ≤ HDD18 < 3800, CDD26 ≤ 90	Taiyuan	112.63° E	37.75° N
			Lanzhou	103.88° E	36.05°N
			Dalian	121.54° E	38.97°N
	Region 2B	2000 ≤ HDD18 < 3800, CDD26 > 90	Beijing	116.59° E	40.08°N
			Shijiazhuang	114.35° E	38.07°N
			Ji'nan	116.98°E	36.68°N

In the simulation process, thermal properties of the building materials are referenced from the IESVE-2018 database. Due to the lack of the data of straw bales in the software, the data are from existing researches. Considering the property differences of straw bales in different researches [40,50,51], the thermal conductivity of straw bale is designated as 0.700 W/mK. Existing researches confirmed the high thermal mass of straw bale walls [29,52] with specific thermal heat capacity ranging from 1388–2000 J/(kgK) [53,54]. As a result, the heat capacity of the PSBC is designated as 2000 J/(kgK) in the simulation process.

Detailed constructions and the requirements of design U-values of the building components are referenced from the standard JGJ-26-2010 [17]. According to the standard, the buildings require highly demanding U-value designs in the climatic regions that feature high HDD and vice versa [17]. Considering the practices of residential buildings in the climatic regions, detailed U-values of reference buildings are listed in the Table 2. There are two walling constructions of the reference building with PSBC walls: The original-thickness construction (OTC) and the reduced-thickness construction (RTC). The OTC features a walling thickness which is similar to that of the existing PSBC, whereas the RTC has a similar walling thickness to that of the typical walling constructions required in the JGJ26-2010.

There are two major aspects in deciding the occupancy factors: The heating and cooling modes and the residential behaviors. As the city levels of district heating systems have been widely used in the climatic regions since the 1950s, the heating periods of the reference buildings are subject to the heating requirements in each city. The heating periods and cooling periods of different cities are developed from the HDD index introduced in the JGJ-26-2010 (Table 3). The cooling requirements in the researched climatic regions are mainly achieved by using separate air conditioning units on a flat basis; there are no fixed cooling periods in this research. The cooling requirements are set 26 °C outside the heating periods in the flats of the reference buildings [17]. For the residential behaviors, the residential periods would be mainly based on the individual life styles of the people living in the flats; this research provides estimations of residential behaviors of working people on average (Table 4). The appliances are considered to be the most essential ones in flats.

Table 2. U-values of reference buildings in different climatic regions.

Climatic Regions	Sub-Regions	U-Value (W/(m²·K))			
		Building Construction	Simulated Standard Construction	PSBC (OTC)	PSBC (RTC)
Severe Cold	Region 1A	Roof	0.199	0.199	0.199
		Wall	0.242	0.150	0.168
		Window	1.600	1.600	1.600
		Door	1.436	1.436	1.436
	Region 1B	Roof	0.243	0.243	0.243
		Wall	0.436	0.150	0.221
		Window	1.600	1.600	1.600
		Door	1.436	1.436	1.436
	Region 1C	Roof	0.243	0.243	0.243
		Wall	0.436	0.150	0.221
		Window	1.600	1.600	1.600
		Door	1.436	1.436	1.436
Cold	Region 2A	Roof	0.389	0.389	0.389
		Wall	0.573	0.150	0.245
		Window	1.600	1.600	1.600
		Door	1.692	1.692	1.692
	Region 2B	Roof	0.389	0.389	0.389
		Wall	0.573	0.150	0.245
		Window	1.600	1.600	1.600
		Door	1.692	1.692	1.692

Table 3. Settings of heating periods in the IESVE-2018.

Sub-Climatic Regions	Region 1A	Region 1B	Region 1C	Region 2A	Region 2B
Heating period	1 October–30 April	10 October–10 April	20 October–31 March	10 November–20 March	15 November–15 March
Heating temperature	20 °C in flat (required 18 °C in JGJ 26-2010) and 12 °C in corridors (required 12 °C in JGJ 26-2010)				

Table 4. Occupancy factors in the IESVE-2018 simulation in this research.

		Bedroom	Living room	Bathroom	Kitchen
Occupied period	Weekday	24:00–7:00	7:00–9:00 and 17:00–24:00	7:00–7:30 and 22:00–22:30	Randomly 1 h between 17:00–19:00
	Weekend	24:00–9:00	0:00–24:00	Randomly 1 h between 0:00–24:00	Randomly 1 h between 0:00–24:00
Lighting		60 W	180 W	40 W	40 W
Occupancy density		2 persons/room	2 persons	2 persons	2 persons
Equipment		300 W	200 W	2000 W	2600 W

3. Research Results and Discussions

The energy-saving potentials of PSBC are estimated through the heating energy requirements of the flats and the heating and cooling energy loads in the reference buildings. The heating energy intensities of residential buildings are standardized in the JGJ-26-2010. To compare the energy intensity of the standard constructed walls and the PSBC walls, the heating energy loads are examined in the simulation processes of the reference buildings with the standard construction. The heating energy demand and the cooling energy demand are based on the results of energy consumption in the simulation processes of the reference buildings across climatic regions in this research. Due to the uncertainty of energy efficiency of the heating and cooling systems in different cities, the energy efficiency levels are set at 100% [55]. The following estimations of energy demand are based on the energy requirements of the reference buildings with the standard construction and the PSBC.

3.1. Heating Load

Detailed heating loads of reference buildings are shown in Table 5. The heating loads of the reference buildings with the standard construction are notably higher than those of the standard requirements. The higher heating loads of the reference buildings would be the result of the higher indoor temperature set-up than the standard requirements. The indoor temperature of the reference buildings is set at 20 °C during the simulation process, whereas the standard (JGJ-26 2010) requires 18 °C for the flat spaces. As the 20 °C indoor temperature is widely applied in existing practices in northern China [56–58], the standard heating load may underestimate the heating load in practices in Mohe, Harbin, and Changchun (Table 5). In the three cities, the gaps of heating load are between 2–4 W/m^2 in the simulation results of reference buildings with standard constructions. Higher heating loads in real practices in these cities may also confirm the gaps of heating demands [59].

Table 5. Heating loads in severe cold regions and cold regions in China.

Climatic Area	Cities	Heating loads (W/m^2)											
		JGJ26-2010 Requirements			Simulated Standard Construction			PSBC (OTC)			PSBC (RTC)		
		6F	12F	18F	6F	12F	18F	6F	12F	18F	6F	12F	18F
Region 1A	Mohe	23.1	20.9	20.6	24.96	23.11	22.71	23.5	21.6	21.2	23.9	21.9	21.51
	Hailar	20.9	18.9	18.8	20.8	19.2	18.90	19.4	17.9	17.6	19.8	18.1	17.8
	Nenjiang	20.7	18.6	18.5	19.8	18.3	18.02	18.5	17.1	16.8	18.9	17.3	17.0
Region 1B	Dunhua	18.0	16.5	15.2	21.6	19.3	18.55	18.0	15.6	14.9	19.0	16.6	16.0
	Qiqihar	19.8	18.1	16.7	21.3	19.0	18.20	17.7	15.3	14.6	18.7	16.3	15.7
	Harbin	20.0	18.3	16.9	21.0	18.7	18.01	17.4	15.1	14.4	18.4	16.0	15.5
Region 1C	Chang chun	19.9	18.6	16.3	22.9	20.5	19.97	19.2	16.7	16.2	20.1	17.7	17.1
	Shen yang	17.2	15.9	13.9	16.9	14.9	14.44	13.8	11.8	11.3	14.6	12.6	12.1
	Wulu muqi	18.7	17.4	15.4	19.8	17.5	16.93	16.4	14.1	13.5	17.2	15.0	14.4
Region 2A	Taiyuan	15.4	14.1	12.5	14.4	12.5	11.99	10.6	8.6	8.1	11.5	9.5	9.0
	Lan zhou	14.4	13.1	11.7	14.6	12.6	12.04	10.8	8.8	8.2	11.7	9.7	9.1
	Dalian	14.3	13.0	11.5	16.8	15.0	14.58	12.6	10.7	10.3	13.6	11.7	11.3
Region 2B	Beijing	15.0	13.4	12.1	14.1	12.3	11.77	10.3	8.4	7.9	11.1	9.2	8.8
	Shijia zhuang	14.6	13.1	11.6	13.6	11.8	11.34	10.0	8.2	7.7	10.8	9.0	8.5
	Ji'nan	13.2	11.7	10.5	11.6	10.0	9.62	8.3	6.7	6.3	9.0	7.4	7.0

□ 5.00–9.99 W/m^2 □ 10.00–14.99 W/m^2 ■ 15.00–19.99 W/m^2 ■ 20.00–24.99 W/m^2.

The simulation results show that the heating loads of the simulated standard constructions are from −14.4% to 26.2% in comparison to the data from the JGJ-26-2010 (Table 6). The simulation results indicate that the standard data overestimate the heating load by around 10% across the cities in the climatic region 2B. The heating load is subjective to the situation of each city in other climatic regions. The heating loads of standard requirements overestimate the heating loads of reference buildings in Taiyuan, Shenyang, and Nenjiang. However, the simulation results of heating loads are higher than those of the standard requirements in other cities in the climatic regions 1A, 1B, 1C, and 2A. Compared to the simulation results, the underestimations of heating loads are more than 15% in Dunhua, Changchun, and Dalian.

Table 6. Comparison of heating loads of simulated buildings with standard construction.

Climatic Area	Cities	Comparison of Heating Requirements with Standard Requirements in JGJ26-2010 (%)								
		Simulated Standard Construction			PSBC (OTC)			PSBC (RTC)		
		6F	12F	18F	6F	12F	18F	6F	12F	18F
Region 1A	Mohe	+8.1	+10.6	+10.3	+1.6	+3.5	+3.1	+3.5	+4.8	+4.4
	Hailar	−0.7	+1.6	+0.5	−7.1	−5.4	−6.5	−5.2	−4.1	−5.1
	Nenjiang	−4.3	−1.5	−2.6	−10.4	−8.3	−9.4	−8.5	−7.0	−8.1
Region 1B	Dunhua	+20.2	+17.1	+22.0	+0.2	−5.6	−2.0	+5.7	+0.4	+5.1
	Qiqihar	+7.6	+4.8	+9.0	−10.4	−15.5	−12.6	−5.5	−10.2	−6.2
	Harbin	+4.9	+2.3	+6.6	−13.0	−17.8	−14.7	−8.0	−12.4	−8.3
Region 1C	Chang chun	+15.1	+10.3	+22.5	−3.7	−10.1	−0.7	+0.9	−5.1	+5.0
	Shen yang	−1.5	−6.1	+3.9	−19.6	−25.9	−18.7	−15.1	−20.9	−13.1
	Wulu muqi	+5.7	+0.7	+9.9	−12.3	−18.9	−12.1	−7.8	−14.0	−6.6
Region 2A	Tai yuan	−6.2	−11.2	−4.1	−31.2	−38.9	−35.2	−25.6	−32.6	−28.1
	Lan zhou	+1.3	−3.9	+2.9	−24.7	−32.9	−29.6	−18.8	−26.2	−22.1
	Dalian	+17.4	+15.4	+26.8	−11.9	−17.7	−10.6	−5.3	−10.1	−2.0
Region 2B	Beijing	−6.2	−8.6	−2.7	−31.6	−37.5	−34.7	−26.1	−31.1	−27.7
	Shijia zhuang	−6.6	−9.6	−2.2	−31.3	−37.7	33.9	−25.9	−31.4	−26.8
	Ji'nan	−12.5	−14.4	−8.4	−37.4	−43.1	−40.2	−31.9	−36.6	−33.1

□ < −20% □ −10% ~ −20% □ −9.9% ~ −0.1% □ 0% ~ +9.9% □ +10% ~ +20% □ > +20%.

Due to relatively low standards of thermal insulative properties of walling constructions in the cold regions, the PSBC has more reductions of heating load than the ones in the severe cold regions. In the region of 2B, the two types of PSBC are estimated to lower the heating loads of reference buildings by more than a quarter compared to the reference buildings with standard constructions. As the PSBC (RTC) has a similar walling thickness to that of the standard constructions, the applications of the PSBC will significantly lower the heating loads of the heating systems and maintain similar levels of indoor space to the reference buildings with standard constructions.

The applications of the two types of PSBC have significant impact on reducing the heating loads of reference buildings in the severe cold regions as well. As the setups of the indoor temperatures are higher than the standard requirements in JGJ-26-2010, the existing heating systems may not meet the heating loads of reference buildings with standard constructions in the severe cold regions. The applications of the two types of PSBC lower the heating loads significantly, and the heating loads meet the existing standards in the buildings in the severe cold regions (Table 6). In Dalian, the heating loads of reference buildings are 14.58–16.78 W/m^2 with standard constructions, whereas the reference buildings require a 10.28–13.55W/m^2 heating load with the two types of PSBC. Compared to the standard heating load in JGJ26-2010, the standard construction fails to meet the standard requirements, whereas the reference buildings reduce the heating load by 2–17% by applying PSBC walls.

Comparing the two types of PSBC, the PSBC (OTC) has advantages in reducing the heating loads of reference buildings, whereas the PSBC (RTC) features maintain similar levels of heating load to those of the PSBC (OTC) with the same internal space as the existing residential buildings. The PSBC (OTC) has 5–8% lower heating load than the PSBC (RTC). Due to the high heating load in the severe cold regions, the advantages of the PSBC (OTC) have a more significant impact on the residential buildings in the regions. In the Region 1C, the heating loads of PSBC (OTC) are 1–1.5 W/m^2 lower than those of the reference buildings with PSBC (OTC). In the cold regions, due to the relatively low requirements of thermal insulative properties of walling constructions in the JGJ26-2010, the standard walling thicknesses are between 200–300 mm, which are notably lower than those for the PSBC (OTC). Compared to the two types of PSBC, the heating loads of the PSBC (RTC) are 0.8–1 W/m^2 more than those of the reference buildings with the PSBC (OTC). Considering the relatively low thickness of the PSBC (RTC), this type of PSBC would be suitable for further construction of residential buildings in the cold regions.

3.2. Heating and Cooling Energy Requirements

The heating and cooling energy demands of the reference buildings across different climatic regions are shown in Figures 6 and 7. Due to the low winter temperatures in the region 1A, the peak energy demands are more than 105 kWh/m^2/year in the six-floor reference buildings, and the energy requirements are around 100 kWh/m^2/year in the reference buildings in the region. Due to the relatively low cooling energy requirements in the Region 2A, the energy demands of the reference buildings are between 48 and 57 kWh/m^2/year, which are the lowest energy demands compared to the ones in all researched climatic regions. The heating and cooling energy demands of the reference buildings are significantly different between the severe cold regions and the cold regions. As the residential periods exclude the daytime (9:00–17:00) in this research, the cooling energy demands exclude the cooling energy requirements to offset the potential overheating periods in the afternoon (12:00–17:00) in the simulation processes. In the severe cold regions, the cooling energy demands contribute less than 5% of the total energy demands and are barely ignorable, whereas the cooling energy demands are much more significant in the cold regions. The features of energy requirements are comparable to the HDD and CDD index in the JGJ26-2010.

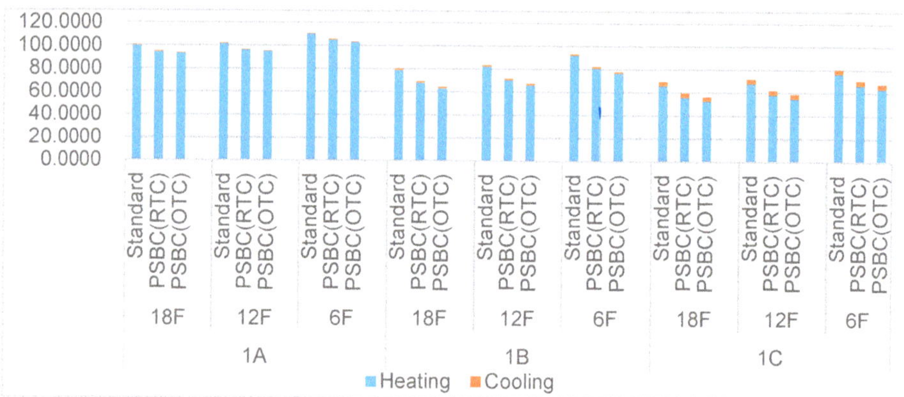

Figure 6. Heating and cooling energy demands (kWh/m^2/year) of reference buildings in severe cold regions.

In the severe cold regions, due to the high demands of heating energy, the high insulative properties of the two types of PSBC have notable advantages compared to the standard walling constructions in reducing energy demand in reference buildings with different building heights (Figure 6). As high thermal insulative properties of building construction are required in the JGJ26-2010, the advantages of the PSBC are not significant in the region 1A. Reductions of heating and cooling energy demands achieved by the PSBC are between 5 and 10 kWh/m^2/year. However, compared to the standard constructions, the two types of PSBC reduce the energy demands by more than 10 kWh/m^2/year in the region 1B and the region 1C. Due to the high energy-saving potentials of the PSBC in the regions 1B and 1C, replacing existing walling constructions with the PSBC would achieve a more significant impact on energy efficiency levels of residential buildings in the severe cold regions.

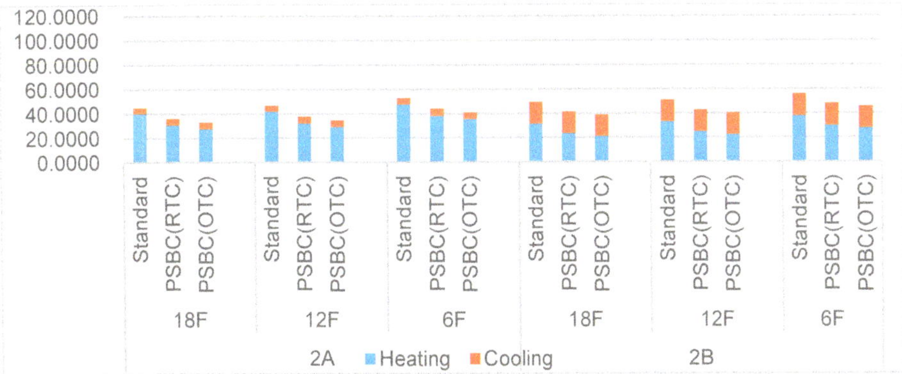

Figure 7. Heating and cooling energy demands (kWh/m^2/year) of reference buildings in cold regions.

As shown in Figure 7, reductions of cooling energy demands are the major concerns for improving the energy efficiency levels of the reference buildings in the cold regions. Compared to the heating and cooling energy demands of the reference buildings in the cold regions, the reference buildings have higher requirements of heating energy demands in the region 2A than the ones in the region 2B. However, due to the significantly higher cooling energy demands of the reference buildings in the region 2B, the overall energy demands of the reference buildings are 4–6 kWh/m^2/year more than those of the reference buildings in the region 2A. In the region 2B, the cooling energy demands are of 33–46% of the total energy demands of the reference buildings with different floor heights. Compared to the energy demands of the standard construction, the PSBC walls have notable advantages in reducing the cooling energy demands of the reference buildings. The cooling energy demands are 17.6–18.3 kWh/m2/year in the reference buildings with different heights in the climatic region. The energy demands of the PSBC (OTC) and the PSBC (RTC) are from 10.4 to 11.1 kWh/m^2/year lower than those of the standard constructions, respectively. As a result, the applications of PSBC will lead to lower energy consumptions of residential buildings by reducing both the heating energy demands and the cooling energy demands in the climatic regions.

Table 7 shows the potential energy saving of reference buildings with the two types of PSBC. Due to the higher walling thickness of the PSBC (OTC) than the PSBC (RTC), the thermal properties of the PSBC (OTC) have advantages in preserving heat inside the reference buildings, and lead to energy demands of the buildings that are lower than those of the PSBC (RTC) in the researched regions. Despite the situation in the region 1A, the PSBC (OTC) has 4–6% reductions of heating and cooling energy demands compared to the PSBC (RTC). The features of the PSBC (OTC) would be suitable for residential buildings that have insignificant requirements for walling thickness, including low-rise residential buildings, terrace-houses, and semi-detached houses. Whereas as the reduced walling thickness of the PSBC (RTC) is at a level similar to that of the existing constructions, the PSBC (RTC) would be more suitable for refurbishment of existing buildings and the buildings. The PSBC (RTC) would be suitable for the construction of new residential buildings in the cold regions as well.

The simulation results (Table 7) indicate that the reference buildings would save more heating and cooling energy by applying the PSBC in the higher-floor ones compared to the low-height ones in the researched regions. Comparing the 18 floor reference buildings to the six-floor reference buildings, the reductions of energy are from 2.9% to 4.6% for the PSBC (OTC) and are from 1.8% to 3.5% for the PSBC (RTC) in the researched regions, excluding the region 1A. As more concentrated surface–volume ratios are in the reference buildings with higher floors, heat losses through walling constructions are notably lower than the ones in the low-rise residential buildings. As a result, the greater potential for reduction of energy consumption would be in constructing and retrofitting of the high-rise residential buildings in northern China with the two types of PSBC.

Table 7. Reduction of heating and cooling energy consumptions achieved by replacing the standard construction with PSBC.

Climatic Area	Cities	Reduction of Heating and Cooling Energy Consumptions					
		PSBC (OTC)			PSBC (RTC)		
		6F	12F	18F	6F	12F	18F
Region 1A	Mohe	6.0%	6.4%	6.5%	4.2%	5.2%	5.3%
	Hailar	6.4%	6.8%	6.8%	4.4%	4.0%	5.5%
	Nenjiang	6.3%	6.8%	6.8%	4.3%	5.4%	5.5%
Region 1B	Dunhua	16.7%	19.3%	19.7%	12.1%	14.2%	13.9%
	Qiqihar	16.1%	18.7%	19.1%	11.6%	13.7%	13.3%
	Harbin	16.6%	19.2%	19.5%	11.9%	14.0%	13.6%
Region 1C	Changchun	15.8%	17.8%	18.3%	11.8%	13.3%	13.7%
	Shenyang	16.9%	19.3%	19.9%	12.6%	14.4%	14.8%
	Wulumuqi	15.4%	17.5%	18.0%	11.4%	12.9%	13.3%
Region 2A	Taiyuan	22.6%	26.1%	27.1%	16.9%	19.5%	20.2%
	Lanzhou	22.7%	26.3%	27.3%	16.8%	19.5%	20.3%
	Dalian	22.3%	25.5%	26.3%	17.0%	19.4%	20.0%
Region 2B	Beijing	19.1%	21.5%	22.1%	14.6%	16.4%	16.8%
	Shijiazhuang	18.3%	20.6%	21.1%	13.9%	15.5%	15.9%
	Ji'nan	18.4%	20.5%	21.1%	13.9%	15.5%	15.9%

□ 0% ~ 10% □ 10.1% ~ 20% □ 20.1% ~30%.

Due to the high thermal insulation demands of walling constructions for the region 1A in the JGJ26-2010, the reductions of energy loads are not significant compared to the ones in other climatic regions. Compared to the standard walling constructions, the heating reductions are 4.2%–6.5% in the reference buildings in the region 1A. The PSBC constructions would be less efficient for replacing the walling constructions in the region compared to applying the PSBC in other climatic regions in northern China.

4. Discussion on Applicability of PSBC

4.1. Environmental Impact of PSBC

The low environmental impact of the straw bales is the capital advantage of using PSBC in building industries. There are three stages, including the materialization stage, the operational stage, and the end-of-life stage, to assess the life cycle assessment (LCA) of buildings [60]. Compared to conventional building materials, the materialization of straw bales has much lower carbon emissions and has notably less environmental impact during the end-of-life stage.

The single research relating embodied carbon of straw bale is the Inventory of Carbon Emission (ICE) done in the University of Bath, and the carbon emission of straw bales is estimated to be 0.01 $kgCO_2/kg$ [61]. However, the CO_2 emission of straw bales is based on the data from the wheat straw in New Zealand [61]. As the growing process of rice produces methane, the carbon emission of rice straw would be notably higher than the one of wheat straw [62]. Due to the limited research on carbon emissions of rice straw and wheat straw regarding the conditions in China, the overall carbon emissions of PSBC walls may not be assessed in detail. However, considering that the straw is a by-product of agricultural activities, the carbon emissions of straw would be significantly lower than those of other building materials [29].

As a bio-based building material, straw will be degradable after the end of the life of PSBC walls. Compared to conventional fossil-fuel-based building materials, straw has a significantly lower degradation period. With the presence of favorable environments of degradation, straw would degrade within four weeks [63]. Due to the lack of a recycling scheme for building materials, existing research did not evaluate the end-of-life stage in the LCA for building industries [55,64]. The advantage of the low end-of-life impact of straw bales may have an insignificant impact on building projects in the short term.

4.2. Hygrothermal Environment and Durability

Due to the high potential of degradation of straw, the designs of controlling hygrothermal environments of straw bales are critical for the durability of PSBC walls. The designs of controlling are subject to the local climatic features of the construction site. Rainscreens have positive impacts on the durability of straw bale walls in the UK [41], whereas similar walling constructions have direct opposite effects on the cases in Japan due to the high temperatures and high humidity in summer months [65]. Existing research on walling constructions of straw bale buildings regarding climatic conditions in the severe cold region of China shows that the lime render maintains enough breathability and favorable hygrothermal conditions for the straw inside straw bale walls [66]. As there is no research on the effects of rainscreens on hygrothermal environment inside straw bale walls regarding the local climatic features, applications of rainscreens on the PSBC walling constructions need further research.

4.3. Material Availability

Both rice production and wheat production are vast in both severe cold regions and cold regions. The rice production contributes more than 15% of total domestic rice production in the three provinces (Heilongjiang, Jilin, and Liaoning) of northeast China [67], and the wheat production is more than 50% in the three provinces (Hebei, Shandong, and Henan) in the cold regions [68]. Both the rice production and the wheat production will keep growing in the future [69] and guarantee the availability of the raw materials of the PSBC walling construction. However, as there are no applications of straw bale walls in existing building projects, the issues relating to logistics, quality control, and prices remain uncertain. As a result, despite the vast availability of the raw materials, the application of PSBC walling construction still needs to be developed.

5. Conclusions

This research provides potential application methods regarding the standards in China for PSBC in multiple-floor and high-rise residential buildings, following an analysis that evaluates the energy-saving potential of the PSBC in the building types across northern China. Overall, this research has extended the understanding of the feasibility of using PSBC and the energy performance of residential buildings to which PSBC is applied, and which are subject to the climatic features of northern China.

The research results indicate that the energy efficiency levels of residential buildings can be significantly improved by applying PSBC in all climatic regions in northern China. In the residential buildings that apply PSBC, heating energy loads are notably lower than in the residential buildings with existing walling constructions in northern China. Compared to the heating loads in reference buildings with standard constructions, the heating loads can be reduced by 2–4 W/m^2 with application of PSBC walling across different cities, and the reductions of heating loads are between 4.8% and 26.1% across different sub-climatic regions in the severe cold and cold regions. The simulation results show that more significant energy saving is achieved by the PSBC by applying the PSBC in high-rise residential buildings rather than the multiple-floor residential buildings. The energy-saving ratios achieved by applying PSBC walling are from 2% to 4% more in the high-rise buildings than in the multiple-floor buildings. Due to the high requirements for thermal insulation properties of existing walling constructions in the existing standards for the region 1A, the advantages of the PSBC are not as significant as the ones in the other climatic regions in northern China. The energy saving achieved by PSBC walling is from 4.2% to 6.5% in the region 1A whereas, the energy reductions are expected to be 11.9% to 27.1% in other regions across the severe cold and cold regions.

Author Contributions: Conceptualization, X.Y. and C.S.; methodology, S.Z.; software, S.Z.; validation, Q.D., J.Y.; formal analysis, X.Y.; investigation, L.H.; resources, Q.D.; data curation, J.Y.; writing—original draft preparation, X.Y.; writing—review and editing, X.Y.; visualization, L.H. and J.Y.; supervision, C.S.; project administration, Q.D.; funding acquisition, C.S. All authors have read and agreed to the published version of the manuscript.

Funding: This research was funded by National Natural Science Foundation of China, grant number 51938003.

Conflicts of Interest: The authors declare no conflict of interest.

References

1. Lin, B.; Liu, X. China's carbon dioxide emissions under the urbanization process: Influence factors and abatement policies. *Econ. Res. J.* **2010**, *2010*, 66–78.
2. Chen, S.; Li, N.; Guan, J.; Xie, Y.; Sun, F.; Ni, J. A statistical method to investigate national energy consumption in the residential building sector of China. *Energy Build.* **2008**, *40*, 654–665. [CrossRef]
3. Hu, S.; Yan, D.; Guo, S.; Cui, Y.; Dong, B. A survey on energy consumption and energy usage behavior of households and residential building in urban China. *Energy Build.* **2017**, *148*, 366–378. [CrossRef]
4. Yuan, T.; Zhu, N.; Shi, Y.; Chang, C.; Yang, K.; Ding, Y. Sample data selection method for improving the prediction accuracy of the heating energy consumption. *Energy Build.* **2018**, *158*, 234–243. [CrossRef]
5. Diao, L.; Sun, Y.; Chen, Z.; Chen, J. Modeling energy consumption in residential buildings: A bottom-up analysis based on occupant behavior pattern clustering and stochastic simulation. *Energy Build.* **2017**, *147*, 47–66. [CrossRef]
6. Yao, R.; Li, B.; Steemers, K. Energy policy and standard for built environment in China. *Renew. Energy* **2005**, *30*, 1973–1988. [CrossRef]
7. Han, K.K.; Golparvar-Fard, M. Potential of big visual data and building information modeling for construction performance analytics: An exploratory study. *Autom. Constr.* **2017**, *73*, 184–198. [CrossRef]
8. Liang, J.Q. Ministry of Housing and Urban-Rural Development. In *China Building Industry Energy Saving Report*; China Architecture & Building Press: Beijing, China, 2014. (In Chinese)
9. Yu, S.; Eom, J.; Zhou, Y.; Evans, M.; Clarke, L. Scenarios of building energy demand for China with a detailed regional representation. *Energy* **2014**, *67*, 284–297. [CrossRef]
10. Chang, C.; Zhu, N.; Yang, K.; Yang, F. Data and analytics for heating energy consumption of residential buildings: The case of a severe cold climate region of China. *Energy Build.* **2018**, *172*, 104–115. [CrossRef]
11. Marszal, A.J.; Heiselberg, P.; Bourrelle, J.S.; Musall, E.; Karsten, V.; Sartori, I.; Napolitano, A. Zero energy building—A review of definitions and calculation methodologies. *Energy Build.* **2011**, *43*, 971–979. [CrossRef]
12. Sartori, I.; Napolitano, A.; Voss, K. Net zero energy buildings: A consistent definition framework. *Energy Build.* **2012**, *48*, 220–232. [CrossRef]
13. Brambilla, A.; Salvalai, G.; Imperadori, M.; Sesana, M.M. Nearly zero energy building renovation: From energy efficiency to environmental efficiency, a pilot case study. *Energy Build.* **2018**, *166*, 271–283. [CrossRef]
14. Sesana, M.M.; Salvalai, G. Overview on life cycle methodologies and economic feasibility for nZEBs. *Build. Environ.* **2013**, *67*, 211–216. [CrossRef]
15. Ministry of Housing and Rural-Urban Development. *Standard of Climatic Regionalization for Architecture*; GB50178-93; Ministry of Housing and Urban-Rural Development: Beijing, China, 1994.
16. Amecke, H. *Buildings Energy Efficiency in China, Germany, and the United States*; Climate Policy Initiative: San Francisco, CA, USA, 2013.
17. Ministry of Housing and Urban-Rural Development. *Design Standard for Energy Efficiency of Residential Buildings in Severe Cold and Cold Zones*; JGJ26-2010; Ministry of Housing and Urban-Rural Development: Beijing, China, 2010; p. 36.
18. Ministry of Housing and Urban-Rural Development. *Design Standard for Energy Efficiency of Public Buildings*; GB50189-2015; Ministry of Housing and Urban-Rural Development: Beijing, China, 2016.
19. King, B. *Design of Straw Bale Buildings: The State of the Art*, 2nd ed.; Aschheim, M., Ed.; Green Building: San Rafael, CA, USA, 2006.
20. Lacinski, P.; Bergeron, M. *Serious Straw Bale: A Home Construction Guide for All Climates*; Chelsea Green Publishing Company: White River Junction, VT, USA, 2000.
21. Steen, A.S.; Steen, B.; Bainbridge, D. *The Straw Bale House*; Chelsea Green Publishing: White River Junction, VT, USA, 1994.

22. ADRA. *What is a Straw Bales Building?* ADRA: Silver Spring, MD, USA, 2006.
23. Gao, X.S. ADRA (Adventist Development Relief Agency). 2008. Available online: http://www.chinadevelopmentbrief.org.cn/news-13062.html (accessed on 27 July 2016).
24. Cao, B.Z.; Yuan, B.; Duan, W.F.; Li, J.; Wen, M. Research and design on load bearing wall with green energy-saving straw bale. In *Applied Mechanics and Materials*; Trans Tech Publications Ltd.: Zurich, Switzerland, 2014.
25. Menet, J.-L.; Gruescu, I.-C. A comparative life cycle assessment of exterior walls constructed using natural insulation materials. *Environ. Eng. Sustain. Dev. Entrep.* **2012**, *1*, 14.
26. Pritchard, M.B.; Pitts, A. Evaluation of strawbale building: Benefits and risks. *Archit. Sci. Rev.* **2006**, *49*, 372–384. [CrossRef]
27. MacDougall, C. Natural building materials in mainstream construction: Lessons from the UK. *J. Green Build.* **2008**, *3*, 3–14. [CrossRef]
28. Goodhew, S.; Carfrae, J.; De Wilde, P. Briefing: Challenges related to straw bale construction. In *Proceedings of the ICE—Engineering Sustainability*; ICE Publishing: London, UK, 2010; pp. 185–189.
29. Sodagar, B.; Rai, D.; Jones, B.; Wihan, J.; Fieldson, R. The carbon-reduction potential of straw-bale housing. *Build. Res. Inf.* **2011**, *39*, 51–65. [CrossRef]
30. Bigland-Pritchard, M. *An Assessment of the Viability of Strawbale Wall Construction in Buildings in Maritime Temperate Climates, in School of Architecture*; University of Sheffield: Sheffield, UK, 2005.
31. Modcell. Technical. 2016. Available online: http://www.modcell.com/technical/ (accessed on 13 December 2016).
32. Florig, H.K. Peer reviewed: China's air pollution risks. *Environ. Sci. Technol.* **1997**, *31*, 274–279. [CrossRef]
33. He, K.; Yang, F.; Ma, Y.; Zhang, Q.; Yao, X.; Chan, C.K.; Cadle, S.; Chan, T.; Mulawa, P. The characteristics of PM 2.5 in Beijing, China. *Atmos. Environ.* **2001**, *35*, 4959–4970. [CrossRef]
34. Mestl, H.; Aunan, K.; Seip, J.M.; Wang, S.; Zhao, Y.; Zhang, D. Urban and rural exposure to indoor air pollution from domestic biomass and coal burning across China. *Sci. Total Environ.* **2007**, *377*, 12–26. [CrossRef]
35. Wang, J.; Zhang, X. Analysis on residential energy conservation for straw-bale building. *J. Build. Mater.* **2005**, *8*, 109–112.
36. Grain, S.A.O. *China Grain Development Report 2016*; China Social Science Press: Beijing, China, 2016.
37. Li, L.; Wang, K.; Zhang, Q.; Li, J.; Yang, X.; Jin, J. Wheat straw burning and its associated impacts on Beijing air quality. *Sci. China Ser. D Earth Sci.* **2008**, *51*, 403–414. [CrossRef]
38. Craig, W.; Wall, K.; Gross, C.; Walker, P.; Mander, T. Development and testing of a prototype straw bale house. In *Proceedings of the ICE—Engineering Sustainability*; ICE Publishing: London, UK, 2012; Volume 165, pp. 377–384.
39. Maskell, D.; Gross, C.; Thomson, A.; Wall, K.; Walker, P.; Mander, T. Structural development and testing of a prototype house using timber and straw bales. In *Proceedings of the ICE—Engineering Sustainability*; ICE Publishing: London, UK, 2015; Volume 168, pp. 67–75.
40. Shea, A.; Wall, K.; Walker, P. Evaluation of the thermal performance of an innovative prefabricated natural plant fibre building system. *Build. Serv. Eng. Res. Technol.* **2013**, *34*, 369–380. [CrossRef]
41. Lawrence, M.; Heath, A.; Walker, P. The impact of external finishes on the weather resistance of straw bale walls. In Proceedings of the 11th International Conference on Non-conventional Materials and Technologies, Bath, UK, 6–9 September 2009.
42. Chatterton, P. Towards an agenda for post-carbon cities: Lessons from LILAC, the UK's first ecological, affordable cohousing community. *Int. J. Urban Reg. Res.* **2013**, *37*, 1654–1674. [CrossRef]
43. International Constructions. LILAC: Low Impact Living Affordable Community. 2014. Available online: http://www.construction21.org/case-studies/h/lilac-low-impact-living-affordable-community.html (accessed on 2 January 2016).
44. Clark, D. *What Colour Is Your Building?* RIBA: London, UK, 2013.
45. Ecococon. Modular Building. 2016. Available online: https://ecococon.eu/ (accessed on 23 April 2020).
46. Yao, S.; Luo, D.; Wang, J. Housing development and urbanisation in China. *World Econ.* **2014**, *37*, 481–500. [CrossRef]
47. Jiang, R.; Mao, C.; Hou, L.; Wu, C.; Tan, J. A SWOT analysis for promoting off-site construction under the backdrop of China's new urbanisation. *J. Clean. Prod.* **2018**, *173*, 225–234. [CrossRef]
48. Ministry of Housing and Urban-Rural Development. *Design Code for Residential Buildings*; GB50096-2011; Ministry of Housing and Urban-Rural Development: Beijing, China, 2011.

49. China Institute of Building Standard Design & Research Co., Ltd. *Standard for Prefabricated Concrete Cladding*; 16J110-2/16G333; China Planning Press: Beijing, China, 2016. (In Chinese)
50. Cascone, S.; Rapisarda, R.; Cascone, D. Physical properties of straw bales as a construction material: A review. *Sustainability* **2019**, *11*, 3388. [CrossRef]
51. Sabapathy, K.A.; Gedupudi, S. Straw bale based constructions: Measurement of effective thermal transport properties. *Constr. Build. Mater.* **2019**, *198*, 182–194. [CrossRef]
52. Garas, G.; Allam, M.; El Dessuky, R. Straw bale construction as an economic environmental building alternative—A case study. *J. Eng. Appl. Sci.* **2009**, *4*, 54–59.
53. Atkinson, C. *Energy Assessment of a Straw Bale Building*; University of East London: London, UK, 2008.
54. Chaussinand, A.; Scartezzini, J.L.; Nik, V. Straw bale: A waste from agriculture, a new construction material for sustainable buildings. *Energy Procedia* **2015**, *78*, 297–302. [CrossRef]
55. Dong, Y.; Cui, X.; Yin, X.; Chen, Y.; Guo, H. Assessment of energy saving potential by replacing conventional materials by Cross Laminated Timber (CLT)—A case study of office buildings in China. *Appl. Sci.* **2019**, *9*, 858. [CrossRef]
56. Ministry of Housing and Urban Rural Development & State Administration for Market Regulation. *Technical Standard for Near Zero Energy Building*; GB/T51350-2019; Ministry of Housing and Urban Rural Development & State Administration for Market Regulation: Beijing, China, 2019.
57. Zhang, W. 'Photovoltaic + Solar Heating' Make Herdman have Warm Life. 2019. Available online: http://www.stdaily.com/index/kejixinwen/2019-12/25/content_847204.shtml (accessed on 6 January 2020). (In Chinese).
58. Zhang, J.L. Spend Less for Warm Home. 2019. Available online: http://news.bjx.com.cn/html/20191231/1032798.shtml (accessed on 6 January 2020). (In Chinese).
59. Zhou, H.; Zhang, L.; Yu, S.; Cao, P.; Jia, B.; Li, M.; Yin, H. *A Study on Institutional Reform and Suggestions of Urban Thermal Supply System*; Department of Social Development Research of DRC, The State Council: Beijing, China, 2016. (In Chinese)
60. Cabeza, L.F.; Rincón, L.; Vilariño, V.; Pérez, G.; Castell, A. Life Cycle Assessment (LCA) and Life Cycle Energy Analysis (LCEA) of buildings and the building sector: A review. *Renew. Sustain. Energy Rev.* **2014**, *29*, 394–416. [CrossRef]
61. Hammond, G.; Craig, J. *Inventory of Carbon & Energy*; University of Bath: Bath, UK, 2008.
62. Van der Gon, H.A.C.D.; Neue, H.U. Influence of organic matter incorporation on the methane emission from a wetland rice field. *Glob. Biogeochem. Cycles* **1995**, *9*, 11–22. [CrossRef]
63. Lawrence, M.; Heath, A.; Walker, P. Determining moisture levels in straw bale construction. *Constr. Build. Mater.* **2009**, *23*, 2763–2768. [CrossRef]
64. Zhang, X.; Wang, F. Assessment of embodied carbon emissions for building construction in China: Comparative case studies using alternative methods. *Energy Build.* **2016**, *130*, 330–340. [CrossRef]
65. Holzhueter, K.; Itonaga, K. The influence of passive ventilation on the interstitial hygrothermal environment of a straw bale wall. *J. Asian Archit. Build. Eng.* **2014**, *13*, 223–229. [CrossRef]
66. Yin, X.; Lawrence, M.; Daniel, M.; Chang, W. Construction and monitoring of experimental straw bale building in northeast China. *Constr. Build. Mater.* **2018**, *183*, 46–57. [CrossRef]
67. USDA. China: Rice Production. 2018. Available online: https://ipad.fas.usda.gov/rssiws/al/crop_production_maps/China/China_rice.jpg (accessed on 18 January 2019).
68. USDA. China: Total Wheat Production. 2018. Available online: https://ipad.fas.usda.gov/rssiws/al/crop_production_maps/China/China_wheat.jpg (accessed on 18 January 2018).
69. Li, Y.; Zhang, W.; Ma, L.; Wu, L.; Shen, J.; Davies, W.; Oenema, O.; Zhang, F.; Dou, Z. An analysis of China's grain production: Looking back and looking forward. *Food Energy Secur.* **2014**, *3*, 19–32. [CrossRef]

© 2020 by the authors. Licensee MDPI, Basel, Switzerland. This article is an open access article distributed under the terms and conditions of the Creative Commons Attribution (CC BY) license (http://creativecommons.org/licenses/by/4.0/).

sustainability

MDPI

Article

Decrepit Building Monitoring Solution for Zero Energy Building Management Using PLC and Android Application

Jun-Ho Huh [1] and Jong Hyuk Park [2,*]

1 Department of Data Informatics, Korea Maritime and Ocean University, Busan 49112, Korea; 72networks@kmou.ac.kr

2 Department of Computer Science and Engineering, Seoul National University of Science and Technology (SeoulTech), 232 Gongneung-ro, Nowon-gu, Seoul 01811, Korea

* Correspondence: jhpark1@seoultech.ac.kr; Tel.: +82-2-970-6702

Received: 15 January 2020; Accepted: 29 February 2020; Published: 5 March 2020

Abstract: Since power line communication (PLC) technology has been considered a platform technology for smart metering, this study introduces an analogous system technology for the PLC-based monitoring systems in zero energy building management. This technology allows a system operator or an operation center to monitor, calculate, or manage power use remotely through the existing power line so that individual suppliers will be able to grasp the present condition of power use and respond to any unexpected incidents. As in any country, many old buildings in Korea are repaired to be used for 40–50 years or even longer since constructing new ones is expensive. Because remodeling old buildings is not only expensive but also generates construction waste, repairing is the preferred alternative in many cases. With the expectation that the proposed platform technology will be one of the promising technologies for sustainable building, its uninterrupted power usage monitoring and remote power control/management functions will be quite useful for economizing power with convenience. A system with such a PLC-based design and algorithm can be quite scalable as well. As the major contribution of this study, a solution suitable for decrepit buildings or apartments has been developed by using the PLC technology, and the test bed experiment was conducted for it. The result showed that the solution worked flexibly and efficiently. Since it is impossible to install a new network especially in the case of decrepit buildings, PLC must be used through the existing power line. Thus, a decrepit building monitoring system has been proposed in this study for a sustainable building. The proposed solution was simulated with OPNET simulation first. Following the mounting of the actual PLC product and development of a monitoring application to complete the platform/solution, it is being tested at the test bed in a decrepit building.

Keywords: smart grid; power line communication; monitoring system; sustainable building; zero energy building; android application

1. Introduction

Smart grids not only require advanced technologies but also contribute to their development, in addition to technical and social innovations. Furthermore, their roles in a transition from the existing energies to sustainable energies require more detailed research on both current and future energy markets and policies, focusing on the analysis of public support for the revised electricity policies involving new pricing models in addition to the application methods of the data obtained from public consumption patterns. Such a transition entails significant changes in the roles played by energy producers and consumers. Consumers are expected to play a more active role in response to future fluid power markets and complex supply/demand networks by claiming more rights and managing their own demands through their superior status in the market, where more local energy companies or some sectors are allowed to participate by opening the systems to the local energy companies or third parties. Ultimately, smart grids are an inevitable consequence in the transitional process during which buildings, transportation, fuels, and other social elements are being integrated consistently. Smart grids do not simply contribute to better sustainability, but also wield an influence on technological innovations, technology-oriented society, energy policies, communication network structures, household energy usage patterns and practices, and indeed, on the research works to be performed. Again, the development of smart grid technology requires comprehensive research on every relevant technology and element to explore the feasibility of a sustainable society supported by smart energy systems [1].

Power Line Communication (PLC) is a technology that communicates voice and data on a frequency signal through a power line. As one of the viable information delivery systems, the PLC technology allows data such as voice or image data to be transmitted over the existing electric power line as high-frequency signals that are then delivered to the terminal devices via a PLC modem [1,2]. However, most of the mediums used in a power grid are not suitable for such communication technology as they have been developed primarily for power distribution rather than communications, so there may be some problems when transmitting data over the power line. The signal attenuation caused by line noise or other factors is the major disadvantage of PLC technology [1,2]. PLC refers to a communication mode using power lines. PLC technology uses high-frequency signals to send information through power lines. These signals are separated from the power line with the PLC modem and subsequently delivered to terminal devices. PLC technology is based on two modes: high-speed PLC and low-speed PLC. The former adopts the frequency bands between 0.5 MHz and 30 MHz to achieve speed ranging from 14 Mbps to 1 Gbps, whereas the latter utilizes the frequency bands between 10 KHz and 450 KHz, offering lower than 9.6 Kbps as the speed that is usually used to control electronic devices [1].

The PLC described in this paper is to achieve communications by loading a frequency signal much higher than AC frequency (50/60Hz) onto the power line supplying AC current. The low-speed PLC carrying a frequency signal from 10 to 450 Hz is being used for the communications between old ships or railroad cars, or for automatic power meter reading. We've validated the use of this technology and the contribution of this research work can be found in the automated meter reading utilizing the PLC system.

Sustainability has become an important concept when constructing a nearly or completely net-zero energy building and is being incorporated into its design process as well as construction and operation processes. Currently, a series of advanced sustainable technological solutions are being introduced to enhance the level of thermal performance of building envelopes or promote the utilization of renewable energies. The sustainable buildings designed in a way to adapt to surrounding environmental factors such as weather conditions or the artificial changes controlled within the building itself to provide and maintain better human comfort. Thus, planning the use of appropriate building materials or technologies is an essential part of the design process in which durability, energy efficiency, maintainability of the building should be considered first followed by recyclability and cost-effectiveness of the materials to be used for the purpose of reducing the negative impact on the environment [3]. Taking these into account, this paper focused on the recyclability and maintainability of decrepit/old buildings.

This paper describes the details of our PLC mechanism, with related protocols implemented with both C and C++ languages. A PLC communication-based solution for decrepit buildings or apartments has been developed along with its test bed experiment. The results from the experiment confirmed that the solution was flexible and suitable for use. The contribution of this study lies in these tasks.

The description of the proposed PLC system and its mechanism is presented in this study along with its protocols developed with either C or C++ language. Developed especially for decrepit buildings or apartments, this PLC-based solution worked flexibly and efficiently during the test bed experiment;

such results obtained through a series of analyses and simulations can be considered to be the major contribution of this study.

In modern society, it has become quite clear that constructing a new building is not economic so most of the old buildings in the Republic of Korea often undergo full renovation or a partial remodeling process to sustain or extend their lifespans. Besides the expense, another major reason for recycling these buildings is to avoid a negative impact on the environment commonly due to various types of wastes generated by construction work. An old steel-reinforced building built in Janurary 1980 and located in the vicinity of a local government office (Geumjeong-gu Office, Busan, Republic of Korea) was selected for this research work. As the building was subjected to reconstruction, it was quite suitable for a testbed experiment to validate the feasibility and stability of the proposed power line communication. The experiment was conducted over a period of two years at an office room where an automatic sprinkler and electric heating system were available. The electricity was shared with another office on a separate floor and the bill was paid equally. The major feature, which can be considered as the contribution of the implemented PLC system is that the power usages are measured individually with the PLC-based power meter which allowed each room's electricity charge to be checked and billed separately on an Android app.

The rest of this paper is organized as follows. Section 2 describes the PLC, Narrowband PLC, G3-PLC modes, and related work. Section 3 presents a PLC-based monitoring system for sustainable building and discussion. Section 4 deals with the simulation of the power system control programs using PLC and RS-232 and discussion. Section 5 presents the implementation method for the Android application of the PLC-based monitoring system for sustainable building and related discussions. Section 6 discusses the analysis result based on the performance evaluation. Finally, Section 7 presents our conclusion.

2. Related Research

PLC can be influenced by future advanced power transmission or distribution systems when achieving end-to-end communications, and is regarded as an innovative, evolutional model compared to conventional power grids. Furthermore, the revised technical/technological regulations in relation to PLC systems have presented some positive and interesting issues in the areas of advanced communication and control technologies. Compared to the existing regulations, these new regulations tend to deal mainly with technical standardization or certification of heterogeneous networks, and such change is expected to contribute to developing a future power system that utilizes renewable or clean energies, creating a more suitable research environment for the development of distributed intelligence systems or demand-response (DR) programs.

A PLC signal of approximately 500Hz was adopted by New York City for the first time [2]; following this technological development, such technology has been used for data transmission applications in the fields of communications, energy, and power system security, conveniently using the existing wired power lines. The effectiveness of the PLC technology has been most evident in the smart metering system [4], which has led to the global utilization of smart metering infrastructure (SMI) for smart grids [5].

The PLC-based telemetry application over High-Voltage (HV) lines (power distribution lines) with frequency ranging from 15 to 500 kHz was first implemented in 1922 [6], followed by the ultra-narrowband power line communication (UNB-PLC) as a bilateral automatic communication system (TWACS) developed by Aclara and Sumitomo in 1930 [7]. The TWACS was developed based on the concept that the power grids have been optimally designed for the 60 Hz power transmission/distribution systems.

The X-10 PLC protocol [8,9] was originally developed by Pico Electronics (Glenrothes, Scotland) in 1975 [8]. This protocol enabled the transmission of binary data by adopting Amplitude Modulation (AM) for light, small-sized appliances. Considered a basic but major technology of modem-based smart home applications, this protocol was highly regarded, securing an advantageous position globally at the time. Nonetheless, the protocol could not overcome problems involving limitations in distance and power phases and unreliable performance. Currently, narrowband power line communication (NB-PLC) systems have been adopted in Japan, North America, and China using a frequency range of up to 500 kHz, offering a wide-range communication bandwidth for PLC-based applications.

NB-PLC systems are usable for a wide variety of applications by supporting low to high voltage systems [10], especially for power utility systems dealing with power and data transmission or detection and control applications (Figure 1). The NB-PLC-based energy management system includes some of the PLC-based (or Internet-based) remote energy management systems [11] that achieve maximum data transmission of 500 Kbps while offering increased reliability.

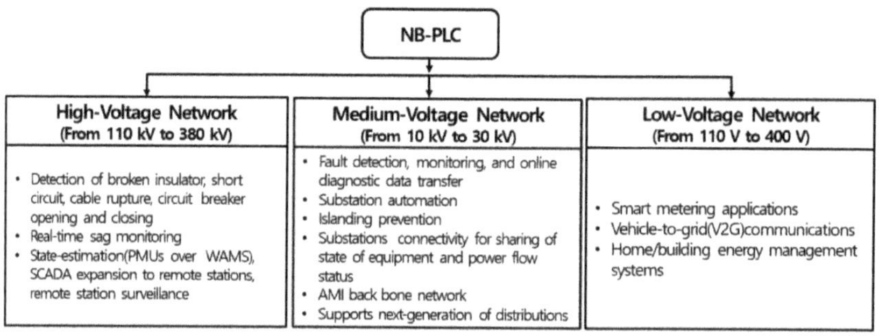

Figure 1. Narrowband Power Line Communication (NB-PLC) for smart power grid applications.

Moreover, NB-PLC is an advanced concept of the existing PLC technology and Internet protocol (IP) backbone for providing an appropriate energy-saving solution in an extensive power/energy distribution network, supporting the network's primary functions such as metering and remote control. Following the modern sophistication of power grids, various types of add-on services or functions such as ground protection, automatic transformer control, load balancing, automatic billing and payment, and fraud detection are being offered through decentralized intelligence (REMPLI system components) [12]. Meanwhile, the DLCþVITP4IP project [13] has facilitated advanced network monitoring/control, intelligent metering, and other innovative services for both power producers and users in a local power distribution network through an IP communication protocol that has been considered quite convenient and adequate for the NB-PLC networks (150–500 kHz).

Although high-impedance faults (HIFs) and their precise localization can be difficult, such is instrumental for SG applications. While an NB-PLC device using a frequency range of 3–95 kHz can track HIFs quite accurately [14], the PLC-access network (PLC-AN) has been proven to be effective for the medium to low-voltage distribution systems used for the SGs [15]. PLC-opportunistic routing (PLC-OR) was introduced as a novel but reliable technology that is able to reduce the packet transmission time significantly, offering much faster performance than conventional sequential routing with an equivalent level of reliability during communications between the substation and users. Depending on the latest revised regulations, standardizations, and technological advances, the NB-PLC protocols achieving higher data transmission rates can be classified into the four categories below.

The PRIME Standard introduced in 2007 performs multi-carrier modulation for a low-voltage network, and it is widely accepted for the development of Smart Grid (SG). It is being supported by the PRIME Alliance consisting of over 50 members and is chaired by Iberdrola, a Spanish multinational power company. From various types of NB-PLC PRIME standards, one with a frequency band between 3 and 95 kHz and a maximum data transfer rate of 130 Kbps has proven to be quite effective and efficient for a large PRIME-based SMI [4]. Australian power company Energex [16] has been evaluating a PRIME-based PLC infrastructure to assess its adequacy for their power network environment. This is to provide flawless services while preparing a next-generation remote metering platform.

Figure 2 shows the SG-oriented communication protocol developed by the G3-PLC Alliance, describing the PHY, MAC, and 6LoWPAN network layers. The G3-PLC operated on five nodes, and the test results were reported at Metering Barcelona (2009). Even at a maximum transformer resonant frequency of 490kHz, the protocol is suitable for transferring PLC signals [17].

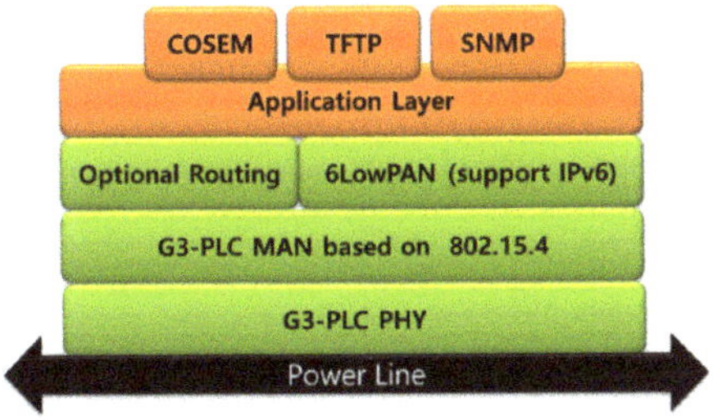

Figure 2. G3-PLC standard communication terminal protocols stack.

In the PLC signal range of 9–500 kHz, however, the noise level will increase as the signals have to go through the distribution transformer and capacitor banks. The G3-PLC standard system [18], which exhibits a higher data transfer rate and high-speed, long-range communication capability on a power line, also undergoes the same process.

In 2009, the IEEE 1901.2–2013 working group began discussion of subjects concerning PLC technology with the companies involved in the automotive standard meetings [18]. Since then, they have developed a low-frequency (max. 500 kHz)-based NB-PLC standard for the global NB-PLC-based systems and applications. The systems and solutions based on the IEEE 1901.2–2013 standard are quite effective for home-networking applications, electric vehicles and their charging stations, and power grids and their meters. They also provide effective performance for communications through medium-to low-voltage transformers. Meanwhile, the ITU-T G.hnem standard [19] provides the definition for the unified NB-PLC OFDM-based technology, targeting several SG applications including smart metering, distributed automation, residential energy management, generic home automation, vehicle recharging, etc., adopting Ipv6 as its primary networking protocol.

Both ITU-T recommendations, G.9955 (G.hnem PHY layer) and G.9956 (G.hnem data link layer), were certified in February 2011. Both G3-PLC and PRIME technologies allow the penetration of electrical devices such as distribution transformers and capacitor banks, and they are found to be effective and applicable to the advanced distribution automation (ADA) systems after being considered to be two major applications for SG infrastructures [20]. These technologies are currently open to the public and industry, so they have been the primary basic technologies for the globally recognized ITU standard called G 9955 narrowband OFDM-based PLC transceivers, a physical (PHY) layer specification [1]. The flexibility requirements for multi-standard systems have been dealt with via the multi-protocol PLC analyzer (MP-PLCA) [21], which has been successful in analyzing industrial power distribution networks, covering a frequency band of up to 400 kHz. The results from several experiments have shown that the PLC signals can cross distances of over 100 m, specifically for industrial use.

Raffaele Carli and Mariagrazia Dotoli published [22] "A decentralized resource allocation approach for sharing renewable energy among interconnected smart homes." Niels Blaauwbroek, et al. [23] published "Decentralized Resource Allocation and Load Scheduling for Multi-commodity Smart Energy Systems," whereas Giovanni Brusco, et al. [24] presented an "Energy Management System for an Energy District with Demand Response Availability" after their respective research activities.

Meanwhile, Raffaele Carli and Mariagrazia Dotoli and João Figueiredo and José Sá da Costa also presented some interesting research results in "Cooperative Distributed Control for the Energy Scheduling of Smart Homes with Shared Energy Storage and Renewable Energy Source [25]" and "A SCADA system for energy management in intelligent buildings [26]," respectively.

Oslo City's successful sustainable energy transition based on some innovative ideas and technological developments is well-described by Lind and Espegren [27], who studied all sorts of previous and present plans or measures taken for energy and climate-related issues and effectively used the results for transforming the city into one of the world's leading low-carbon cities.

Meanwhile, Sun et al. [28,29] collectively studied the RES's role in China's sustainable energy transition policy and concluded that it is essential to develop renewable energies to achieve the "Intended Nationally Determined Contribution (INDC)" in the post-2020 era after examining China's renewable energy status and potential in achieving the goal set by INDC. The eight different scenarios assuming 40% of total electric power output to be generated with renewable energies were established under the "Energy Plan," and the simulation results revealed that renewable energy sources were able to replace up to 20% of fossil energy sources such as oil or natural gases and could be used for electric power production for 40–50% of the current fossil energy-based generation systems.

A number of countries such as India [30,31], Thailand [32], Japan, and the Republic of Korea are also seeking the most efficient ways to achieve successful sustainable energy transition [33] in their respective territories. Under these conditions, Steg et al. [34] studied the human factor that could be involved in such process, proposing a general framework that can recognize and understand human behaviors that encourage/promote/advocate sustainable energies [35,36].

Most of the studies on PLC technology reveal that this technology can be an efficient platform technology for a smart metering network. Borrowing such idea, this study attempts to construct a PLC-based monitoring system that allows the central control facility or system operator to monitor, estimate, and manage all power usages remotely and provide subscribers with some important information that could help them understand or control their usages or use patterns, including helping them prepare for future incidents that may occur due to system failures. C or C++ programming language was used to create the protocols for the system operation algorithm. The completed platform can be quite useful for the sustainable building process, and all of its efficient functions will enable the user or the power system itself to conserve power systematically. The system is also scalable, and its design and algorithm are applicable to other network systems including Zero Energy Building Management, for example. One of the significant contributions of this study is that, unlike before when new buildings were usually considered for the application of new technologies, the proposed system targeted decrepit buildings or apartments. Prior to the actual application, a Test Bed experiment was conducted, from which the flexibility and efficiency of the system have been confirmed.

Jae-Shik Yoon, et al. [37] devised a power monitoring system by developing a low-power PLC modem with power monitoring and abnormal power-blocking function (Figure 3) to detect power consumption and block unusual power activities with their test program (Figure 4).

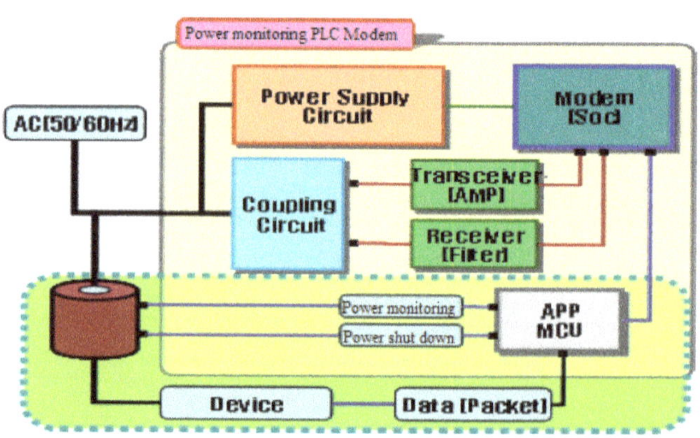

Figure 3. Power Line Communication (PLC) power consumption monitoring system.

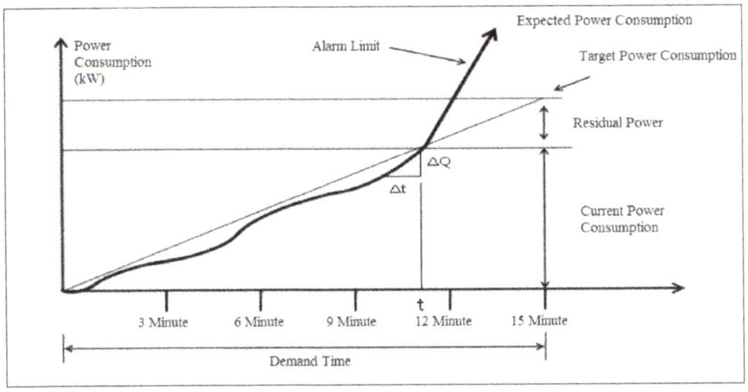

Figure 4. Power consumption measurement by the Power Consumption Monitoring System.

Seon-Tae Yoon et al. [38] implemented a PLC-based human disaster prevention and location information monitoring system by using PLC communication and Zigbee. The system consists of a fixed-node system using a PLC + Zigbee board, a mobile-node system using a Zigbee sensor board, and a monitoring system that allows real-time monitoring of the data received from each node (Figure 5). The fixed nodes are installed on the power lines laid out in each work zone, whereas the mobile nodes are carried by the workers or attached to their work hats. The range of these nodes is set at tens of meters.

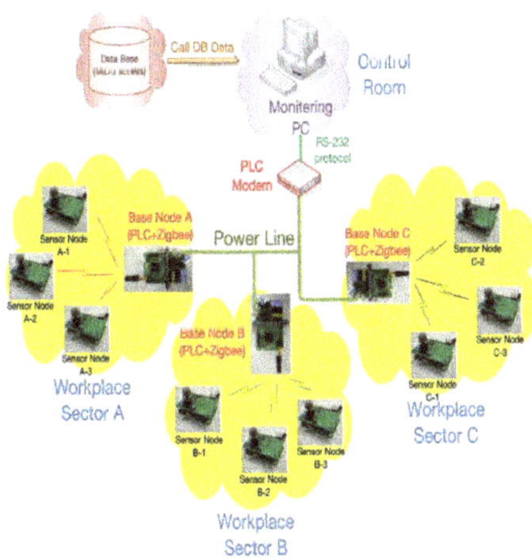

Figure 5. Configuration of the proposed System.

Keun-Soo Park [39] implemented a system equipped with a PLC modem and a DCU to monitor the data from the household meter via PC. The collected data includes each household's previous and current months' power usage so that the user can conveniently manage his/her electricity rate or meter troubles by checking them on the server PC.

It shows the required values on the PC after simultaneously storing seven different pieces of information through the DCU. Although the IC used in this experiment guarantees a maximum of 1024 nodes allocated to a single DCU, it is commonly specified at the actual sites that up to 20 units should be linked to establish reliability. Still, such a requirement can vary depending on the company utilizing the system.

A private line had to be laid for the existing meters to send data to a PC or a mobile device, but the system implemented in this study allows the serial data to be checked through a direct connection to the power line. Moreover, this system can be modified to convert the power usage into the power rate based on the calculation method set by Korea Electric Power Corp. (KEPCO) or visualize the power consumption level by indicating the hourly power usage with a graph. Such effort can be helpful to consumers in using electricity sensibly. The additional modification can include adding a remote-control system through a mobile device or the Internet.

Keun-Ho Park and Young-Ho Lee [40] implemented a PLC modem and an Internet map-based data collection device with a Graphical User Interface (GUI) monitoring system to monitor household meter data on a PC. The device can conveniently monitor the previous and current months' power usage from the household electricity meters through visualization by utilizing the Internet map information. This has made it easier for the public (non-experts) to manage power usage and meter troubles by checking their power use information on the PC.

Figure 6 shows the experiment conducted for the system configured in Figure 7. The load used for each meter is the two metal halide lamps connected serially (220 V, 70 W per lamp). The light and thermal energies consumed by the lamps are converted into data values by the meter, and then the changed values are finally checked by the GUI tool. The information of all the meters is stored in the DCU and subsequently shown on a display when the relevant position is clicked.

Figure 6. View of the experimental set.

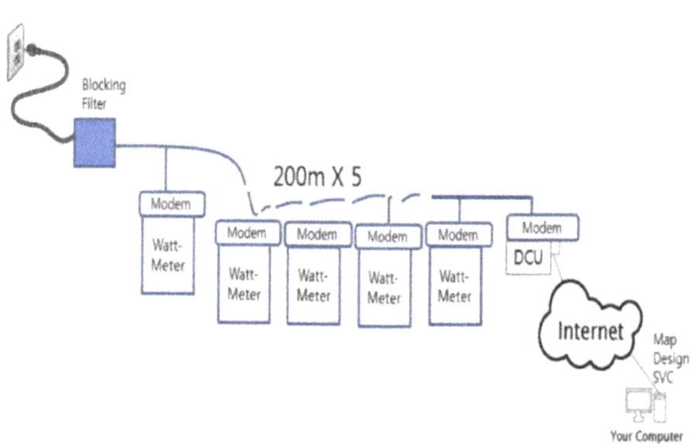

Figure 7. Block diagram for experimental setting.

The infrastructure introduced by Paulo Régis C. De Araújo et al. [41]. in their work 'Infrastructure for Integration of Legacy Electrical Equipment into Smart-Grid Using Wireless Sensor Networks' proposes a system where individual legacy electrical devices are integrated to be connected to a sensor node and the sink node links it with a Smart Grid using an appropriate communication protocol. Also, the research work 'OpenZmeter: An Efficient Low-Cost Energy Smart Meter and Power Quality Analyzer' by Eduardo Viciana et al. [42]. specifically points out that consumers usually are not provided with the information about their energy consumption status on a real-time basis as well as the quality of power they are using due to the cost of available electric energy meters that are often so complex for them to understand or operate. The authors proposed an alternative affordable but innovative home-use electric energy meter that was supposed to be more precise, manageable, and reliable while offering better convenience. The system performance was quite satisfactory when it was tested in a household for a period of two weeks and the performance data also supported the system proposed in this paper.

Additionally, González et al. [43] introduced a systemical approach that allows PLCs and Arduino platforms to be integrated into a SCADA system with the connectivity offered by Open Platform Communications (OPC). In this system, the open-source device focuses on sensing tasks exclusively. The validity of this approach was proven by applying its concept to an experimental automated SMG which was intended to monitor the temperature level of a specific PV panel, presenting satisfactory results [44].

Meanwhile, Maria-Angeles, Sanchez-Hidalgo, and Maria-Dolores Cano [45] pointed out in their work 'A survey on visual data representation for Smart Grids control and monitoring' that all the collected/processed data in a smart grid are inherently heterogeneous because of the intrinsic nature of smart grids, and it is essential to devise effective ICT tools or platforms that can provide a broader spectrum of multi-dimensional control perspective that could effectively support modern energy industry. They emphasized that data visualization could be an effective solution when monitoring/analyzing/responding to the events efficiently within a Smart Grid. For this study, a wide range of relevant research works was studied to investigate previously available visualization methods and propose an optimal one for the control and monitoring tasks in a Smart Grid.

Today, people and things are being connected by IoT systems regardless of time or space owing to the number of networks and services getting larger and larger every day. In this context, IoT can be regarded as a dynamic network infrastructure connecting worldwide internet users with online services. IoT technology/system has become an essential part of a Smart Grid where data is exchanged through a communication network connected to a power grid to autonomously collect and analyze the data obtained by monitoring transmission lines, substations distributing power, or household/business users.

3. Design of Monitoring System for Target Building

As in any country, many old buildings in Korea are repaired to be used for 40–50 years or even longer since constructing new ones is expensive. Because remodeling old buildings is not only expensive but also generates construction waste, repairing is the preferred alternative in many cases.

The structure cited for the testing of this paper was a steel-concrete building constructed in January 1980. The building located in front of Geumjeong-gu Office in Busan was facing reconstruction in the near future, and we used it as a testbed for two years and verified the safety of power line communication.

The office room we selected for the test had no city gas service, only the electrical heating system As such, it had the spring cooler that is activated automatically in case of fire caused by an electrical short circuit. It shared the electricity with an accountant's office on the third floor and paid half of the total electricity bill. The contribution of this paper would be the implementation of a system (power metering) that metered the power line communication and automatically divided the electricity bill for payment with an Android app.

3.1. Design of PLC between Electronic Ammeter and Main Server

An example of PLC implemented between an electronic ammeter and the main server is presented in Figure 8, where the MCU (Micro Controller Unit) controls the PLC module along with the DCU (Data Concentration Unit) that collects the meter age and sends its value to the main server regularly. The values delivered to the main server are then transmitted to the user's portable device or other display devices by the PLC communication module. At this time, RS-232 communications are performed between the PLC module, the MCU, and the main server with a serial interface connecting the communication modem and serial port. Such a communication type requires an adequate communication protocol.

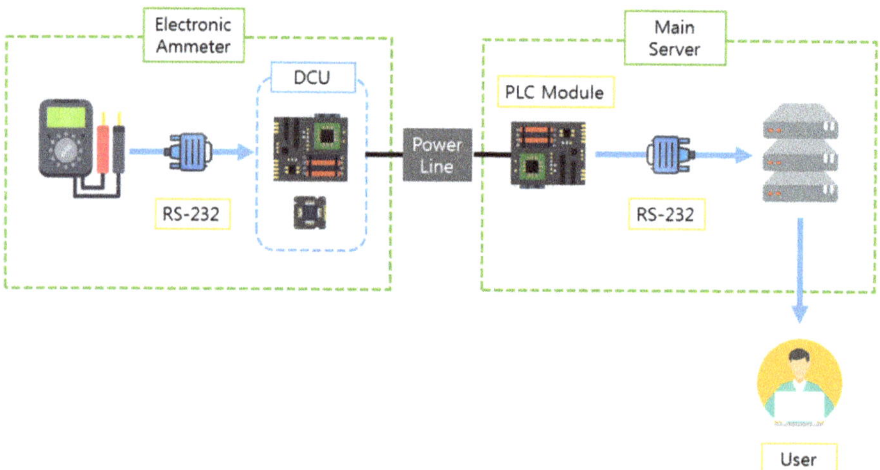

Figure 8. Power Line Communication (PLC) between an electronic ammeter and the main server.

3.2. Target Building's System Connections

The connections between the digital watt-hour meter, field terminal, and DCU are shown in Figure 9, where the modem communication line extended from the watt-hour meter connects the meter with the field unit. The PLC communication module is connected to the RS-232 communication unit, and then to the field connection unit, to perform communications with the main server over the power line. In this structure, optical communication components are used to establish the connections. They should be supplied or designated by the supplier/manufacturer of the digital meters to avoid any communication errors due to "out-of-spec" components. Figure 9 shows the connections between the watt-hour meter, field terminal, and DCU.

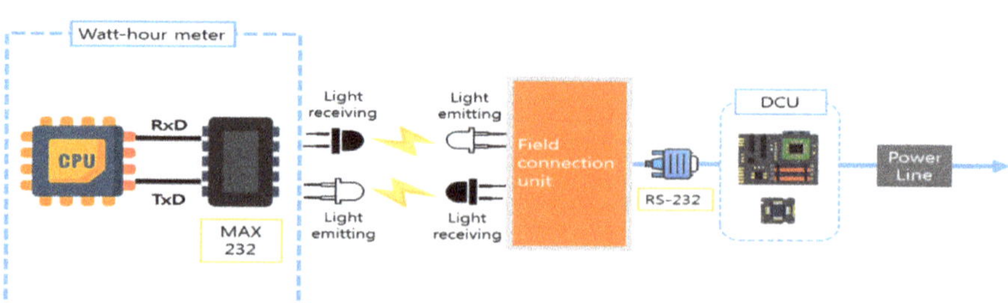

Figure 9. Connections between the watt-hour meter, field terminal, and DCU.

3.3. Design of Mail Box and Port

A queue that could be designated by one or more senders or receivers is called a "mail box." Figure 10 shows the related technique. Since anyone among the senders(n) can generate a message that can be taken away by anyone among the receivers(m), mail boxes offer more common communication equipment. Unfortunately, in distributed environments, receivers who see the same Mail box could be located at different computer sets, so the implementation of the "receive" computing operation is costly. As such, it is implemented with a limited mail box form mainly described as a "port." The port involves only one receiver. The messages coming in from other processes and the designated port are sent to a location associated with the receiver, as shown in Figure 11.

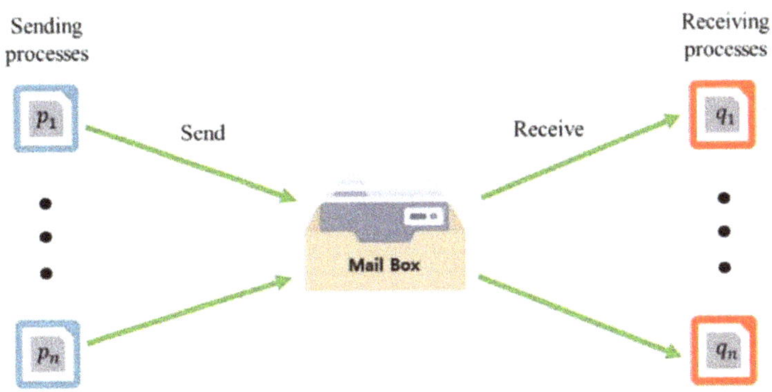

Figure 10. Mechanism of the Mail Box.

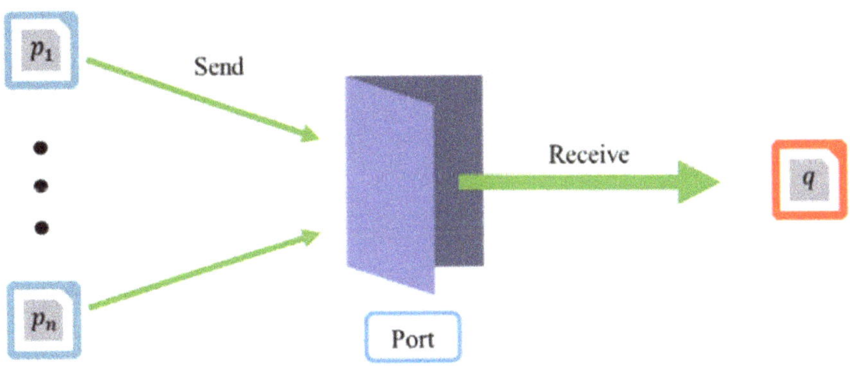

Figure 11. Communication mechanism via a port.

3.4. Design of RS-232-based Protocol

PLC requires a suitable protocol when establishing connections between the main server and other individual devices. Since the PLC module executes a broadcasting function, individual devices require their own unique numbers, and the sign "Unused" has to be included as well. The protocol usually follows the TCP/IP model where the plus (+) sign should be attached at the end, meaning the communication has ended. Another requirement is that the minus (–) sign should be inserted between words. The commands used among the devices and their execution flow are presented in Figure 12.

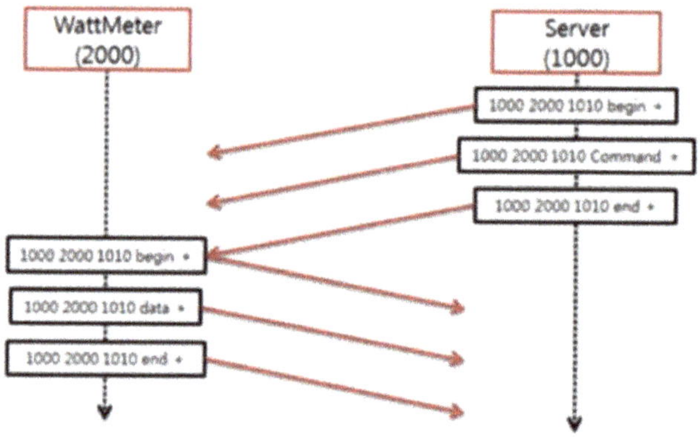

Figure 12. RS-232-based protocol design.

As the server plays the main role, synchronization is not required; the connected devices or equipment will be in idle state until they are to start communicating. Likewise, for the purpose of controlling the individual device group, a broadcasting number will be given to each group to control/manage all the power sockets installed at home simultaneously. At the very first device group number is the broadcasting number, which is 3000 for the general device group, for example.

Meanwhile, communication using PLC technology usually causes the occurrence of noise, so we have tried to reduce them by developing technology such as intelligent agent [1,6]. When it comes to reducing noise, however, RUDP showed better results than this agent.

For the research, OPNET-based simulations have been performed first, and then some actual PLC products were mounted to the system, followed by the development of a monitoring application to complete an appropriate platform/solution that is undergoing the testbed experiment at a decrepit building.

4. OPNET Simulation of the Power System Control Programs Using PLC and RS-232

Considerable installation cost was needed for the smart grid-based monitoring at a decrepit building, so an OPNET simulation was performed first in this research, after which a number of PLC products were mounted together with the monitoring application to complete the appropriate platform and solution for testing.

In this section, the contents to be implemented in the next section are simulated using OPNET 14.5 PL8. When the results are examined as shown in Figure 13, the transmission data and the received data can be confirmed to be identical in both communications. Likewise, Figure 14 shows PLC and RS-232 implementation using OPNET.

As one of the serial interfaces, RS-232 is generally called a serial port. It uses a small number of line cables, and it is cost-effective. It is an asynchronous serial communication method wherein the meaning of serial is to send only one bit at a time for the information to transmit. RS-232 has this feature and allows regular data transmission and reception. In the case of parallel communications, multiple bits can be sent simultaneously. This means that one line is required for each bit, which is cost-inefficient compared to serial communication. In the case of PLC, which is parallel communication using a power line, irregular data transmission/reception occurs compared with RS-232. This is due to the inherent limitation of PLC, i.e., it is not designed for communication purposes, compared to RS-232, which transmits 1 bit at every moment over a short distance due to the difference in transmission environment. By transmitting power and data simultaneously, it has high load interference and noise due to the fact that AC signals are a worse transmission environment than RS-232, a serial communication. In addition, due to the characteristics of AC power, it is difficult to control the signal attenuation, which constantly changes according to the geometry or load of the transmission line. In contrast, RS-232 PLC does not require difficult construction.

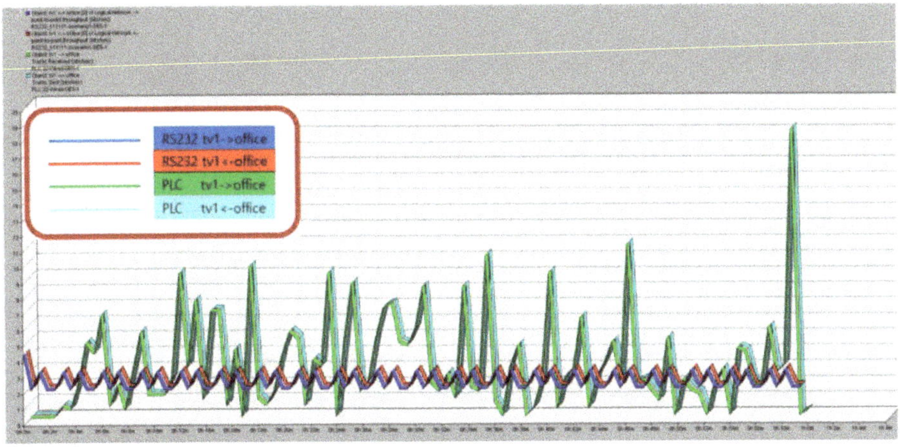

Figure 13. Traffic Send/Receive comparative simulation results for PLC and RS-232 using OPNET Simulation.

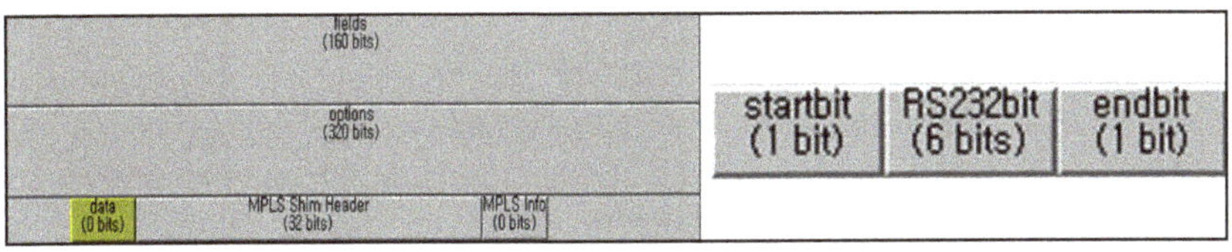

(**a**) (**b**)

Figure 14. PLC and RS-232 implementation using OPNET Simulation: (**a**) Packet Format of ip_dgram_v4 used in PLC; (**b**) RS-232 Packet Format used in RS-232.

In RS-232, it is necessary to transmit in the form of bit units through a physical channel. For this purpose, information must be broken up into units of data words of various lengths. In the PC, a length between 5-bit and 8-bit is basically selected, and this length is quantitative. Additional bits have sync or error-check purposes, and they are added to the data bits. The important point here is that the number of bits in the sender and receiver must be coordinated in advance and must be the same. Otherwise, the data may be misinterpreted, or the information broken up by data words may not be fully recognized.

The disadvantage of RS-232 lies in the selection of the receiving time of the receiver. If the receiver incorrectly chooses the moment of information reception, resynchronization is required to recover it, and all data sent prior to resynchronization is lost, resulting in inefficiency.

Start bit and End bit can be used to avoid this disadvantage. When the start bit is received, the receiver recognizes the start of the transmission of information from that point and recognizes the end of the information by receiving the End bit. In the case of a Packet format of RS-232 configured by OPNET, an asynchronous characteristic is implemented by allocating 1 bit each of the start bit and end bit. No separate parity bit is allocated for communication efficiency. For RS-232, which is not in transit, the TD line is set to 1. When initiating data transmission, the sender sends a start bit that lasts for the time frame of the baud rate with a logical 0. By sending the start bit, the data communication device (DCE) simply waits for the TD line of the data terminal device (DTE) to change from the logical 1 state to the logical 0 state, enabling the synchronization implementation of the transmission. Just as the start bit specifies the start of a character, the stop bit specifies the end of a character. The stop bit transmitted by the serial transmitting device is always a logical 1, and it also sets the TD line. When the stop bit is received, it remains idle. Table 1 shows the 9-pin RS-232 connector configuration.

Table 1. 9-pin RS-232 Connector Configuration.

Pin	Abbreviation	Name	Function
1	DCD	Data Carrier Detect (DCD)	Set to logic state 1 if the data link is in progress.
2	RD	Received Data (RD)	Set to logic state 1 s in the data terminal equipment (DTE) when ready to communicate with the data communication device (DCE).
3	TD	Transmit Data (TD)	Data Terminal Equipment (DTE) is used to deliver data to a Data Communication Device (DCE).
4	DTR	Data Terminal Ready (DTR)	Set to logic state 1 s in the data terminal equipment (DTE) when ready to communicate with the data communication device (DCE).
5	-	Signal Common	Signal Common
6	DSR	Data Set Ready (DSR)	Set to logic state 1 in the data communication device (DCE) when ready to communicate with the data terminal equipment (DTE).
7	RTS	Request to Send (RTS)	Set to logic state 1 for the data terminal equipment (DTE) when data is to be sent to the data communication device (DCE).
8	CTS	Clear to Send (CTS)	Set to logic state 1 in the data communication device (DCE) when ready to receive data.
9	RI	Ring Indicator (RI)	Set to logic state 1 in the data communication device (DCE) when a telephone signal is received.

This is the same as the data format used in IPv4. It consists of hiding the information in the upper layer and adding header information to the encapsulated data. The header information includes all information required to transmit a packet to the other party as well as information such as the version of the Internet protocol, length of the header, DSCP, ECN, length of the entire IP packet, and identification number for fragmentation.

In the simulation result of Figure 15, when the packet format is considered, the ip_dgram_v4 used in the PLC is big, and RS-232 is small. Intuitively, the PLC queuing delay is long, and RS-232 queuing delay is short. In the case of PLC, however, it is possible to transmit a plurality of bits by using parallel communication. In the case of RS-232, by using serial communication, only one bit is transmitted at a time. It can be seen that the queuing delay is longer, which indicates the difference until the transmission from the entry to the transmit channel queue.

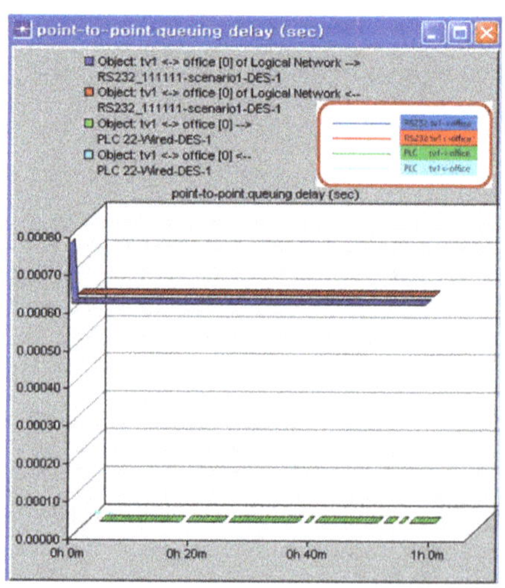

Figure 15. Comparison of results for PLC and RS-232 communications using OPNET simulation.

5. Implementation of Android Application for Monitoring System for Sustainable Building Management

5.1. Implementation of Power System Control Programs Using PLC

The implemented system can be largely divided into the power usage reception part and device control part. In the power usage reception part, the power consumption data from a wattmeter is received and processed by the MCU, and the server can receive the data via communications between PLC modules. In the device control part, the server that has received a command from a client delivers a command (i.e., device control, etc.) to the MCU via the PLC module connected with a power line, and then the MCU controls the power supply of a light bulb. A light bulb was used for the experiment in this study to enable use for other devices.

C and C++ languages were used for the implementation, with the implementation flow described with Lines 1 to 23 in Table 2. The indicators are represented as $\mathbf{A_p}$ (total number of previous access addresses) and $\mathbf{A_n}$ (total number of present access addresses).

Table 2. Implementation algorithm.

1	f(x) =($\sum$ $Number(A_p)$ − $\sum$ $Number(A_n)$)
2	**If** f(x) = 0
3	**If** $\sum$ $Time(A_n) \geq 7Days$
4	**Then** Warning Message transmission to the User's family
5	**Not** Return(f(x))
6	**Not** (f(x)≠0)
7	**If** $\sum$ $Number(A_n) \geq 2\ (= g(x))$
8	**Then** Return(g(x))
9	**Not**
10	**If** $\sum$ $Time(A_n) \geq 3Days$
11	**Then** transmit a visitation request message to the volunteer center
12	**Not** Return(f(x))
13	P(x) = Wi-Fi system identifies the number of connecting users
14	H_{IP} = Home IP address
15	① P(x) = Wi-Fi connection ② SSID search ③ N_{IP} = IP address of accessing Wi-Fi router
16	**If** $H_{IP} = N_{IP}$
17	**Then** P(x) = C(x) + 1
18	Wi-Fi disconnection
19	**If** $H_{IP} = N_{IP}$
20	**Then** P(x) = C(x) − 1
21	End
22	**Not** end
23	**Not** Return P(x)

The description of each formula is as follows: Line 1: A comparison between previous and current access addresses is made; Line 2: When the current and previous access addresses are the same; Line 3: When the access time of the current address remains the same over a week; Line 4: A warning is sent to the user's family; Line 5: Return to the comparison stage f(x); Line 6: When the current *and previous* access addresses are not the same; Line 7: When it has been calculated that the number of current access addresses exceeds 2(g(x)); Line 8: Return to g(x) comparison; Line 10: When the access time of the current access address remains the same for over 3 days; Line 11: A "visit request" is transmitted to the volunteer center; Line 12: Return to comparison f(x); Line 13: The number of users being connected to Wi-Fi is estimated; Line 14: Home IP addresses are allocated; Line 15: Wi-Fi connections are checked along with the SSIDs, and then the current addresses are allocated; Line 16: Determine whether the home address is the same as the current access address or not; Line 17: Determine if the number of users being connected has increased; Line 18: Wi-Fi connection is disconnected; Line 19: Determine whether the home address is the same as the current access address or not; Line 20: The number of users being connected is reduced; Line 22: The function stops; Line 23: Return to P(x).

5.2. *Estimation of Power Use Status: Graphical User Interface Using Java*

The key element in the proposed system is that an actual PLC product has been installed in a decrepit building along with an efficient platform/solution complete with a monitoring application with which customers will be able to interact intelligently concerning their power usages.

The power consumed for a three-minute period is estimated, and the corresponding strings have been processed by splitting them into Carriage Return (\n) units. The aggregated power use amounts entered from the Arduino circuit are shown in Figure 16. A desk lamp was used for simplicity, and its power consumption amount was measured every five seconds.

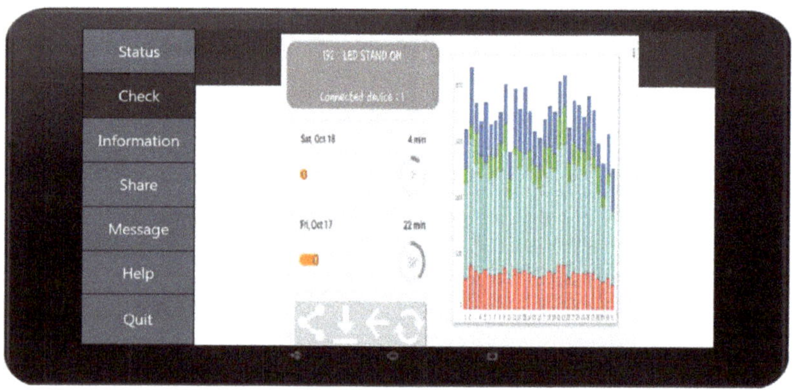

Figure 16. Application for PLC-based monitoring.

As shown in Figure 17, the proposed application is for the monitoring system to harvest energy. "Connected Device" refers to the current number of connected devices, with the left part of the picture indicating the date, hours spent to collect electricity, and its volume (%). Likewise, with the menu bar at the bottom left, Sharing, Download, Back, and Device Change can be performed. This is a system that intelligently reports the electricity usage to the user by monitoring it based on PLC communication and compiling the statistics to allow the manager to save electricity.

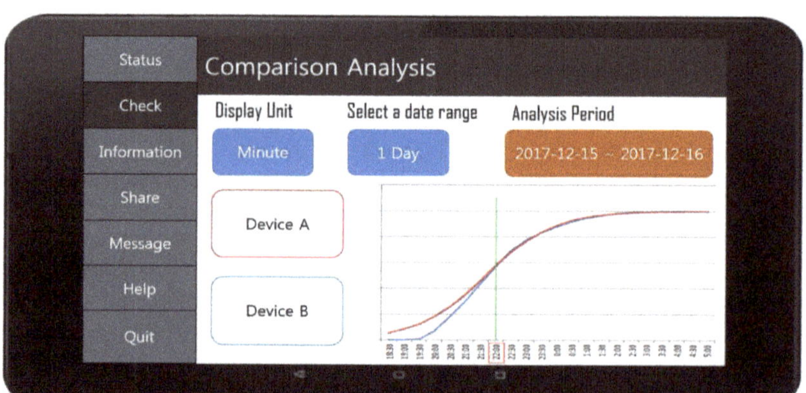

Figure 17. Java Android-based user interface comparing power usage.

Figure 17 shows the Java-Android-based user interface that compares the power usages. The menu consists of Status, Check, Information, Share, Message, and End from the top. The screen in the picture demonstrates the result when the Check function has been selected to perform a comparative analysis. The expression unit, period setting, and date setting are shown in the first line.

The screen unit was set as a per-minute rate, whereas the period was a single day from 15/12/2017 to 16/12/2017. The period can be adjusted by the user. Nevertheless, to minimize energy consumption, the data is transmitted once every minute. Household appliances such as refrigerators have to be connected to a power source for continuous use. This way, any Big Events (uses) can be monitored when detected. As shown in the picture, the minimum unit was set as a per-minute rate, and the period, as a day, for the graph so that the user can check the usage for one day.

The interface below is for comparative analysis. Equipment A and B have been selected, with the result of the analysis shown with the adjoining graph. The accumulated energy of equipment A is shown with a red graph line, and that of equipment B is drawn in blue.

Figure 18 shows the time graph user interface implemented with Java Android, which can represent the energy by comparing it with the time graph. In the picture, the top screen indicates how energy has been collected over time. The red line represents the total amount, and the yellow line indicates the energy harvested. The blue line represents the lost energy. By subtracting the amount represented by the blue line from the amount represented by the yellow line, the total amount represented by the red line can be obtained. The graph shows the progress depending on the time graph directly below. There is a typical user interface at the bottom left where the Menu, Message, Share, and Help functions are located. At the bottom right is an indication of the harvested amount of energy. On the left, the aggregated energy level in the ESS is visually indicated, and the pie chart on the right shows the energy capacity in percentage with various colors.

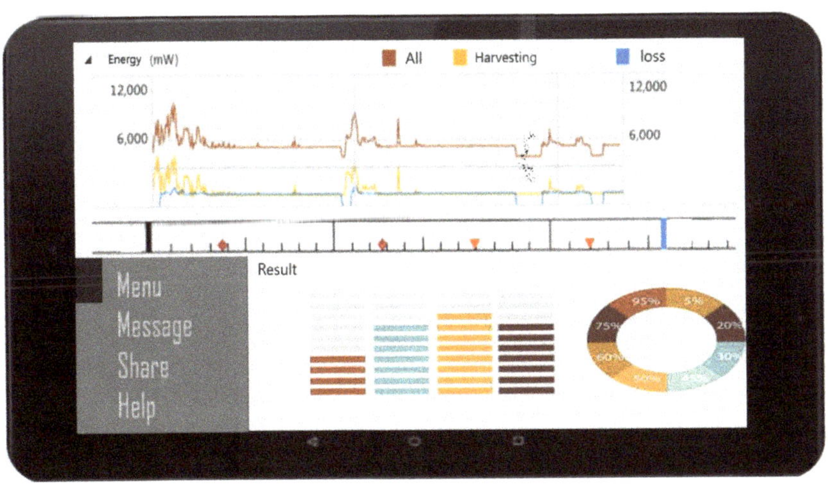

Figure 18. Time graph user interface implemented with Java Android.

The system where the data of a digital watt-hour meter can be checked with a smartphone in the UI through the PLC system installed in a decrepit smart home environment has been developed (Figures 19 and 20). As shown in the picture of the test bed, the user can check that the measurement of the digital watt-hour meter is 22.4 kW.

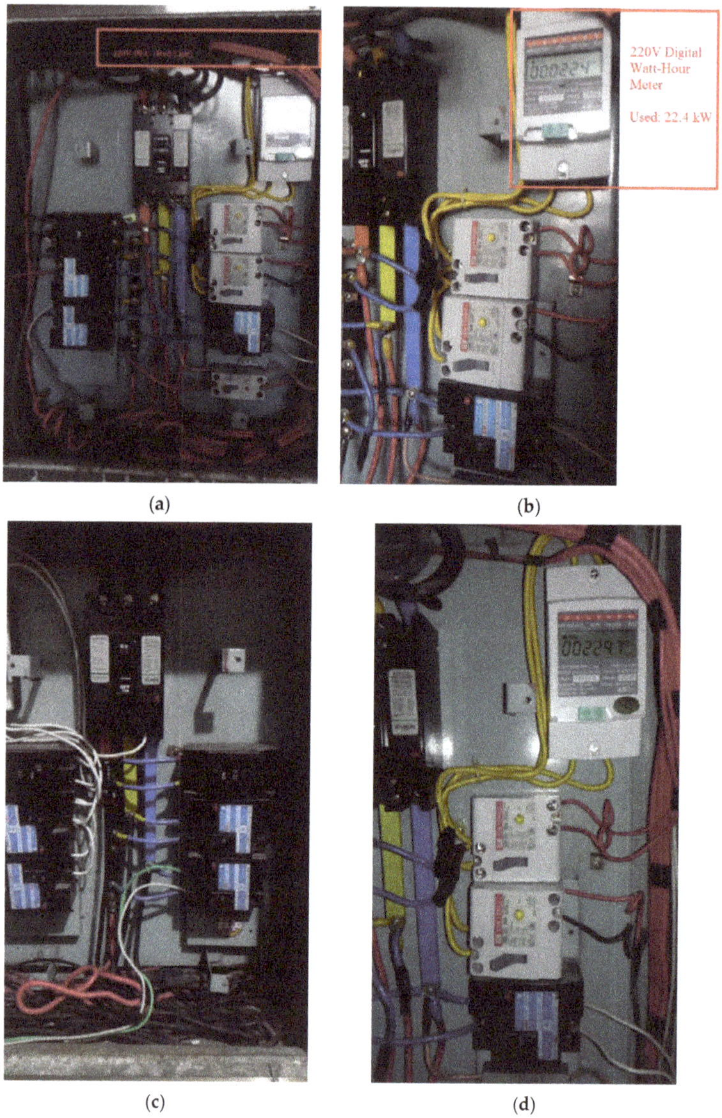

Figure 19. PLC System and Digital Watt-Hour Meter installed in an old home environment. (**a**) PLC Centric Monitoring System (Main part); (**b**) Digital Watt-Hour Meter; (**c**) Before Installation (PLC and Digital Watt-Hour Meter); (**d**) Digital Watt-Hour Meter (1 July 2018–30 June 2019).

Figure 20. User Interface for power usage monitoring.

6. Analysis of Performance Evaluation

6.1. Comparison with Other Systems

PLC communication is mainly used in infrastructures where the power line network is systemized [46]. The PLC communication system allows the remote control of electronic equipment and household appliances through mobile devices or Internet as well as security measures like intrusion detection or safety measures including remotely shutting off gas valves. Such a system is also used for the control of air-conditioning units, home network systems, streetlights, airport runway guide lights, golf course sprinklers, train lights, and product management systems that use a small-scale wireless communication system such as Zigbee. Table 3 shows a comparison of the results with other systems.

Table 3. Comparison with other Systems.

Authors	Specification	Technology Used
Jae-Shik Yoon, et al. [37]	PLC Power Consumption Monitoring System	Low-power PLC Modem
Seon-Tae Yoon, et al. [38]	PLC-based human disaster prevention and location information monitoring system	PLC + Zigbee board
Keun-Soo Park [39]	Watt Hour Meter Monitoring System	PLC modem, DCU
Young-Ho Lee [40]	PLC modem and Internet map-based data collection device with a GUI monitoring system	PLC modem, Internet map-based data collection device, GUI monitoring
Ours (proposed system)	Decrepit Building Monitoring System	PLC modem, DCU, Digital Watt-Hour Meter, Application

One of the success cases of using PLC technology for old buildings can be first found in the PLC testbed service carried out by SK Broadband (Republic of Korea) for 100 households in 2017. Their service allowed the users to enjoy an average speed of 100 Mbps just by plugging the PLC equipment into a power outlet and it was validated that the speed up to 800 Mbps can be achieved with this technology, which is not much different from the 100 Mbps high-speed or Max. 1 Gbps Giga-internet service currently available. Although the PLC system seemed to be promising after the ROK's Ministry of Commerce, Industry and Energy and Ministry of Information and Communication revised the Radio Waves Act and prepared the PLC National Standard in 2007 to promote and spread power line-based home networks, it was not actually adopted for the domestic high speed internet services due to the dispute over its stability and oppositions by most of the communication service providers; only a few companies including Samsung and LG Electronics provided the PLC-based home network services. At the same time, Korea Electric Power Corp. (KEPCOM) also attempted to enter the local home network service business with their own PLC system but failed to achieve its commercialization.

SK Broadband explained at that time that they were able to provide their testbed service after securing enough speed and stability close to a gigabyte-based service, which was not possible earlier.

Most of the buildings constructed before 2000 had some limitations when trying to use high-speed internet, internet TV/telephone or Wi-Fi so that local area network (LAN) cables had to be installed additionally. Also, the demand for LAN cables is increasing recently as the number of households using a multiple number of these devices/equipment is growing rapidly, supporting the economic feasibility of PLC-based systems. SK Broadband is expecting that these systems will play a big role when they attempt to penetrate the niche market targeting old areas/buildings where installation of LAN cables can be costly or difficult [47].

Meanwhile, the ROK's Geumcheon-gu Office conducted a testbed service for an apartment complex in 2019 to replace its old mechanical power meters with smart meters using PLC equipment. The main point of this project was to allow the residents of Doksan Newtown Branew Apartment Complex to check their own power usages on a real-time basis on an app anytime anywhere through newly installed smart meters. The task of replacing old meters was carried out by an energy company Energynet specializing in the PLC business that focuses on voice-call and data communication systems through existing power lines as a medium. Geumcheon-gu Office expected that the accuracy and efficiency would increase much through remote meter reading without requiring a lot of manpower once the testbed service was implemented. They are also expecting that the data collected from the real-time transformer and power use monitoring systems will be used as big data in the future through an integrated management system. All of these systems will be helpful for the residents to save their power usages which in turn will be returned to them in a form of Ecomileage [48].

6.2. Performance Evaluation

For the performance evaluation of mail box, its functional diagram has been analyzed and examined. First, the User creates Mail on the server when a request is made and puts Mail in the Inbox of the Device Manager (1st process). The Device Manager then checks the Inbox and, if there is any Mail, sends out the command to the relevant device and gets the response from the device (2nd and 3rd processes). Finally, the Device Manager puts the response content in the Outbox, and the User checks it to get his mail if there is any. Here, as a performance evaluation, a load test that checks how fast the Device Manager would process simultaneous mail-requests from multiple users, including a robustness test to determine whether the normal operations are being performed by discarding wrongfully sent mail from the multiple users properly, was conducted. Figure 21 shows a functional diagram of the mail box.

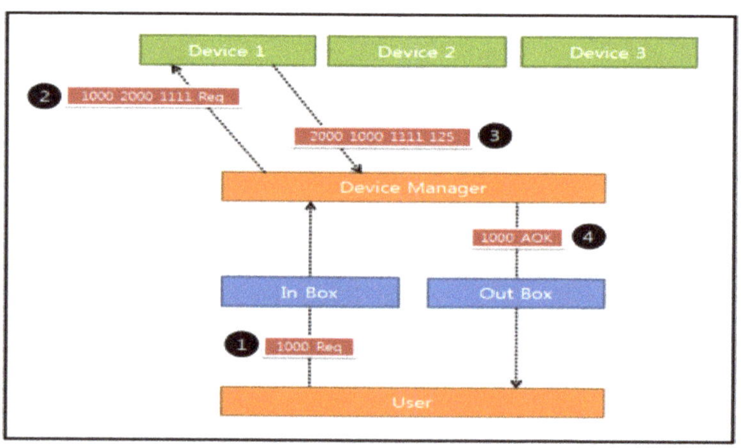

Figure 21. Functional diagram of Mail Box.

A "User" has been created with the program, and he sends out 50 requests to the server continuously. After transmitting the messages, the user terminates its action immediately without waiting for the response. The transmission failures were analyzed through the recorded "Logs" on the server, and the direct device-to-device link method was adopted to reduce failures caused by PLC interference. Figure 22 presents the results obtained from the load test and the robustness test.

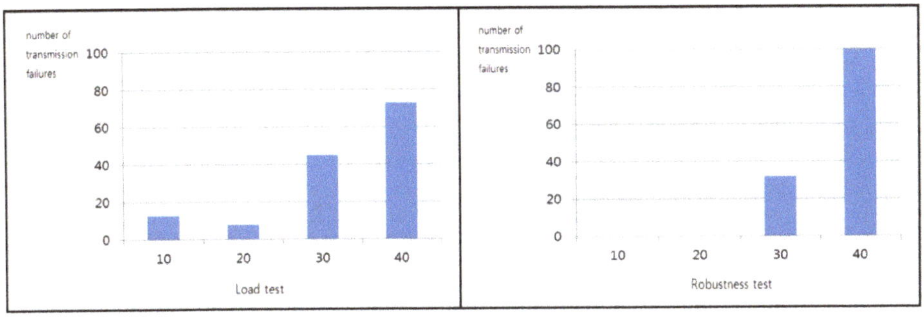

Figure 22. Load test and Robustness test.

First, the load tests were conducted by adjusting the number of users from 10 to 40 people. Transmission failure rapidly increased when the number of users exceeded 29 people, and all of the times taken to complete the response showed the same increase. Second, similar to the load test, robustness tests were conducted by adjusting the numbers of users within the range of 10–40 people. Likewise, the number of wrong replies also increased rapidly when the number of users exceeded 29 people. It was possible to determine both the robustness and the load resistance level of the mail box through these two tests and of the entire program itself. If the mail box works properly, and the failures do not originate from PLCs, we can assume that all functions are in good order. Thus, even though the performance of the program declined rapidly with over 29+ users, and considering the fact that the program is intended for ordinary families in our country (i.e., families of fewer than 10 people), we do not anticipate much of a problem.

The analysis result above revealed that the PLC-based method did not show an adequate level of reliability, and it could be seen that the failure rate increased with greater interference. If the reliability is low, more retransmission requests would be transmitted, so the efficiency of the entire system could decline. Due to the nature of PLC technology and the fact that it would be difficult to increase the reception success rate without replacing the module, the simplest form of communications is preferable. Moreover, a mechanism that deals with re-requests and re-responds when the original messages have not been processed should be designed and added to the system. As the key element in the proposed system, an actual PLC product has been installed in a decrepit building along with an efficient platform/solution complete with a monitoring application with which customers will be able to interact intelligently concerning their power usage.

7. Conclusions

The issues discussed in this study in relation to the proposed system can be classified as follows. First is the time element, specifically the time required for the network connection. A delay of a few seconds occurs when the proposed system is used to connect with Wi-Fi, determine the present number of residents, and then initiate control of the devices installed at home via PLC. Although this type of delay may not be much of a problem, it could actually cause some inconvenience for the system users or operators. Second, due to the nature of PLCs, the vulnerability to noise is inevitable, and the instability caused by such vulnerability limits the expected capacity of the system. These issues can be viewed as the problems originating from the limitations of the present PLC and network technologies, not from the system design and implementation method presented in this study. Nevertheless, such time and noise-related issues are expected to be solved in the near future following the development in research works on advanced PLC technologies. Additionally, the proposed system has been designed for use in an environment where the users are actively utilizing smart devices, which gives rise to the issue of generality. Considering the rapid distribution and use of smart devices including smartphones/tablets or wearables, however, it is safe to say that the problem of generality is almost ignorable at this stage. It is also possible to develop a simpler form of such devices for senior citizens or handicapped persons who are not familiar or who are having some difficulties with smart devices.

A PLC-based solution for decrepit buildings or apartments has been developed along with its testbed. The results from the experiment confirmed that the solution is flexible and suitable for use. As the main contribution of the study, the research has shown that there are great potentials for the PLC technology when it is converged with some of the advanced communication technologies.

The content presented by the first author Jun-Ho Huh when he was a senior research engineer at SUNCOM Company was about the three-phase three-line power line communication system [49] which was installed at a shadow area when building a ship and successfully resolved the communication problem.

Another reason for the transmission rate to drop during actual navigation is due to the noise occurring while conducting power line communications [49]. When the transmission rates were compared between office, land-based testbed, and actual ship [49,50], the rate measured in the office was a little lower than that of the testbed but higher or almost similar to the rate measured aboard. Different from the actual ship the walls of the office were made of cement. This shows that the fading effect can occur in the office as well and the attenuation of electromagnetic waves can be higher when compared with testbed surrounded by steel walls.

This study had continued to develop a monitoring solution for the old homes and completed the testbed experiment, securing the system stability.

Thus, this paper introduced ship PLC-based legacy-line communication (LLC) system construction technology and tested its performance level by conducting a data transmission experiment.

LLC system is a technology to secure a communication line without damaging the exterior of the facility by utilizing the conductive media already being installed in a ship [49], railroad [6] or a building [51–53]. Thus, this technology has the merit of largely reducing the structural changes as well as the time and expense required to design and construct additional lines. One of the typical examples of LLC systems is the power line (PL)-based system proposed in this study, but the existing telephone lines or copper cables that have been used as a communication line can be used as well [54–56]. Also, the technology utilizing air-conditioning pipes installed all cross the building can be a good candidate even though they are not used as a typical transmission media [52,57]. Therefore, we expect that the solution proposed in this study would lay a foundation for sustainability research.

Meanwhile, the PLC system requires at least two communication equipment per line to set the master-slave and from the second line, slaves can be added as many as necessary lines so that it is quite economic. In some of the foreign nations where it is difficult to install LAN systems for their houses, many leading network equipment companies manufacture PLC modems and sell the products using Wi-Fi. The PLC systems are useful in the areas where wireless systems are not suitable or installing Ethernet cables is difficult and can be considered as the easiest way of constructing a wired communication network in the decrepit building.

This study is also expected to contribute to improving the working environment at construction sites, especially in shadow areas. The platform technology designed here could be foundational for the future sustainable energy strategy, and functions such as uninterrupted power usage monitoring and remote power control/management will be quite useful for saving energy in both present and future worlds. The secondary technologies that should be studied in detail are not described here, but they will be introduced in future extended research work after completing patent registration.

Our future research plan is to develop a monitoring system for the smart grid equipment in Southeastern countries and it will be essential to develop a faster PLC-based communication solution at the same time.

Author Contributions: Conceptualization, J.-H.H.; Data curation, J.-H.H.; Formal analysis, J.-H.H.; Funding acquisition, J.-H.H.; Methodology, J.-H.H.; Project administration, J.-H.H. and J.H.P.; Resources J.-H.H. and J.H.P.; Software, J.H.P.; Supervision, J.H.P.; Validation, J.H.P.; Visualization, J.H.P.; Writing—original draft, J.-H.H. and J.H.P.; Writing—review & editing, J.H.P. All authors have read and agreed to the published version of the manuscript.

Funding: This research was supported by Energy Cloud R&D Program through the National Research Foundation of Korea (NRF) funded by the Ministry of Science, ICT (NRF-2019M3F2A1073385). Also, this work was supported by the National Research Foundation of Korea (NRF) grant funded by the Korea government (MSIT) (No.2017R1C1B5077157).

Acknowledgments: The part of first draft of this Congress Proceeding [46] was presented in at The 2015 World Congress on Information Technology Applications and Services Proceedings of the "Advanced Mobile, Communications, Security, Multimedia, Vehicular, Cloud, IoT, and Computing" (World-IT 2015), February 24-26, 2015, Jeju, Republic of Korea. We are grateful to Tae-Woo Park, president of Samsung Renault, for allowing us to use the building as a test bed.

Conflicts of Interest: The authors declare no conflict of interest.

Abbreviations

PLC:	Power Line Communication
SMI:	Smart Metering Infrastructure
DR:	Demand-Response
HV:	High-Voltage
UNB-PLC:	Ultra-narrowband Power Line Communication
NB-PLC:	NarrowBand Power Line Communication
SG:	Smart Grid
DCD:	Data Carrier Detect
RD:	Received Data
TD:	Transmit Data
DTR:	Data Terminal Ready
DSR:	Data Set Ready
RTS:	Request to Send
CTS:	Clear to Send
RI:	Ring Indicator
HIFs:	High-Impedance Faults
PLC-AN:	PLC-Access Network
PLC-OR:	PLC-Opportunistic Routing
MP-PLCA:	Multi-Protocol PLC Analyzer
GUI:	Graphical User Interface
MCU:	Micro Controller Unit
DCU:	Data Concentration Unit
LLC:	Legacy-Line Communication
PL:	Power Line
LAN:	Local Area Network

References

1. Jun-Ho, H.; Otgonchimeg, S.; Seo, K. Advanced Metering Infrastructure Design and Test Bed Experiment Using Intelligent Agents: Focusing on the PLC Network Base Technology for Smart Grid System. *J. Supercomput.* **2016**, *72*, 1862–1877.
2. Konark, S.; Saini, L.M. Power-line communications for smart grid: Progress, challenges, opportunities and status. *Renew. Sustain. Energy Rev.* **2017**, *67*, 704–751.

3. Mazzeo, D.; Oliveti, G. Advanced Innovative Solutions for Final Design in Terms of Energy Sustainability of Nearly/Net Zero Energy Buildings (nZEB). Available online: https://www.mdpi.com/journal/sustainability/special_issues/Advanced_Innovative_Solutions (accessed on 14 January 2020).
4. Lopez, G.; Moreno, J.I.; Amaris, H.; Salazar, F. Paving the road toward Smart Grids through large-scale advanced metering infrastructures. *Electr. Power Syst. Res.* **2015**, *120*, 194–205. [CrossRef]
5. Konark, S.; Mohan, S.L. Performance analysis of smart metering for smart grid: An overview. *Renew. Sustain. Energy Rev.* **2015**, *49*, 720–735.
6. Jun-Ho, H. Smart Grid Test Bed Using OPNET and Power Line Communication. In *Advances in Computer and Electrical Engineering*; IGI Global: Hershey, PA, USA, 2017; pp. 1–425.
7. Mak, S.T.; Moore Thomas, G. TWACSTM, A New Viable Two-Way Automatic Communication System for Distribution Networks. Part II: Inbound Communication. *IEEE Trans. Power Appar. Syst.* **1984**, *103*, 2141–2147. [CrossRef]
8. Build Your Smart Home. Available online: http://buildyoursmarthome.co/home-automation/protocols/x10/ (accessed on 14 January 2020).
9. Power Line Technologies in Home Networking. Available online: http://www.xilinx.com/esp/consumer/home_networking/pdf_files/ch_7_plc/complete.pdf (accessed on 14 January 2020).
10. Galli, S.; Lys, T. Next generation Narrowband (under 500 kHz) Power Line Communications (PLC) standards. *Commun. China* **2015**, *12*, 1–8. [CrossRef]
11. Bumiller, G.; Sauter, T.; Pratl, G.; Treytl, A. Secure and reliable wide-area power line communication for soft-real-time applications within REMPLI. In Proceedings of the International Symposium on 2015 Power Line Communications and Its Applications, Vancouver, BC, Canada, 6–8 April 2005; pp. 57–60.
12. Real-Time Energy Management via Power-Lines and Internet. Available online: http://www.iad-de.com/Rempli_Flyer.pdf (accessed on 14 January 2020).
13. Haidine, A.; Portnoy, A.; Mudriievskyi, S.; Lehnert, R. DLC VIT4IP project high-speed NB-PLC for smart grid communication design of field trial. In Proceedings of the IEEE Power Line Communications and Its Applications (ISPLC), Beijing, China, 27–30 March 2012; pp. 88–93.
14. Milioudis, A.N.; Andreou, G.T.; Labridis, D.P. Enhanced protection scheme for smart grids using power line communications techniques -part II: Location of high impedance fault position. *IEEE Trans. Smart Grid* **2012**, *3*, 1631–1640. [CrossRef]
15. Sung-Guk, Y.; Jang, S.; Yong-Hwa, K.; Bahk, S. Opportunistic Routing for Smart Grid With Power Line Communication Access Networks. *IEEE Trans. Smart Grid* **2014**, *5*, 303–311.
16. PRIME Alliance. Available online: http://www.prime-alliance.org/?page_id492 (accessed on 14 January 2020).
17. DLMS/COSEM Solution for Power-Line Communication. Available online: http://www.st.com/web/en/resource/sales_and_marketing/presentation/product_presentation/dlmscosem_marketing_pres.pdf (accessed on 14 January 2020).
18. G3-PLC Overview. Available online: http://www.g3-plc.com/content/g3-plc-overview (accessed on 14 January 2020).
19. Oksman, V.; Zhang, J.G. HNEM: The new ITU-T standard on narrowband PLC technology. *IEEE Commun. Mag.* **2011**, *49*, 36–44. [CrossRef]
20. Quang-Dung, H.; Chon-Wang, C.; Derakhshani, M.; Le-Ngoc, T. An analysis on throughput and feasibility of Narrow-band Power Line Communications in Advanced Distribution Automation scenarios. In Proceedings of the IEEE International Conference on Communications (ICC), Sydney, NSW, Australia, 10–14 June 2014; pp. 4233–4238.
21. Stefano, R.; Paolo, F.; Alessandra, F.; Mattia, R.; Emiliano, S.; Angelo, V. Performance analysis of power line communication in industrial power distribution network. *Comput. Stand. Interfaces* **2015**, *42*, 9–16.
22. Carli, R.; Dotoli, M. A decentralized resource allocation approach for sharing renewable energy among interconnected smart homes. In Proceedings of the 2015 54th IEEE Conference on Decision and Control (CDC), Osaka, Japan, 15–18 December 2015; pp. 5903–5908.
23. Blaauwbroek, N.; Nguyen, P.H.; Konsman, M.J.; Shi, H.; Kamphuis, R.I.; Kling, W.L. Decentralized resource allocation and load scheduling for multicommodity smart energy systems. *IEEE Trans. Sustain. Energy* **2015**, *6*, 1506–1514. [CrossRef]
24. Brusco, G.; Burgio, A.; Menniti, D.; Pinnarelli, A.; Sorrentino, N. Energy management system for an energy district with demand response availability. *IEEE Trans. Smart Grid* **2014**, *5*, 2385–2393. [CrossRef]

25. Carli, R.; Dotoli, M. Cooperative distributed control for the energy scheduling of smart homes with shared energy storage and renewable energy source. *IFAC Pap. Line* **2017**, *50*, 8867–8872. [CrossRef]
26. Figueiredo, J.; da Costa, J.S. A SCADA system for energy management in intelligent buildings. *Energy Build.* **2002**, *49*, 85–98. [CrossRef]
27. Lind, A.; Espegren, K. The use of energy system models for analyzing the transition to low-carbon cities—The case of Oslo. *Energy Strat. Rev.* **2017**, *15*, 44–56. [CrossRef]
28. Sun, X.Y.; Zhang, B.A.; Tang, X.; Benjamin, C.M.; Mikael, H. Sustainable Energy Transitions in China: Renewable Options and Impacts on the Electricity System. *Energies* **2016**, *9*, 980. [CrossRef]
29. Sun, X.Y.; Zhang, B.A.; Zou, L.P.; Wang, R.Z. The Role of Renewable Energy in China's Sustainable Energy Transition. In Proceedings of the 2nd International Conference on Sustainable Development (ICSD 2016), Xi'an, China, 2–4 December 2016; pp. 327–332.
30. Amrutha, A.A.; Balachandra, P.; Mathirajan, M. Model-based approach for planning renewable energy transition in a resource-constrained electricity system—A case study from India. *Int. J. Energy Res.* **2018**, *42*, 1023–1039. [CrossRef]
31. Thapar, S.; Sharma, S.; Verma, A. Local community as shareholders in clean energy projects: Innovative strategy for accelerating renewable energy deployment in India. *Renew. Energy* **2017**, *101*, 873–885. [CrossRef]
32. Mah, N.Y.; Wu, Y.Y.; Ip, C.M.; Hills, P.R. The role of the state in sustainable energy transitions: A case study of large smart grid demonstration projects in Japan. *Energy Policy* **2013**, *63*, 726–737. [CrossRef]
33. Steg, L.; Perlaviciute, G.; van derWerff, E. Understanding the human dimensions of a sustainable energy Transition. *Front. Psychol.* **2015**, *6*, 805. [CrossRef]
34. Becchio, C.; Corgnati, S.P.; Delmastro, C.; Fabi, V.; Lombardi, P. The role of nearly-zero energy buildings in the transition towards Post-Carbon Cities. *Sustain. Cities Soc.* **2016**, *27*, 324–337. [CrossRef]
35. Wang, H.; Di Pietro, G.; Wu, X.; Lahdelma, R.; Verda, V.; Haavisto, I. Renewable and sustainable energy transitions for countries with different climates and renewable energy sources potentials. *Energies* **2018**, *11*, 3523. [CrossRef]
36. Liu, J.; Zhao, L.; Geng, Z. Current situations and future developments of PLC technology in China. In Proceedings of the IEEE Power Line Communications and Its Applications (ISPLC), Beijing, China, 27–30 March 2012; pp. 60–65.
37. Jae-Shik, Y.; Jung-Chu, W.; Chung-Ha, P.; Yong-Jae, S.; Jae-Heon, J. Development of Low Power PLC Modem for Monitoring of Power Consumption and Breaking of Abnormal Power. *Trans. Korean Inst. Electr. Eng.* **2009**, *58*, 2281–2285. (In Korean)
38. Seon-Tae, Y.; Ki-Won, P.; Bong-Jin, K. Implementation of the Man-made Disaster Prevention and Localization Information Monitoring System based on PLC. *J. Adv. Navig. Technol.* **2010**, *14*, 384–390. (In Korean)
39. Keun-Soo, P. Implement of a Watt-Hour Meter Monitoring System using Power Line Communication. *J. Korea Inst. Electron. Commun. Sci.* **2013**, *8*, 1143–1148. (In Korean)
40. Keun-Soo, P.; Young-Ho, L. Implement of watt-hour meter monitoring system by internet map based GUI using power line communication. *J. Korea Inst. Electron. Commun. Sci.* **2013**, *8*, 1435–1441. (In Korean)
41. de Araújo, P.R.C.; Filho, R.H.; Rodrigues, J.J.P.C.; Oliveira, J.P.C.M.; Braga, S.A. Infrastructure for integration of legacy electrical equipment into a smart-grid using wireless sensor networks. *Sensors* **2018**, *18*, 1312. [CrossRef]
42. Viciana, E.; Alcayde, A.; Montoya, F.G.; Baños, R.; Arrabal-Campos, F.M.; Zapata-Sierra, A.; Manzano-Agugliaro, F. Openzmeter: An efficient low-cost energy smart meter and power quality analyzer. *Sustainability* **2018**, *10*, 4038. [CrossRef]
43. González, I.; Calderón, A.J. Integration of open source hardware Arduino platform in automation systems applied to Smart Grids/Micro-Grids. *Sustain. Energy Technol. Assess.* **2019**, *36*, 100557. [CrossRef]
44. Ghasempour, A. Internet of Things in Smart Grid: Architecture, Applications, Services, Key Technologies, and Challenges. *Inventions* **2019**, *4*, 22. [CrossRef]
45. Hidalgo, S.; Angeles, M.; Ma, C. A survey on visual data representation for smart grids control and monitoring. *Sustain. Energy Grids Netw.* **2018**, *16*, 351–369. [CrossRef]
46. Jun-Ho, H.; Dong-Geun, L.; Kyungryong, S. Design and implementation of the basic technology for realtime smart metering system using power line communication for Smart Grid. *Adv. Comput. Sci. Ubiquitous Comput.* **2015**, *373*, 663–669.

47. SK Broadband. Available online: https://www.mk.co.kr/news/business/view/2017/03/151149/ (accessed on 14 January 2020). (In Korean)
48. Available online: http://m.electimes.com/article.php?aid=1566520662184954010 (accessed on 14 January 2020). (In Korean)
49. Jun-Ho, H.; Taehoon, K.; Kyungryong, S. Design of a Shipboard Outside Communication Network and Its Testbed Using PLC: For Safety Management during the Ship Building Process. *Processes* **2018**, *6*, 67.
50. Huh, J.-H.; Seo, Y.-S. Understanding Edge Computing: Engineering Evolution with Artificial Intelligence. *IEEE Access* **2019**, *7*, 164229–164245. [CrossRef]
51. Kim, H.-S.; Kim, J.R.; Lee, H.Y.; Kim, K.U.; Huh, J.S.; Lee, J.H.; Oh, Y.W.; Byun, W.B.; Gwak, K.Y.; Ju, S.H. Signal transmission properties of the inductive coupler using the high permeability magnetic materials. *J. Korean Inst. Electr. Electron. Mater. Eng.* **2006**, *19*, 339–343. [CrossRef]
52. Kim, H.S.; Byun, C.H. Development of IBS Network System Using Copper Line of Air Conditioner. Mattron Corp, Gyeongnam at Republic of Korea. Available online: http://may.ulsan.ac.kr/muju2015/sogae_files/%E1%84%8E%E1%85%A9%E1%84%85%E1%85%A9%E1%86%A8%E1%84%8C%E1%85%B5%E1%86%B8%20V5.pdf (accessed on 15 February 2020).
53. Kim, H.-S.; Park, S.H.; Kang, S.G. Development of communication joint tools for implementing a legacy-line communication system in a train. *J. Korean Inst. Inf. Commun. Eng.* **2015**, *19*, 877–887. [CrossRef]
54. Dostert, K. *Powerline Communications*; Prentice Hall: Upper Saddle River, NJ, USA, 2001.
55. IEEE Std. *IEEE Standard for Broadband over Power Line Networks: Medium Access Control and Physical Layer Specifications*; IEEE: Piscataway, NJ, USA, 2010.
56. Duche, D.; Gogate, V. Signal attenuation in power-line communication channel. *Int. J. Emerg. Trends Technol. Comput. Sci.* **2014**, *3*, 123–130.
57. Hyun-Sik, K.; Kang, S.G. A powerline-based legacy-line communication system for implementation of a communication network in ship. *J. Korea Inst. Inf. Commun. Eng.* **2015**, *19*, 1831–1838.

© 2020 by the authors. Licensee MDPI, Basel, Switzerland. This article is an open access article distributed under the terms and conditions of the Creative Commons Attribution (CC BY) license (http://creativecommons.org/licenses/by/4.0/).

sustainability

MDPI

Article

A Integrated Dedicated Outdoor Air System to Optimize Energy Saving

Yew Khoy Chuah * and Jun Jie Yang

Department of Energy and Refrigerating Air-conditioning Engineering, National Taipei University of Technology, Taipei 10608, Taiwan; k19800575@gmail.com

* Correspondence: yhtsai@ntut.edu.tw; Tel.: +886-2-27712171; Fax: +886-2-87733713

Received: 25 November 2019; Accepted: 21 January 2020; Published: 2 February 2020

Abstract: Outdoor air supply is required to maintain good indoor air quality (IAQ). For tropical or subtropical regions, warm and humid outdoor air would cause excess air-conditioning energy use. This study has proposed an integrated dedicated outdoor air system (IDOAS), which integrates the enthalpy exchange and outdoor air cooling into a unitary system. IDOAS could operate independently of central air-conditioning systems thus saving tremendous piping cost and energy needed to deliver chilled water to outdoor air unit in a conventional centralized system. An experimental unit of IDOAS was built to prove this novel concept. Enthalpy exchange efficiency was tested to be about 44%. The test results show that about 44% of energy needed to condition the outdoor air can be saved. A reverse Rankine refrigeration cycle was integrated to cool the outdoor air. Due to this integrated configuration, the air passing through the condenser would be at a lower temperature. The consequent lower refrigerant condensing temperature would improve the cooling cycle efficiency. The cooling coefficient of performance (COP) was improved by about 46%. In addition, the outdoor air could be conditioned to a lower humidity before being supplied to space, which would improve the thermal comfort. The test results of this novel IDOAS show that it could provide good air quality at lower energy use.

Keywords: outdoor air; IAQ; energy saving; enthalpy heat exchange; green building

1. Introduction

Indoor air quality has become an indoor environment problem worldwide. The Indoor Air Quality Act of Taiwan [1] was enacted in 2012 and it became one of the few countries which imposed a regulation on indoor air quality. The indoor air quality standards imposed by the Indoor Air Quality Act of Taiwan are given in Table 1. Public health was the major concern in the enactment of this regulation. There are nine items of air quality standards. In Table 1, 8 h means 8 h average value, so for 1 h and 24 h. The standards were first imposed in public spaces such as public transport stations, hospitals, and later all levels of education institutions. Spengler and Sexton [2] discussed the risk of indoor air pollution with a public health perspective. They mentioned that most people stay indoors about ninety percent of the time and the exposure to indoor pollution was a health issue.

Table 1. Taiwan indoor air quality standards.

Item		Standard	Unit
CO_2	8 h	1000	ppm
CO	8 h	9	ppm
HCHO	1 h	0.08	ppm
TVOC	1 h	0.56	ppm
Bacteria	Maximum	1500	CFU/m^3

Table 1. *Cont.*

Item		Standard	Unit
Fungi	Maximum	1000 or ratio of indoor to outdoor ≤1.30	CFU/m^3
PM_{10}	24 h	75	μg/m^3
$PM_{2.5}$	24 h	35	μg/m^3
O_3	8 h	0.06	ppm

Monitoring all the items in Table 1 would be very costly. In the guiding principles, only items identified with potential sources are needed to be monitored. The selection is also based on the risk assessment of public health. In general, fresh air ventilation is a mean to maintain good air quality. It is often said a solution to pollution is dilution.

South Korea had also imposed regulations on indoor air quality. The revised Act in 2003 [2] increased the number of pollutant standards. In some countries or regions, the IAQ Certification Scheme is promoted with the same purpose to improve indoor air quality, such as that carried by the Hong Kong government [3].

It is generally known that fresh outdoor air is important to maintain good indoor air quality (IAQ). Early in 1992, Levenhagen [4] proposed using a dedicated outdoor air system (DOAS) to ensure a proper ventilation rate. Mumma [5] in 2001 proposed a design of DOAS. Other related references suggested using DOAS to handle air-conditioning latent load [6–9]. Mumma [10] gave an overview of integrating dedicated outdoor air systems with parallel terminal systems. Shank and Mumma [11] further proposed using DOAS to handle the latent cooling load, working with distributive parallel systems to handle sensible cooling. Chien et al. [12] presented experimental results of radiant floor cooling. They proposed that a parallel system such as DOAS be used to handle latent load. Li et al. [13] considered a multistage cooling design for DOAS for the purpose of supplying lower air temperature.

DOAS with heat recovery was considered in some research [14,15]. Chuah and Lin [14] analyzed the air-conditioning load for a typical office building in Taiwan. The results showed that about 27% of air-conditioning energy was used to condition the outdoor air to the indoor conditions. Lower CO_2 concentration means more outdoor air has to be supplied to the space. Then, more energy would be needed to cool and dehumidify the outdoor air. James [16] also mentioned that a significant portion of commercial air-conditioning load was due to outdoor air.

Currently, the common practice in central air-conditioning systems is by using an outdoor air handling unit (OAU) to bring in sufficient outdoor air. Outdoor air will be cooled to indoor conditions of temperature and humidity. Such practices have the disadvantage of coupling to central air-conditioning system operation. In such practices, pumping energy to transport chilled water from the central plant to OAU is costly. As shown in Figure 1, chilled water from the central system has to be pumped to the OAU via piping. Normally, the piping construction cost is quite high. The indoor air quality standard of ASHRAE in 2013 [17] further required for fresh air to be effectively distributed to all parts of a building. ASHRAE later published a technical guide for designing DOAS [18]. The above discussion indicates that conventional OAU has the disadvantage of coupling to the central system. The current outdoor air unit (OAU) is generally designed with the purpose to ensure proper outdoor air supply, often not considering the piping construction cost and energy cost of OAU. As seen in Figure 1, OAU is normally designed to distribute outdoor air to each space that would further cause ducting cost and fan energy. Due to climate change, warmer outdoor air could only result in more energy use by an air-conditioning system, especially that to condition the outdoor air.

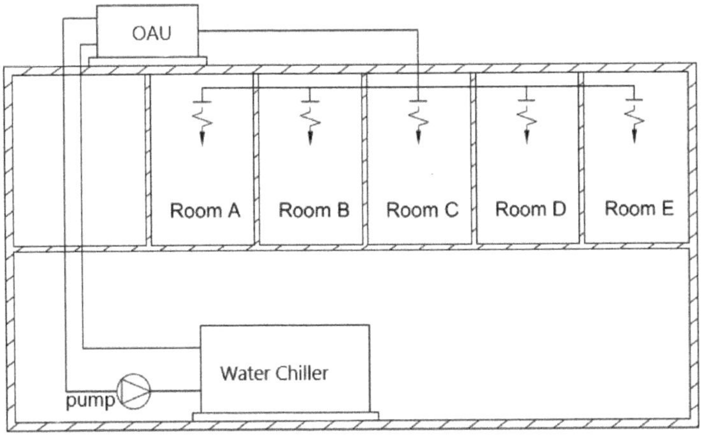

Figure 1. Outdoor air supply unit of central air-conditioning system.

Mui et al. [19] attempted to analyze the air-conditioning energy use in some Hong Kong office buildings. They concluded the energy use was strongly related to the CO_2 concentration. CO_2 concentration is strongly related to the outdoor air supply rate. This further shows that more fresh air ventilation would result in higher air-conditioning energy use.

Chaeand and Strand [20] attempted using computer modeling to understand the energy performance of DOAS. Mumma [21] further studied the feasibility of using DOAS to maintain a building at a positive pressure preventing excess leakage of outdoor air. Xiao et al. [22] applied a liquid desiccant method to dehumidify the outdoor air but the liquid desiccant method needed cooling to maintain the liquid temperature. There were also research studies on applying solid desiccant to remove moisture from fresh air such as that discussed by Liu et al. [23]. Either using liquid or a solid desiccant would have to obey the thermodynamic law. The heat of dehumidification has to be removed by some means of cooling. Therefore, the advantage of desiccants would be constrained.

The review of the above literature indicates the problem of reducing energy use while maintaining good IAQ. Global warming effects are becoming more eminent as days pass. This study was then an attempt to investigate a mean of saving air-conditioning energy while maintaining good indoor air quality. A novel integrated unitary outdoor air system (IDOAS) was studied, an experimental unit was fabricated, and then tested in this study.

2. The Advantage of Integrated Unitary Dedicated Ventilation Unit

Due to the large enthalpy difference of indoor and outdoor air, it would cause excessive air-conditioning energy use to cool and dehumidify the outdoor air to room air conditions. Enthalpy exchange between room exhaust air and outdoor air intake is generally known to be effective in reducing the enthalpy (energy state) of outdoor air. This study thus proposed a novel design that integrated enthalpy exchange and air cooling into a unitary system. Such a design has the advantage of distributive unitary systems instead of centralized outdoor air supply. A unitary system can be individually installed and decoupled from the central system. Therefore, tremendous cost of piping, ducting not the least fan, and pump energy can be saved. As mentioned, and explained in Figure 1, a conventional centralized outdoor air unit (OAU) would be coupled with the operation of a central air-conditioning system and cannot operate independently when fresh air is required but not cooling.

The ventilation unit proposed was a unitary integrated dedicated outdoor air system as shown in Figure 2, hereafter referred to as IDOAS. It can be seen in Figure 2 that an IDOAS consists of an enthalpy exchange unit and a refrigeration cycle cooling unit in a unique configuration. This configuration is important so that optimal energy saving can be achieved. It is shown in Figure 2a that the refrigerant compressor, refrigerant tubes, expansion valve, condenser, and evaporator are all housed in a unit with a total heat exchanger (enthalpy exchanger).

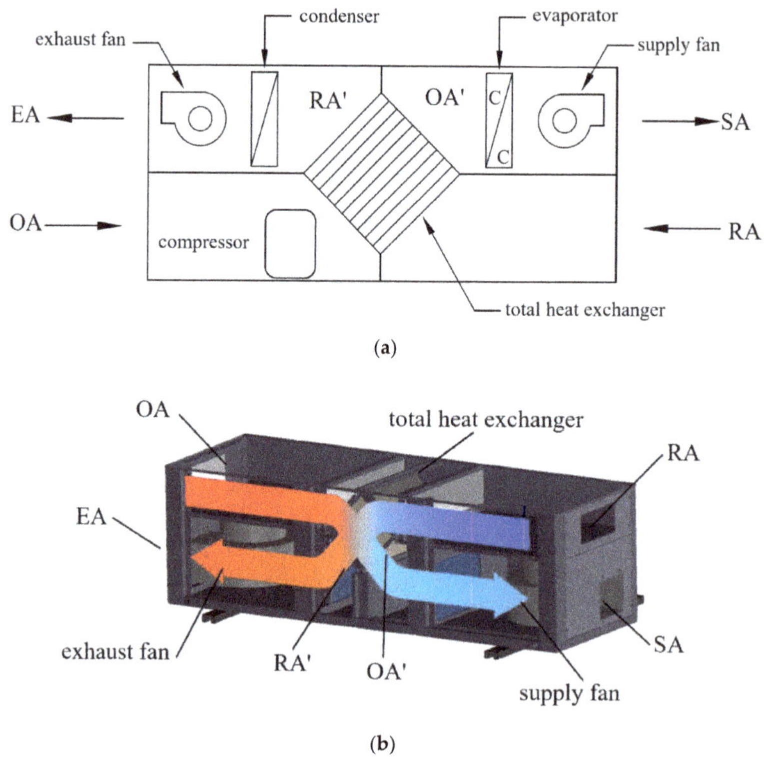

Figure 2. The configuration of an integrated dedicated outdoor air system (IDOAS). (**a**) The major components of IDOAS. (**b**) The air flows and heat exchange through the total heat exchanger.

Figure 2b shows that room air (RA) would undergo total heat exchange with outdoor air (OA) and becomes RA′. RA′ will then flow through the refrigerant condenser before being exhausted as exhaust air (EA). With this configuration, the condenser would be cooled by a lower temperature RA′ (compared to OA for an air-cooled system). Consequently, the system will have a lower refrigerant condensing temperature. In terms of the laws of thermodynamics, the refrigerant cycle efficiency will be higher at a lower refrigerant condensing temperature. This explains the basics that IDOAS could operate at a higher refrigeration cycle efficiency.

It is also noted in Figure 2 that OA becomes OA′ at a cooler and drier state after total heat exchange with RA. OA′ would be further cooled by a refrigerant evaporator and passing out as supply air (SA). In the process heat, and moisture will be further removed from OA′. Therefore, the outdoor supply air (SA) will be at a state of lower temperature and humidity. Temperature and humidity are important factors of thermal comfort. This explains that IDOAS could provide outdoor air at states that would enhance the thermal comfort for the air-conditioned space. Besides the advantages as mentioned, IDOAS can be installed at the point of use as unitary systems, thus can be more effectively distribute outdoor air to air-conditioned spaces.

A model unit of IDOAS was built for experimental tests. A picture of the test unit of IDOAS is shown in Figure 3. Figure 3 shows that IDOAS has all components housed in a single unit with the configuration shown in Figure 2. The refrigerant condenser and the evaporator are integrated into IDOAS. The refrigerant condenser is air-cooled by design but not cooled by the outdoor air, and instead, it will be cooled by exhaust room air at a much lower temperature. As for the evaporator, the refrigerant undergoes direct expansion, absorbs heat, and removes moisture from OA′. On the contrary, for OAU, as shown in Figure 1, chilled water (say at 7 °C) has to be pumped to the OAU to cool the air.

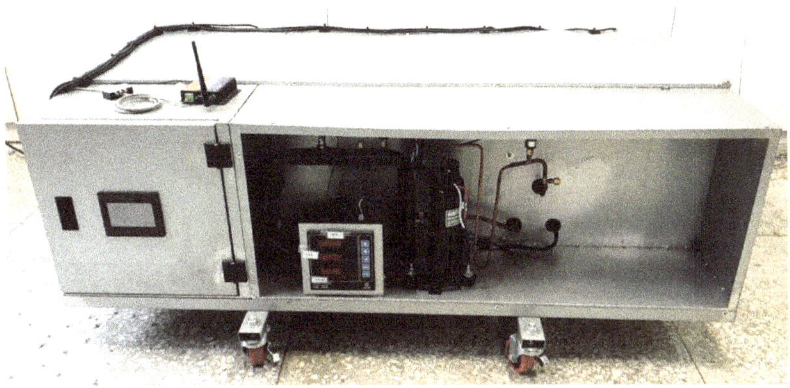

Figure 3. The IDOAS fabricated.

The energy saving principles of IDOAS can be explained using a psychrometric chart as shown in Figure 4. The ordinate of the psychrometric chart is the humidity ratio given as kg vapor in a kg of dry air (or g vapor/kg dry air), often referred to as the absolute humidity of air. The abscissa is the air temperature. Enthalpy is the slant axis on the psychrometric chart with the unit as kJ/kg of dry air. In the psychrometric chart, outdoor air (OA) after the total heat exchanger is brought to OA′ with much lower enthalpy. RA′ after the total heat exchanger will be at higher enthalpy. The increase of enthalpy for RA in principle will be equal to that of the enthalpy decrease of OA. Therefore, total exchange is often referred to as enthalpy exchange. In air-conditioning, the energy state of air is evaluated as enthalpy as indicated in Equation (1). In Equation (1), temperature T is evaluated as °C and enthalpy h at kJ/kg.

$$\mathrm{h} - c_{pa}T + \omega\left(c_{pw}T + h_g\right) \tag{1}$$

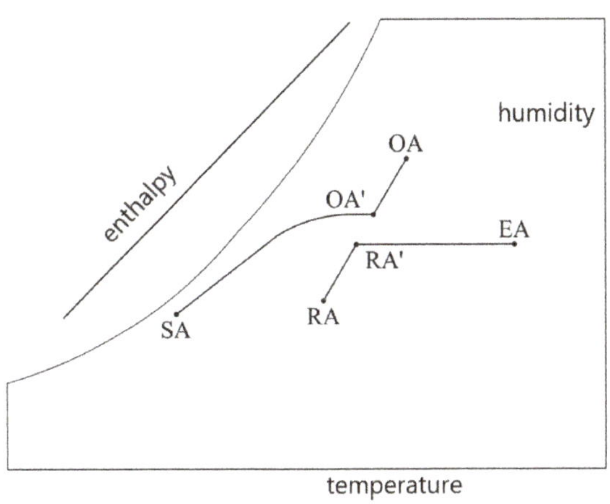

Figure 4. Psychrometrics of IDOAS.

In Equation (1), ω is the humidity ratio (kg vapor/kg dry air), c_{pa} and c_{pw} are, respectively, the specific heats of dry air and water vapor; h_g is the water vapor enthalpy at 0 °C.

A cross-flow-type total heat exchanger was used in this study. A typical configuration of a cross-flow type total heat exchange is shown in Figure 5. The cross-flow of two air streams flow through alternate permeable layers of the total heat exchanger. Total heat exchange between OA and RA would reduce the energy state change of OA to OA'. For a total heat exchanger, the corrugated structure is impregnated with some kind of desiccant, for example silica gel. The humidity ratio of OA is higher than that of the equilibrium value at the desiccant interface. The humidity ratio represents the concentration of moisture in air. In principle, the moisture in OA will be adsorbed by the desiccant. The adsorbed moisture would then permeate to the other side of the permeable layers. RA has a lower humidity ratio than the equilibrium value at the desiccant surface. Therefore, moisture will be desorbed from the desiccant into the RA air stream. This explains the latent heat exchange. The sensible exchange is by the principle of the heat exchange of air streams between the alternate layers. The above explains the principles of total heat exchange (sensible plus latent) or the enthalpy exchange that designed into the IDOAS.

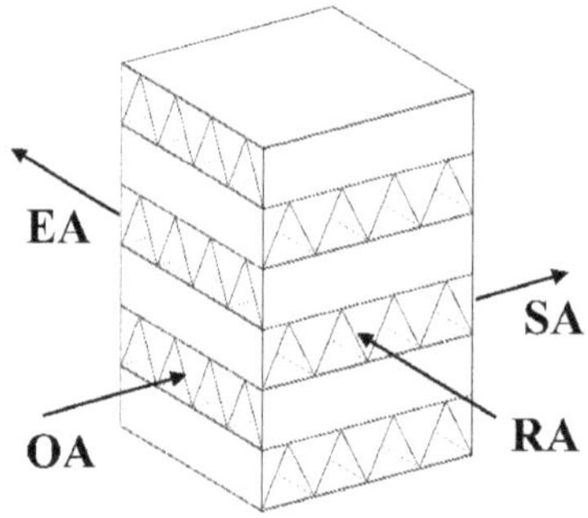

Figure 5. A total heat exchanger.

Taiwan is situated in the subtropical region located about 23 degrees north in latitude. During the summer months in Taiwan, outdoor air (OA) temperature is often higher than 35 °C, especially in the afternoon. Air-cooled air-conditioners are generally rated at low energy efficiency. This is due to the condenser refrigerant cooled by the high ambient temperature. The air temperature is the key factor for energy efficiency. For IDOAS, as indicated in Figures 2 and 4, a lower temperature of RA' instead is used to cool the refrigerant in the condenser. The energy state change from RA' to EA is the cooling of the refrigerant in the condenser. The energy state change from OA' to SA represents the removal of heat and moisture from OA'. This explains the advantages of this novel ventilation unit, which can provide sufficient high-quality fresh air that is low in energy consumption.

3. Experimental Tests

The measurements in the experimental unit are shown in Figure 6. The measurements included the air flow rates separately for exhaust room air and outdoor air intake. The calculation of total heat exchange efficiency requires the measurements of temperature and humidity (T, RH) before and after the total heat exchanger for the two air streams. Enthalpy change for the two air streams (room exhaust and outdoor air) were calculated from temperature and humidity measurements as indicated in Equation (1). Theoretically, enthalpy exchange efficiency ε can be calculated from the enthalpy change of *OA* or the enthalpy change of *RA* as shown in Equation (2). The calculation of enthalpy exchange efficiency is based on the theoretical maximum enthalpy exchange from *OA* to *RA* $\dot{m}_{min}(h_{OA} - h_{RA})$.

$$\varepsilon = \frac{\dot{m}_{OA}(h_{OA} - h_{OA'})}{\dot{m}_{min}(h_{OA} - h_{RA})} = \frac{\dot{m}_{RA}(h_{RA'} - h_{RA})}{\dot{m}_{min}(h_{OA} - h_{RA})} \tag{2}$$

The air flow rates were calculated as the average velocity multiplied by the flow areas of IDOAS. Three velocities (test point v in Figure 6) were measured for each flow area to calculate the average velocity. The power consumption of the compressor was measured by a multi-function power meter shown in Table 2. Temperature, humidity, and velocity probes were connected to the testo multi-function analyzer to measure and collect the measurement data.

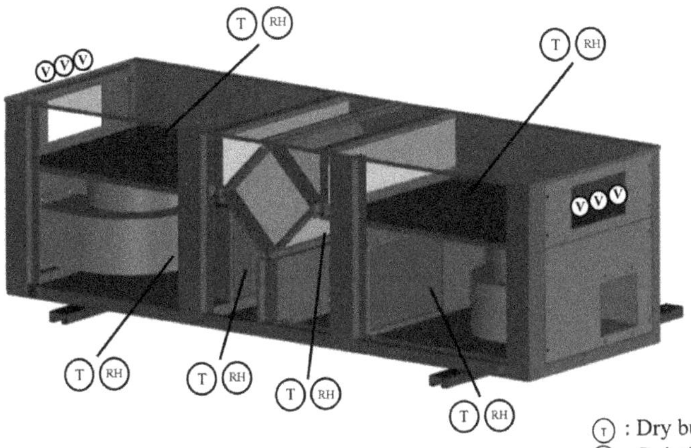

Figure 6. The measurements in the experimental tests.

Table 2. Instruments used in the experiments after calibration.

Instrument	Made	Specifications and Accuracy
Multi-function analyzer	testo 480	Probes made by testo can be connected to 480 for measuring temperature, humidity, and air velocity
Vane velocity probe	testo	Diameter 60 mm, range from 0.25 to 20 m/s, accuracy ±0.1 m/s
Temperature probe	testo	Range from −20 to +70 °C, accuracy ±0.1 °C
Humidity probe	testo	Range from 0% to 100% RH, ceramic capacity type, accuracy 2% RH
Multi-function power meter	TAIK S6–300	Measure voltage, current, and power, accuracy ±0.2%

The cooling rate (W) of IDOAS was the air cooling rate by the evaporator calculated using Equation (3).

$$\text{Cooling rate } \dot{Q} = \frac{1}{3}\sum\nolimits_{1}^{3} \rho v_i \, x \left[A \, x \, (h_i - h_o)\right] \quad (3)$$

The first term $\frac{1}{3}\sum_1^3 \rho v_i$ is the mass flux (kg/(s.m^{-2}) and v_i is the velocities measured. h_i and h_o are the enthalpy before and after flowing through the evaporator. The energy efficiency of the cooling cycle is represented by the coefficient of performance (COP), which is calculated using Equation (4).

$$\text{COP} = \frac{\dot{Q}}{W} \quad (4)$$

In Equation (4), $\dot{Q}$ is the cooling rate as calculated using Equation (3) and W is the power measured for the refrigerant compressor. The instruments used to measure the performance of IDOAS are described in Table 2. Power use of the refrigerant compressor was measured by a multi-function power meter.

A small experimental test unit of IDOAS was designed, fabricated, and tested. The specifications of the experimental test unit are described in Table 3. The experimental unit was designed with a ventilation rate of about 450 m^3/h, sufficient for a space of 15 persons [15]. The compressor was supplied by a Taiwanese compressor manufacturer. The total heat exchanger was made in Japan.

The refrigerant condenser and the evaporator were sized by the authors and supplied by a local coil company. The objectives of the experiments were to prove the novelty of IDOAS, to understand the effectiveness of enthalpy exchange, and the refrigerant cycle cooling performance. The energy saving potential of IDOAS was also evaluated.

Table 3. The experimental IDOAS unit.

Item	Specifications
Compressor	Constant speed type, R410A refrigerant, rated cooling capacity 1900 W, and COP 2.90
Total heat exchanger	300 mm × 175 mm, cross-flow type
Condenser	3423 W at 35 °C air temperature
Evaporator	2867 W at 26 °C air temperature
Air volume	450 m^3/h or 0.125 m^3/s

4. Test Results and Discussion

The IDOAS had two variable speed controlled fans. Several tests were carried out to understand the fan performance. The designed supply air volume was 0.125 m^3/s. The test results of fan frequency and the corresponding air volume are given in Table 4. Graphical plots of frequency with air flow volume are given in Figure 7. The two fans were connected to a single power source thus operated at the same frequency. The frequency chosen was 1280 Hz, and the flow volumes at this frequency were 0.132 m^3/s and 0.1206 m^3/s, respectively, for the exhaust and supply fans.

Table 4. Tested air volume at different frequency.

Frequency, Hz	Exhaust Fan		Supply Fan	
	Air Volume (m^3/s)	Power (kW)	Air Volume (m^3/s)	Power (kW)
1600	0.1576	0.118	0.1749	0.108
1280	0.1137	0.062	0.1288	0.058
960	0.0830	0.030	0.0860	0.026

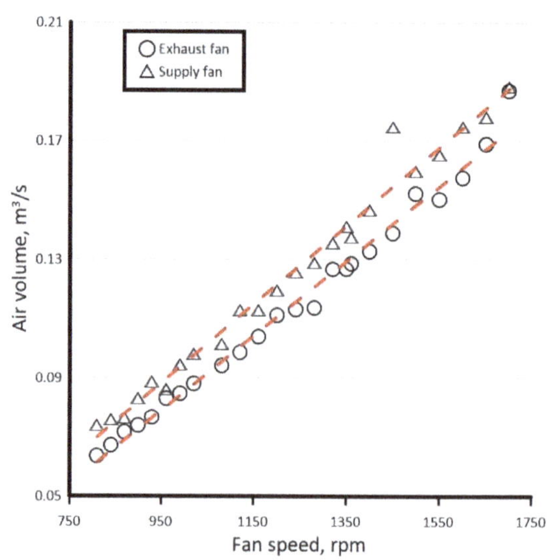

Figure 7. Graphical plots of frequency with air flow volume for fans.

5. Total Heat Exchange Performance

The test results of total heat exchanger performance are given in Tables 5 and 6. In Table 5, it is noted that OA′ was about 6 °C lower than OA. The humidity ratio of OA′ was lower than that of OA by about 2.29 g/kg. More significantly, enthalpy of OA′ was lower than OA by 15.33 kJ/kg. It would mean that 15.33 kJ of cooling can be saved per kg of outdoor air. The reduction of about 2.29 g/kg of moisture from OA is also significant. It shows that OA had been cooled and at the same time dehumidified without mechanical means but just through the total heat exchange with the room exhaust air. The cooler and drier air after flowing through the evaporator will be conditioned to an even lower temperature and lower humidity supply air (SA) would, in effect, improve the air quality, and at the same time increase the thermal comfort.

Table 5. The total heat exchange for outdoor air.

OA States				**OA′ States**			
Temp, °C	RH%	Humidity ratio g/kg	h, kJ/kg	Temp, °C	RH%	Humidity ratio g/kg	h, kJ/kg
33.1	57.8	18.57	81.97	27.2	72.2	16.49	69.41
33.0	58.7	18.77	81.25	27.1	72.0	16.34	68.93
33.2	58.2	18.82	81.59	27.3	71.7	16.47	69.46
Average OA enthalpy = 81.60 kJ/kg Average OA humidity ratio = 18.72 g/kg				Average OA′ enthalpy = 66.27 kJ/kg Average OA′ humidity ratio = 16.43 g/kg			

Table 6. The total heat exchange for exhaust room air.

RA States				**RA′ States**			
Temp, °C	RH%	Humidity ratio g/kg	h, kJ/kg	Temp °C	RH%	Humidity ratio g/kg	h, kJ/kg
21.5	59.5	9.55	45.87	29.7	44.3	11.61	57.88
21.6	59.3	9.58	46.04	28.9	45.5	11.38	57.73
21.4	60.0	9.57	45.73	28.5	46.6	11.39	57.86
Average RA enthalpy = 45.88 kJ/kg Average RA humidity ratio = 9.57 g/kg				Average RA′ enthalpy = 57.82 kJ/kg Average RA′ humidity ratio = 11.46 g/kg			

Similar analyses of the test results are given for the exhaust indoor room air (RA) as shown in Table 6. It is noted in Table 6 that an enthalpy exchange of about 11.94 kJ/kg was obtained, closely in agreement with the enthalpy exchange of outdoor air. This congruence in energy balance indicates the reliability of the test results.

6. Energy and Efficiency Analysis

The heat exchange efficiency was calculated using the average values in Tables 5 and 6. The total heat exchange efficiency was calculated based on enthalpy, which was calculated from the measurements of temperature and humidity using Equation (1). Equations (5) and (6) were used to calculate the enthalpy exchange efficiency, respectively, for OA and RA.

$$\varepsilon_{OA\rightarrow OA'} = \frac{\Delta h}{\Delta h_{max}} = \frac{\dot{m}_{OA}(h_{OA} - h_{OA'})}{\dot{m}_{min}(h_{OA} - h_{RA})} = 37.82\% \tag{5}$$

$$\varepsilon_{RA\rightarrow RA'} = \frac{\Delta h}{\Delta h_{max}} = \frac{\dot{m}_{RA}(h_{RA'} - h_{RA})}{\dot{m}_{min}(h_{OA} - h_{RA})} = 33.43\% \tag{6}$$

The calculated results are also shown in Equations (5) and (6), which indicate again the accuracy of the measurements. It is to be noted here that total heat exchange (enthalpy exchange) efficiency is related to the dimensions. The experimental unit was of small size and the total heat exchange efficiency tested agrees well with the supplier's data.

The heat exchange efficiency could be different for sensible heat and latent heat. Sensible means effect of temperature change and latent means humidity change. The sensible heat exchange efficiency for OA and RA were calculated and given, respectively, in Equations (7) and (8). It is noted that sensible heat exchange is higher than total heat exchange as given in Equations (5) and (6).

$$\varepsilon_{SH,OA\rightarrow OA'} = \frac{\Delta T}{\Delta T_{max}} = \frac{\dot{m}_{OA}(T_{OA} - T_{OA'})}{\dot{m}_{min}(T_{OA} - T_{RA})} = 55.7\% \quad (7)$$

$$\varepsilon_{SH,RA\rightarrow RA'} = \frac{\Delta T}{\Delta T_{max}} = \frac{\dot{m}_{RA}(T_{RA'} - T_{RA})}{\dot{m}_{min}(T_{OA} - T_{RA})} = 64.94\% \quad (8)$$

The latent heat exchange efficiency was also computed from the test results, given, respectively, for OA and RA in Equations (9) and (10). It is noted that latent exchange efficiencies of 27.39% and 20.68% are much lower than that of sensible heat exchange.

$$\varepsilon_{LH,OA\rightarrow OA'} = \frac{\Delta \omega}{\Delta \omega_{max}} = \frac{\dot{m}_{OA}(\omega_{OA} - \omega_{OA'})}{\dot{m}_{min}(\omega_{OA} - \omega_{RA})} = 27.39\% \quad (9)$$

$$\varepsilon_{LH,RA\rightarrow RA'} = \frac{\Delta \omega}{\Delta \omega_{max}} = \frac{\dot{m}_{RA}(\omega_{RA'} - \omega_{RA})}{\dot{m}_{min}(\omega_{OA} - \omega_{RA})} = 20.68\% \quad (10)$$

Latent heat exchange is important to regions of high humidity. Higher latent heat exchange efficiency would be beneficial to regions with high humidity. Global warming has resulted in warmer summers in high latitude countries, where sensible heat exchange could be more important.

The cooling and heat rejection rate of IDOAS were also measured. The cooling rate is the heat removal at the evaporator Q_{evap}, and the heat rejection rate Q_{cond} is the heat rejected by the cooling unit. Q_{evap} and Q_{cond} were measured to be 2.454 kW and 3.361 kW given, respectively, in Equations (11) and (12). The calculations were based on OA′ and RA′, at states after the total heat exchanger. Q_{evap} is the cooling rate required to condition the air to the supply conditions SA.

$$Q_{evap} = \dot{m}_{OA}(h_{OA'} - h_{SA}) = 2.454\ kW \quad (11)$$

$$Q_{cond} = \dot{m}_{RA}(h_{EA} - h_{RA'}) = 3.361\ kW \quad (12)$$

Due to the effects of total heat exchange, the cooling capacity can be significantly reduced. The result shown in Equation (13) was calculated using the average values of OA and OA′ in Table 5. Energy saving is quite significant, otherwise Q_{evap} would be 4.41 kW instead (2.454 + 1.956), a saving of 44.3% (1.956/4.410).

$$Q_{THX,OA\rightarrow OA'} = \dot{m}_{OA}(h_{OA} - h_{OA'}) = 1.956\ kW \quad (13)$$

The standard rating of outdoor air temperature for the performance test of air-conditioners for cooling is 35 °C [24]. For IDOAS, the air temperature entering the condenser was about 29 °C in this study, significantly lower than 35 °C. This 6 °C temperature difference would reduce the condensing temperature and increase cooling efficiency.

Isentropic efficiency of the compressor was measured to be 0.78, obtained by comparing the compressor power to that of isentropic compression. The compressor was of constant speed with the rated COP at 2.90. The coefficient of performance COP is defined as in Equation (3), calculated as the ratio of cooling rate against the power input into the compressor. In this study, COP was measured to be 4.25. Therefore, the increase in COP was 46.6%. This improvement in COP is very significant, as about 46% of compressor energy can be saved.

7. Conclusions

Indoor air quality has become an important public health issue worldwide. Past studies have shown that more fresh air supply could result in more energy use of air-conditioning. However, increasing energy use would cause adverse effects on global warming. The IDOAS proposed in this study might provide a part of the solution to the above problem. In comparison to the outdoor air unit that coupled to a central system, IDOAS was configured in a unique unitary system to integrate enthalpy recovery and a cooling system in a single housing. An experimental unit of IDAOS was designed and tested. The test results showed that, after enthalpy exchange, outdoor air could be brought to a lower temperature and lower humidity, with the advantages of reducing the energy needed to condition the outdoor air. Moreover, a lower temperature and lower humidity of outdoor air could be supplied to the spaces for better thermal comfort. The refrigerant condenser was cooled by lower temperature of exhaust air. It resulted in higher cooling energy efficiency. The test results indicate that about 44% of cooling can be saved due to enthalpy exchange. The cooling efficiency in terms of COP was improved by about 46%. This novel unitary IDOAS could be installed at the point of use and operated independently to the central air-conditioning system. Therefore, the application of IDOAS could save tremendous piping costs and energy needed to deliver chilled water compared to a conventional outdoor air unit. Saving energy and resources is becoming more important than ever as global warming is more severe as days pass. The test results obtained for the proposed IDOAS were encouraging and are worth further research and development.

Author Contributions: Conceptualization, Y.K.C.; methodology, Y.K.C.; formal analysis, Y.K.C. and J.J.Y.; investigation, Y.K.C. and J.J.Y.; resources, Y.K.C.; writing—original draft preparation, J.J.Y.; writing—review and editing, Y.K.C.; project administration, Y.K.C.; funding acquisition, Y.K.C. All authors have read and agreed to the published version of the manuscript.

Funding: This research was funded by Ministry of Science and Technology, grant number MOST 108-2221-E-027-042.

Acknowledgments: This research was funded by the Ministry of Science and Technology of Taiwan by research grant MOST 108-2221-E-027-042, it is acknowledged hereby gratefully.

Conflicts of Interest: The authors declare no conflict of interest

References

1. Taiwan Indoor Air Quality Act, Administered by Environmental Protection Administration. Available online: https://www.roc-taiwan.org/us_en/post/4612.html (accessed on 1 February 2020).
2. Spengler, J.D.; Sexton, K. Indoor air pollution: A public health perspective. *Science* **1983**, *221*, 9–17. [CrossRef] [PubMed]
3. The Indoor Air Quality Management Act in the Public Using Facility. Available online: http://seaisi.org/thumbnail/32d419d10e65d65060e4b898af5f8942.pdf (accessed on 1 February 2020).
4. Hong Kong Environmental Protection Department. *IAQ Certification Scheme*; Hong Kong Environmental Protection Department: Hong Kong, China, 2003.
5. Levenhagen, J.I. Control systems to comply with ASHRAE standard 62–1989. *ASHRAE J.* **1992**, *34*, 40–44.
6. Mumma, S.A. Designing outdoor air systems. *ASHRAE J.* **2001**, *43*, 28–31.
7. Dieckmann, J.; Rothand, K.W.; Brodrick, J. Dedicated Outdoor Air Systems. *ASHRAE J.* **2003**, *45*, 58–59.
8. Murphy, J. Selecting DOAS Equipment with Reserve Capacity. *ASHRAE J.* **2010**, *52*, 30–40.
9. Larrañaga, M.D.; Beruvides, M.G.; Holder, H.W.; Karunasena, E.; Strauss, D.C. DOAS & Humidity Control. *ASHRAE J.* **2008**, *50*, 34–40.
10. Mumma, S.A. Overview of integrating dedicated outdoor air systems with parallel terminal systems. *ASHRAE Trans.* **2001**, *107*, 545–552.
11. Shank, K.; Mumma, S.A. Selecting the supply air conditions for a dedicated outdoor air system working in parallel with distributed sensible cooling terminal equipment. *ASHRAE Trans.* **2001**, *107*, 562–571.
12. Chien, K.C.; Chuah, Y.K.; Chang, W.T. An experimental study on the performance of radiant floor cooling modules. *Int. J. Air-Cond. Refrig.* **2013**, *21*. [CrossRef]

13. Li, H.; Lee, W.L.; Jia, J. Applying a novel extra-low temperature dedicated outdoor air system in office buildings for energy efficiency and thermal comfort. *Energy Convers. Manag.* **2016**, *121*, 162–173. [CrossRef]
14. Zhang, Z.Y.; Cao, X.; Yang, Z.; Shao, L.L.; Zhang, C.L.; Yub, Y. A new dedicated outdoor air system with exhaust air heat recovery. In Proceedings of the 12th IEA Heat Pump Conference, Rotterdam, Netherlands, 15–18 May 2017.
15. Chuah, Y.K.; Lin, C.Y. Health Building. In Proceedings of the Performance Assessment of Dedicated Outdoor Air Systems (DOAS) for Commercial Buildings, Tainan, Taiwan, 2–5 September 2017.
16. Energy Consumption Characteristics of Commercial Building HVAC System, Volume 3 Energy Savings Potential. Available online: https://www1.eere.energy.gov/buildings/publications/pdfs/commercial_initiative/hvac_volume3_final_report.pdf (accessed on 1 February 2020).
17. The American Society of Heating, Refrigerating and Air-Conditioning Engineers. *Standard 62.1 Ventilation for Acceptable Air Quality*; ASHRAE: Atlanta, GA, USA, 2013.
18. The American Society of Heating, Refrigerating and Air-Conditioning Engineers. *Design Guide for Dedicated Outdoor Air Systems (DOAS)*; ASHRAE: Atlanta, GA, USA, 2017.
19. Mui, K.W.; Wong, L.T.; Law, L.Y. An energy benchmarking model for ventilation systems of air-conditioned offices in subtropical climates. *Applied. Energy* **2007**, *84*(1), 89–98. [CrossRef]
20. Chaeand, Y.T.; Strand, R.K. Development of a Dedicated Outdoor Air System (DOAS) Module for a Whole Building Annual Energy Simulation Program. In Proceedings of the Fourth National Conference of IBPSA-USA, New York, NY, USA, 11–13 August 2010; pp. 458–465.
21. Mumma, S.A. Dedicated Outdoor Air Systems and Building Pressurization. *ASHRAE J.* **2010**, *52*, 42–52.
22. Xiao, F.; Ge, G.M.; Niu, X.F. Control performance of a dedicated outdoor air system adopting liquid desiccant dehumidification. *Appl. Energy* **2011**, *88*, 143–149. [CrossRef]
23. Liu, W.W.; Lian, Z.W.; Radermacher, R.; Yao, Y. Energy consumption analysis on a dedicated outdoor air system with rotary desiccant wheel. *Energy* **2007**, *32*, 1749–1760. [CrossRef]
24. The American Society of Heating, Refrigerating and Air-Conditioning Engineers. *ANSI/ASHRAE/IES standard 90.1 Energy Standard for Buildings Except Low-Rise Residential Buildings*; ASHRAE: Atlanta, GA, USA, 2016.

© 2020 by the authors. Licensee MDPI, Basel, Switzerland. This article is an open access article distributed under the terms and conditions of the Creative Commons Attribution (CC BY) license (http://creativecommons.org/licenses/by/4.0/).

sustainability

MDPI

Article

Numerical Study of Balancing between Indoor Building Energy and Outdoor Thermal Comfort with a Flexible Building Element

Choul Woong Kwon [1], Kang Jun Lee [2,*] and Soolyeon Cho [3]

1 SUSB Centre, Hanyang University, Gyeonggi-do 15588, Korea; hottrinity@gmail.com
2 School of Architecture, Hanyang University, Gyeonggi-do 15588, Korea
3 School of Architecture, College of Design, NC State University, Raleigh, NC 27695, USA; soolyeon_cho@ncsu.edu
* Correspondence: leekangjun@gmail.com; Tel.: +82-(0)10-2988-6902

Received: 22 October 2019; Accepted: 22 November 2019; Published: 25 November 2019

Abstract: This study analyzed the environmental role of a flexible canopy as a microclimate modifier in balancing indoor energy demands and outdoor thermal comfort. Flexible building elements are often installed in traditional buildings, depending on the local climate in southern Europe. The architectural performance of a canopy was analyzed using several environmental software packages (Ecotect, Rayman, WinAir, DaySim, and EDSL TAS). Coupling methods were applied to determine the environmental influence of the attached building element, a canopy with fixed and operable panes in different orientations and locations. The results showed that the flexible canopy played a crucial role in reducing indoor energy demands (heating and electricity for lighting) and increasing outdoor thermal comfort under the canopy area. Outdoor thermally comfortable conditions ranging between 13 and 29 °C in the canopy space could be enhanced by 56.3% over the entire year by manipulating a flexible canopy, compared with a fixed canopy with 90% transparency in London. The flexible canopy with higher transparency helped increase outdoor thermal comfort in Glasgow, while one with lower transparency showed better performance during summer in London. The findings of this research will help broaden the range of architectural elements used in buildings.

Keywords: flexible canopy; outdoor thermal comfort; daylighting; heating energy demand; balancing energy; coupling method

1. Introduction

This study investigated the environmental performance of a flexible canopy as an alternative to generally used fixed ones. Flexible use of building elements could maximize their environmental potential, which influences both the energy consumption of the building and indoor environment. Flexible architectural elements such as blinds or canopies are often used in regions with Mediterranean climates [1–3]. However, there is limited research regarding the impact of flexible building elements in terms of building energy consumption. The focus is mostly on finding new engineering elements for controlling building environments with very limited information on handling or controlling common building elements.

Efforts to develop environment friendly building designs have evolved in various ways to overcome global warming. These include using passive design approaches or considering engineering aspects (e.g., materials and renewable energy) to design environment-friendly buildings. Designers tend to apply passive design techniques in the early design stage as it is related to the building form and orientation rather than each element of the building. However, some building design elements found frequently in historic buildings work effectively to control environmental factors such as humidity and solar radiation [4–6]. In particular, flexible use of building elements (e.g., shutters and canopies) can be beneficial for controlling the building environment, in terms of both design and engineering [7,8]. However, as there is relatively little research in this regard, a combined approach comprising architecture and engineering is needed. Hence, the purpose of this study was to analyze the environmental performance of a flexible building element and determine whether it can be utilized in the current global environmental scenario.

Flexible building elements such as sunshades have been used for a long time to adapt to the external environment (seasonal, daily environment). However, the engineering approach involving the development of new building materials and equipment is mostly preferred for constructing environmentally friendly buildings. Therefore, through this research, we intend to analyze the effectiveness of flexible building elements in reducing the energy consumption of existing building elements (materials).

This research attempted to develop a flexible design strategy that can help overcome the seasonal limitations resulting from the use of a fixed canopy in school buildings with a particular transparency percentage and capacity for allowing airflow. The main idea behind proposing a flexible canopy is to overcome the limitations in its environmental performance. A flexible canopy was tested with computer simulation software using the same conditions as those for commonly used fixed canopies. The findings of this study can contribute to architectural design in terms of energy efficiency methods and design options beneficial to both designers and engineers.

2. Background Information

Changing school curriculums present challenges for everyone involved in education; hence, the role of outdoor spaces is becoming more important as an extension of indoor classrooms and as a framework for future expansion [9–11]. We analyzed the effectiveness of using a flexible canopy by examining the following performance parameters.

2.1. Thermal Comfort

Perceptions of outdoor thermal sensations are different from indoor thermal sensations, and it is even assumed that indoor thermal comfort standards are not applicable to outdoor settings. A study conducted by Hoppe in 2002 shows that outdoor thermal sensations are perceived differently from indoor thermal sensations, and indoor thermal comfort standards do not apply to outdoor settings [12,13]. Various studies have been conducted using the thermal comfort index specified for outdoor settings in urban studies. Methods for improving urban microclimates include the use of more appropriate materials, increased use of green spaces, cooling sinks for heat dissipation, and proper layout of urban canopies [14]. These methods are for outdoor spaces such as those in schools [15] to improve microclimate conditions. By changing urban structures and architectural elements, designers can create thermally comfortable outdoor spaces that are also appropriate for school environments [16]. A similar study on urban environments shows the importance of architectural and thermal diversity in urban settings. Architectural diversity comes from various spatial characteristics identified in urban space, such as geometry, orientation, urban structure, and materials [17]. Spatial diversity is also accompanied by thermal diversity, defined in terms of various microclimatic conditions related to temperature and air circulation in hot environments [18,19].

One of the main parameters influencing comfort in both outdoor and indoor spaces is solar radiation. The amount of incident solar radiation is affected by the surroundings, i.e., the presence of open or closed spaces. In winter, direct solar radiation is considered as a crucial environmental influencing thermal comfort in temperate climate zones.

2.2. Airflow

Airflow is an essential factor affecting the thermal comfort in outdoor spaces. Owing to changing global and regional conditions, it is difficult to predict and control airflow. Therefore, it is crucial to understand that there are significant differences in the wind environment from one part of the city to another, ranging from macroscale to microscale differences. Analyzing the effect of wind speed on people, the effect of building height on common flow patterns, and combining simulated wind conditions with measured thermal parameters is essential to predict outdoor thermal comfort [20,21].

The personal perception of wind speed associated with airflow was examined by Szokolay et al., (1997) [22]. However, the thermal comfort associated with airflow effects varies depending on the climate [23,24]. Winds can cause discomfort during winter cold climates; however, in humid conditions, it can be comfortable because wind promotes cooling of the human body by evaporation. For example, during periods of overheating in hot and humid areas, higher wind speeds can alleviate physiological heat stress caused by higher temperatures. In addition, the tendency of urban temperatures to exceed regional levels (e.g., heat islands) decreases as urban wind speeds increase. Installing windbreaks to avoid cold winds in cold weather will help reduce the cooling effect while it may exacerbate thermal discomfort in summer. Therefore, when designing windbreaks, it is desirable to incorporate flexibility to meet the needs during different seasons [25–27].

2.3. Applied Bioclimatic Index

The physiological equivalent temperature (PET) is a general and useful bioclimate indicator because it is expressed in °C and widely known as an indicator of heat stress. This makes it convenient for potential users unfamiliar with modern human biometeorology terms, including planners, decision makers, and the general public, to understand the concept of heat stress. For evaluating thermal conditions in a physiologically relevant manner [28], PET is defined as the temperature at which the human energy balance for an assumed indoor condition is balanced by the same skin temperature and sweating rate as in the actual complex outdoor condition being evaluated [29], which is based on the heat balance equation:

$$M + W + R + C + ED + E_{re} + E_{sw} + S = 0 \quad (1)$$

where m = metabolic rate (internal energy produced by food oxidation)

W = the physical work

R = the net radiant balance of the body

C = the convective heat flow

ED = the latent heat flow evaporate water through skin (perspiration)

E_{re} = heat flows for respiration (air heating and humidifying)

E_{sw} = heat flow for sweat evaporation

S = heat flow accumulated in body.

PET allows different users to compare their experience indoors with the essential effects of complex outdoor thermal conditions [30]. Moreover, PET can be used in different climates throughout the year [31]. Weather parameters such as temperature, humidity, wind speed, and shortwave/longwave radiation affect human energy balance. These parameters should be determined at a human biometeorological height of 0.5 m above the ground; this corresponds to the average height of the center of gravity of a standing child. Significant differences between air temperature and mean radiant temperature (Tmrt) (and PET) occur during windy winter days and under mild and sunny conditions. In these cases, extreme cold and heat stress can occur. In this study, we used solar radiation and PET to characterize thermal comfort conditions and evaluate human bioclimatological comfort employing flexible canopies with different transparency levels in locations with different climates.

As some studies [32–34] revealed, external shading devices are advantageous to prevent solar radiation from entering internal spaces in summer, while allowing low-angle sunlight to enter in winter. However, permanently fixed canopies cannot fully meet the requirement of outdoor thermal comfort and decreased indoor energy demand in both summer and winter. Therefore, an adjustable canopy system is recommended to address the difficulty in preventing sunlight from entering internal spaces in summer and allowing sunlight to enter in winter.

2.4. *Flexible/Fixed Canopy*

Providing sheltered spaces outdoors would be helpful for pupils to use outdoor spaces throughout the year regardless of weather conditions. Even though there are no specific guidelines for using outdoor spaces and regarding the duration for staying outdoors, sheltered spaces such as canopies are frequently used to enable the utilization of outdoor spaces for as much of the year as possible. Easy access to outdoor spaces enables teachers to use them as often as possible. For young pupils, free movement between indoor and outdoor spaces is essential. However, even for older pupils, direct access to outdoor spaces from their classrooms could mean frequent use of these outdoor spaces. Different types of canopies have been developed diversely and applied in schools.

There are different types of flexible canopies, namely foldable, retractable, and rotatable, which are adjustable in terms of controlling solar radiation according to sun angles, and are found in modern buildings [35,36] as seen in Table 1. These types of canopies can provide ideas for designing an optimal canopy to enhance outdoor thermal comfort in schools and reduce indoor energy consumption [37]. The main objective behind designing flexible canopies is to overcome the limitations of fixed canopies in terms of ensuring outdoor thermal comfort and reducing indoor energy consumption.

Table 1. Examples of canopies in existing buildings.

Types	Fixed Canopy	Flexible Canopy		
		Retractable	Ratable	Foldable

Locations	Lee-on-the-Solent Infant School, UK	Urban street, open space	Ponzano Primary and Nursery School, Empoli, Italy	Asilo Infantile Antonio Sant' Elia, Como, Italy

3. Methodology

3.1. *Concept of Flexible Building Components*

For explaining the concept of flexible building components, the canopy is selected, which is mostly found in school buildings in southern Europe and in the UK. The idea behind using a flexible component is that it allows for adjusting major environmental parameters (solar radiation and wind speed) according to user requirements. It can improve thermal comfort during summer and winter either through manual or automatic operation. Simulations involving flexible canopies were performed following the same methods as those used for fixed canopies, examining daylighting, outdoor thermal comfort, and energy demand.

The main factors considered for designing the canopy are as follows:

- Summer: Increase wind speed and avoid direct solar radiation.
- Winter: Decrease wind speed and allow as much direct solar radiation as possible.

3.2. *Component Design*

The proposed design for a flexible canopy comprises two divided panels where each panel is operable according to the season for maximizing performance, as illustrated in Figure 1. The closed canopy shown in (a) allows solar radiation to enter as the glass pane at the front is transparent while the rear glass pane has different transparencies of 0%, 50%, and 90%. The rear pane can fold toward the front pane, as illustrated in (b) according to the environmental conditions. Once the pane is folded, there are open holes to facilitate free airflow between the area under the canopy and upper open space; this would increase the wind speed under the canopy, as indicated in Figure 1c.

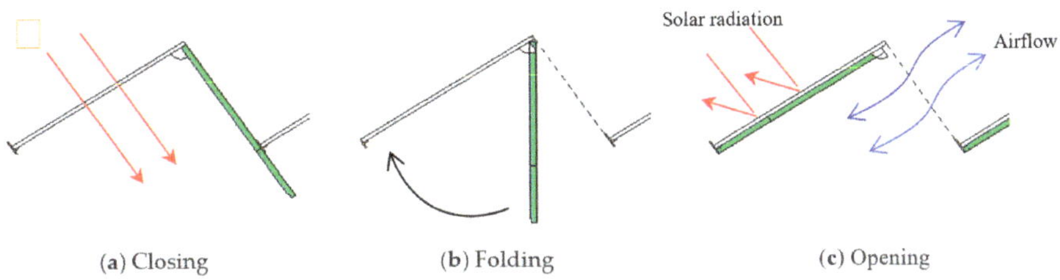

Figure 1. (**a**) Closed rear pane, (**b**) folding toward front pane, and (**c**) opening of rear pane to allow airflow.

In the prototype of the canopy unit, each jigsaw-shaped unit comprises two 1 m long and 0.4 m wide rectangular panes with the rear pane being foldable toward the front pane.

This canopy can be partially or fully folded and can be operated either manually or automatically. This flexible canopy can overcome the functional drawbacks of a fixed canopy without any influence from environmental changes. The schematic in Figure 2 shows the annual use of the flexible canopy as an open canopy in summer and closed canopy in winter. This canopy is designed for blocking solar radiation while promoting airflow to enhance thermal comfort in summer. In winter, the opposite approach is used by closing panes to reduce the wind speed under the canopy while allowing direct solar radiation to enter.

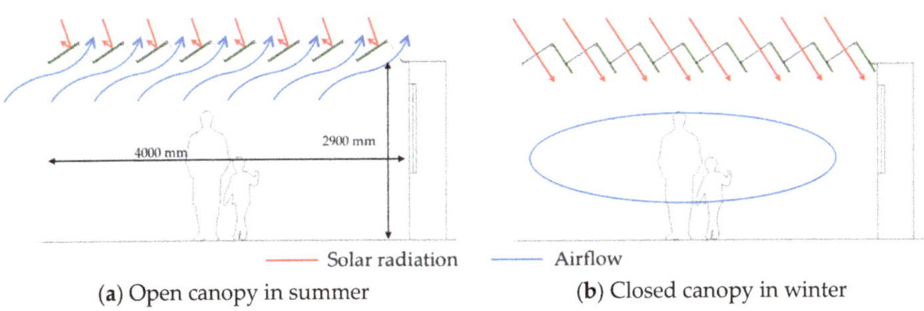

Figure 2. Schematic diagrams showing a sectional view of the flexible canopy unit in (**a**) summer (**b**) winter.

Using a fixed canopy is not beneficial considering environmental requirements in different seasons for both indoor and outdoor spaces in schools. Each flexible canopy unit can be automatically operated; manual operation is also possible whenever teachers or students want to change the canopy to suit their needs.

Modeling of the canopy for simulation studies was performed using EDSL TAS version 9.1.4, which is a dynamic thermal simulation software used to calculate the heating energy demand according to climatic conditions [38,39].

The calculated values were compared with those for fixed canopies with different transparencies, namely 0%, 50%, and 90%. The school classroom unit was set up based on typical classroom sizes in the UK, as seen in Figure 3. Diverse environmental parameters were applied considering four main orientations, thermal transmittance/*U*-values (0.25 and 0.35 W/m^{2}K for wall and roof respectively), 40% window ratio on the front wall, and three different transparencies of the fixed canopy, as presented in Table 2.

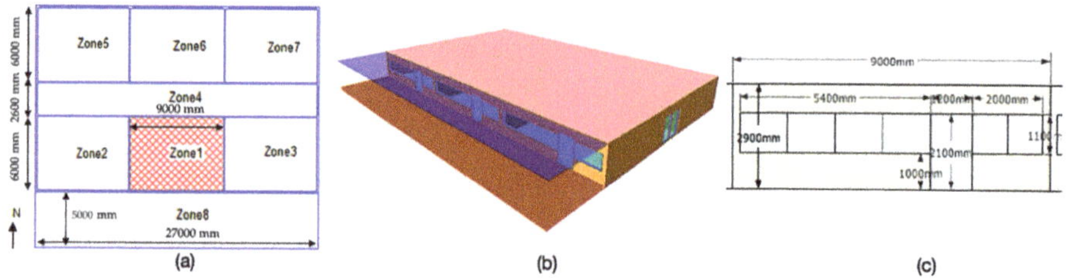

Figure 3. EDSL TAS numerical model: (**a**) Plan, (**b**) perspective, and (**c**) front façade of the classroom.

Table 2. Input parameters for modeling.

Design Variables	Classroom Built Recently	Design Variables	Classroom Built Recently
Orientation	S, N, E, W	Window U-value	U = 2.974 W/m^{2}K
Wall insulation	U = 0.25 W/m^{2}K	Room dimensions	9 (w) × 7 (d) × 2.9 m (h)
Roof	U = 0.35 W/m^{2}K	Canopy dimensions	9 (w) × 5 m (d)
Infiltration	0.5 ACH	Material of the canopy space	Glass material with 0%, 50%, and 90% transparency
Ventilation	Natural ventilation via openable window	Reflectance	Ground 0.2 Canopy 0.5
Glazing ratio	40%	Wall material	Brick

In the calculation of outdoor thermal comfort, wind speeds under the canopy are calculated with Equation (2) with WinAir4. For the wind speed input data, a prerequisite simulation for calculating the wind speed exponent (α) was performed, and the results are presented in Table 3.

$$\frac{V_{zex}}{V_{Z10}} = \left(\frac{Z_{ex}}{Z_{10}}\right)^{\alpha} \tag{2}$$

where V_{zex}: the mean wind speed at height Z (m/s)

V_{z10}: the mean wind speed at reference height Z_{10} (m/s)
Z_{ex}: the height above ground level (m)
Z_{10}: the reference height (m)
α: the exponent characteristic of terrain roughness.

Table 3. Wind speed at height z (V_{zex}) with exponent (α) for flexible canopy (calculated by WinAir 4).

Wind Direction	Flexible canopy	
	Vzex	Exponent (α)
N	0.64	1.04
NW/NE	0.72	0.99
W	3.47	0.31
SW/SE	2.13	0.52
S	0.81	0.94

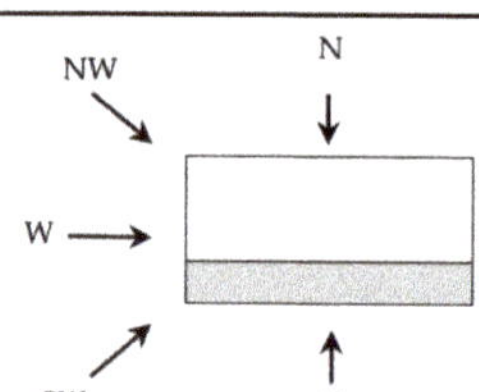

V_{10} : 7 m/s; Zex : 1 m; Z_{10} : 10 m. Wind directions (every 45°): N(337.5–22.5°), NE(22.5–67.5°), E(67.5–112.5°), SE(112.5–157.5°), S(157.5–202.5°), SW(202.5–247.5°), W(247.5–292.5°), and NW(292.5–337.5°).

In Manchester, flexible canopies facing east or west work better than those with other orientations as the funnel effect caused by the increased wind speed under east- and west-facing canopies reduces thermal comfort significantly in winter. As the transparency of the flexible canopy can be changed from 0% to 90% at any time, it can provide spaces with improved thermal comfort.

In summer, thermal comfort can be achieved by controlling solar radiation to maintain comfortable outdoor thermal conditions where the thermal index, PET, is in the range of 13%–29 °C (equivalent to slightly cold to slightly warm). Thermal comfort can vary depending on exposure to solar radiation; the extent of variation in thermal comfort deduced from the simulations was 8 °C in summer and 4 °C in winter. Other than solar radiation, controlling wind speed in winter plays an important role as there is little solar radiation during winter in the UK. Cold winds with high speeds are detrimental to outdoor thermal comfort; therefore, decreasing the wind speed is necessary to avoid the wind chill effect [40].

3.3. Thermal Comfort

The extent of thermal comfort enhancement is mainly dependent on the operation rather than the material used in fabricating the canopy. The performance would vary primarily according to the local climate and operation. In particular, if there were a high incidence of direct solar radiation, the efficiency of the flexible canopy would be higher than that in a climate with more diffused light rather than direct solar radiation. The results showed that using a flexible canopy could improve thermal comfort in flexible canopy spaces and reduce indoor overheating and daylight energy demands. The monthly thermal comfort levels plotted in Figure 4 help understand outdoor thermal performances under different canopy transparencies (0, 50, and 90) and without a canopy. The green curves indicate comfort bands; PET values in the range of 13–29 °C correspond to comfortable conditions ranging from slightly cool to slightly warm. Dark green indicates neutral conditions, i.e., neither hot nor cold, corresponding to comfortable temperatures in the range of 18–23 °C.

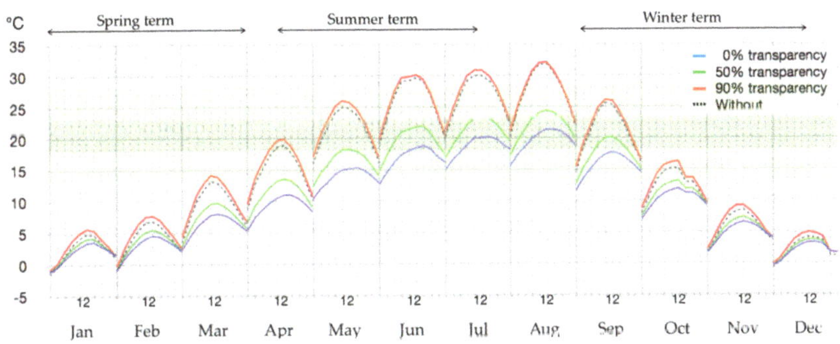

Figure 4. Monthly average physiological equivalent temperature (PET) on an hourly basis with different canopy transparencies (Comfort bands: 13–29 °C; 18–23 °C).

Figure 5 shows the outdoor comfort levels in south-facing canopy spaces. PET in the 13–29 °C range that makes occupants feel thermally comfortable is achieved to a high extent over the whole year when using flexible canopies; comfort levels of 56.3% in London, 51.2% in Manchester, and 49.2% in Glasgow were achieved. Opaque glass (0%) shows the worst performance in Manchester and Glasgow where solar radiation is relatively rare compared to London. Thermal comfort in the outdoor canopy space is enhanced by 12.6% (compared to that achieved using a canopy with 90% transparency) and by 11.7% (compared to a canopy having 0% transparency) in London, as observed in Figure 5. Yearly PETs were calculated considering a south orientation for London, Manchester, and Glasgow and are presented in Table 4. It presents the improvement observed for each orientation and in each city when using a flexible canopy over the whole year during school hours (09:00 am–4:00 pm).

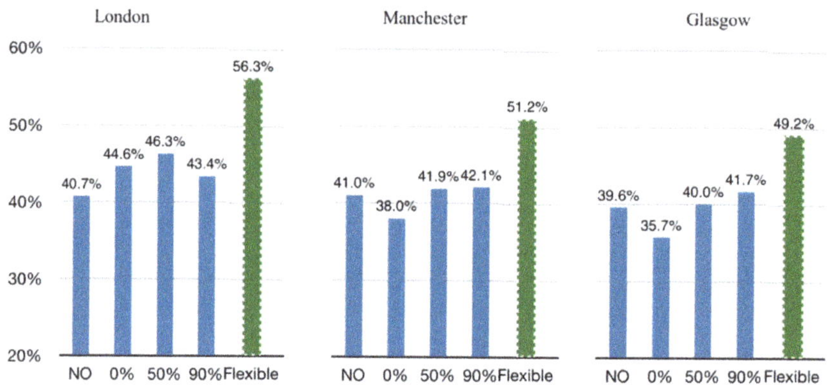

Figure 5. Outdoor thermal comfort levels throughout the year in the temperature range of 13–29 °C in outdoor south-facing canopy spaces in London, Manchester, and Glasgow.

Table 4. Outdoor thermal comfort levels in the temperature range 13–29 °C achieved using a flexible canopy with facing south orientation during the daytime (09:00 am–4:00 pm).

	London	Manchester	Glasgow
North	55.4%	50.4%	47.6%
West	54.9%	51.5%	49.6%
East	55.8%	52.4%	50.3%
South	56.3%	51.2%	49.2%

The performance efficiency of the flexible canopy was considerably more significant in London compared to Glasgow as the canopy plays a vital role in controlling solar radiation. Wind speed is also affected when the canopy is opened or closed; increased wind speeds when the canopy is open positively influenced thermal comfort in summer while having a negative influence in winter. According to the results of this simulation, it is best to keep the canopy open in warm seasons from June to September but closed in cold seasons from October to May. The improvement in thermal comfort clearly shows that the flexible canopy performs well in different cities compared to a fixed canopy by just varying the transparencies. The flexible canopy shows a substantial improvement of 16.6% in Glasgow in an east-facing canopy space compared to a canopy with 0% transparency in Glasgow. The thermal comfort improvement was in the range of 10%–12.6% when using the flexible canopy compared to the base-case simulations using canopies with 0%, 50%, and 90% transparency.

Most improvements can be achieved during the warm season as there is a profound change in thermal comfort when blocking solar radiation, as observed in the simulation involving a fixed canopy. If there is a considerable amount of solar radiation, there is also excellent potential for improving thermal comfort through manipulation when using a flexible canopy.

3.4. Daylighting

Figure 6 presents the results of daylighting simulations conducted using the flexible canopy. The primary method of reducing electric lighting is to minimize any glare problems while providing enough diffuse light in the classroom. A canopy can function as a shading device to protect an indoor area from direct radiation and maximize diffuse light, thus improving the visual conditions for the students. Blinds were included in the simulation as most schools generally use blinds operated manually by the occupants of the classroom, which block direct solar radiation entering through the classroom window. With blinds in use, 25% of the diffuse light can still enter the indoor classroom. The use of light switches was also considered in the simulations.

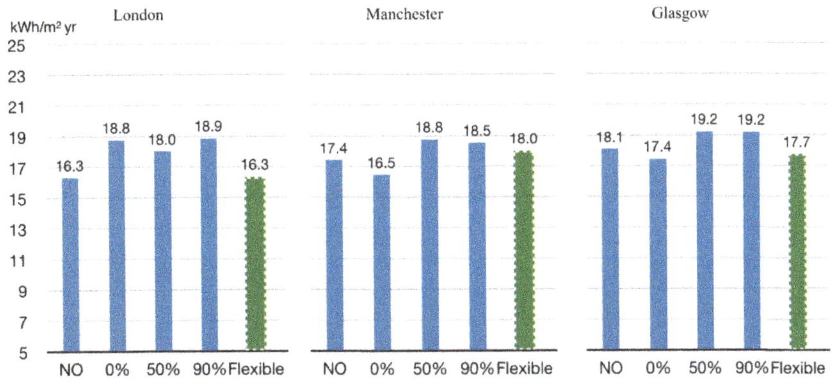

Figure 6. Annual indoor lighting energy demand for a south-facing canopy area.

The physical conditions used for the numerical model in the daylight simulations were the same as those used for the fixed canopy. The illuminance level on the working plane had to be no less than 300 lux in the teaching area [41]. The simulation period was the same as that for the fixed canopy, i.e., from Monday to Friday between 09:00 am and 4:00 pm. The measured desk height was 900 mm from the floor and reflectance values of 0.3 for the floor, 0.5 for wall, and 0.7 for the ceiling were set. At a minimum illuminance level of 300 lux, the minimum allowed level of illuminance at the desk level was used for calculating the lighting energy demand. As illustrated in Figure 6, the daylighting performance of the flexible canopy shows an improvement of approximately 20% achieved compared to that without a canopy in the case of a south facing canopy in London. Even though there is no software yet available to calculate indoor lighting demands when using a flexible type canopy, simplified simulations were conducted. In the simulation, the method with which the blinds at the window would be operated automatically, set by the DaySim software. As direct sunlight can cause a glare problem, blinds will be drawn automatically which is considered as the same mechanical process with that of outdoor canopy system.

As shown in Figure 6, in a London classroom, the maximum potential annual saving in energy used to light the classroom when using a flexible canopy is 2.6 kWh/m^2 compared to a canopy with 90% transparency, this amounts of a saving in electric lighting energy of approximately 10%–14% per square meter.

In all cities, considerable amounts of savings in electric lighting energy can be made when using a flexible canopy rather than one with 0% transparency in which there is little available daylight. One of the advantages of using a flexible canopy is that it can minimize the use of blinds that block direct sunlight and then lead to the use of electric lights during daytime. The reduction in daylighting energy demands associated with different classroom orientations and fixed canopy transparencies were calculated through simulations using the DaySim software and are presented in Table 5. In general, there is an enhancement in the daylighting level in the classroom, thus reducing the use of electric lights. However, in the south-facing classrooms in Manchester and east-facing ones in Glasgow, more electric light is used than when using a flexible canopy. Glaring is the main reason for daytime light usage in schools. However, using a flexible canopy that can be automatically adjusted according to light conditions would potentially increase the electrical energy demand. Therefore, it is essential to consider how a flexible canopy would be operated and what its running costs would be before installation. Daylight in the classroom is influenced by the reflectance of both internal and external surfaces, size of the glazing, and architectural reflectors such as light shelves and reflectors. This simulation showed how a flexible canopy can be manipulated in the same way as an external blind, thus helping to minimize the usage of both indoor electric lighting and blinds. The simulation showed that the canopy could perform well in this respect in most orientations and cities but especially in north-facing transitional outdoor spaces.

Table 5. Annual indoor lighting energy demand (kWh/m^2 yr) when using a flexible canopy.

	London	Manchester	Glasgow
North	14.64	14.97	18.50
West	15.13	15.13	16.57
East	16.26	16.65	18.55
South	16.33	17.99	17.75

Computer simulations of internal daylighting through DaySim showed that using a flexible canopy can reduce the need for artificial lighting and allow students to benefit from daylight almost throughout the year. Even though it does not always achieve the recommended daylighting level of over 300 lux for a classroom, it can help minimize glare problems associated with direct solar radiation; moreover, illuminance levels can be increased by opening the canopy on a cloudy day. As there are several environmental factors associated with an outdoor canopy space, the solution to one problem should not compromise the environmental performance of the canopy in terms of its other functions. Therefore, the simulations involving all the other factors such as outdoor thermal and heating energy demands should be considered at the same time to optimize the canopy design.

3.5. Heating Energy Demand

The heating energy demand of the classrooms was analyzed using the same physical conditions as those when using a flexible canopy in the canopy space. Both internal and external gains influence the heating energy demand. The window size and orientation, as well as external shading devices, play essential roles in admitting sunlight into the classroom. With the same building settings as in the simulations, the heating energy demand of a classroom was determined when using a flexible canopy.

The annual heating demands of classrooms using flexible canopies are given in Figure 7 and Table 6. Using the EDSL TAS software and a design that follows current building regulations, the annual heating energy consumption was calculated to be generally low, but slightly higher than when using a canopy with 90% transparency.

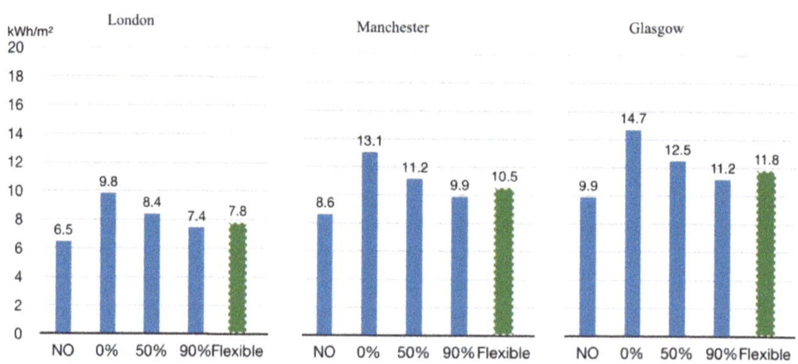

Figure 7. Annual heating energy demands (kWh/m^2 yr) of south-facing classrooms.

Table 6. Annual indoor heating energy demands (kWh/m^2 yr) when using a flexible canopy.

	London	Manchester	Glasgow
North	9.94	13.77	16.01
West	8.61	11.98	13.80
East	9.47	13.05	15.34
South	7.80	10.54	11.84

The simulation modeled seasonal operations of the canopy in which a canopy with 0% transparency was used during summer (June–July) and one with 90% transparency for the rest of the year. The canopy transparency should be selected so as to maximize the incident solar radiation in cold/cool seasons and to minimize outside heat in warm seasons. The internal heating temperature was set at 18 °C according to the recommendations of BB87. According to the simulations, Glasgow showed a higher heating energy demand than either London or Manchester, requiring a canopy transparency slightly higher than 90%, as illustrated in Figure 7.

The heating energy demand of a classroom needs to be minimized as the winter school holidays are relatively shorter than the summer holidays, meaning that students spend more time in school in winter than in summer. As contemporary buildings are built according to stringent building regulations (e.g., using high-performance building materials and airtightness standards), the mean value of the annual heating energy demand of the classroom ranged between 7.8 and 9.9 kWh per square meter in London and 11.8 and 16.0 kWh per square meter in Glasgow. Being at a higher altitude, the lower temperatures and limited solar radiation experienced by the Glasgow classroom caused an increase in heat gain. The impact of an external canopy in a transitional outdoor space on the heating energy demand of the neighboring classroom(s) must be evaluated for each school term. Using a flexible canopy can potentially reduce both the heating energy demand in winter and internal heating problem in summer (June–July) as it blocks direct sunlight by adjusting its panes to reduce overheating. Such canopy devices are popular in southern Europe. During cold periods (November–February), as much solar radiation (direct and diffuse) needs to penetrate the indoor classroom as possible to provide external heat gain. Therefore, when designing a flexible shading device, it is crucial to consider the solar altitude and azimuth angles during the hot seasons to prevent direct sunlight from passing through. The extent to which direct sunlight passes through windows depends on the orientation of the windows, the sun position at the time under consideration, and the operating mechanism of these devices, all of which need to be considered to maximize the environmental performance of the canopy.

It is useful to analyze how energy consumption and outdoor thermal comfort vary with different classroom orientations as classrooms can be allocated differently in school planning. The stacked column charts in Figure 8 present the energy consumed for lighting (colored in blue) and heating (colored in yellow) while outdoor thermal comfort is indicated using dotted gray lines. Energy consumed for lighting and heating is the highest when using a canopy with 0% transparency in most orientations, except for the east-facing classroom. The canopy with 50% transparency requires 28.9 kWh/m^2 per year owing to the glare problem during early morning hours when blinds are used to block direct sunlight. The flexible canopy shows the best performance among all the canopies. Among the fixed canopies, the one with 0% transparency should not be installed in front of classrooms while that with 90% transparency is the most suitable. In particular, energy consumption in north-facing classrooms with opaque canopies (0% transparency) increases sharply owing to the relative shortage in daylight and solar radiation.

However, it is hard for stakeholders and designers to understand energy consumption and outdoor thermal comfort with a balanced view. Therefore, it is required to use a simplified graph to select the appropriate canopy unit (based on factors such as enhancement percentage) easily; hence, it is essential to understand the simulation results in the early stages of designing. The simplified graphs developed during this study are shown in Figures 9–11. Energy and thermal comfort are combined with the enhancement percentage. Therefore, stakeholders can effectively understand the performance of each canopy considering the enhancement percentage. The average performance enhancement values considering all the cases (no canopy, transparencies of 0%, 50%, and 90% and flexible canopy) were calculated and are given in each of the figures for comparison.

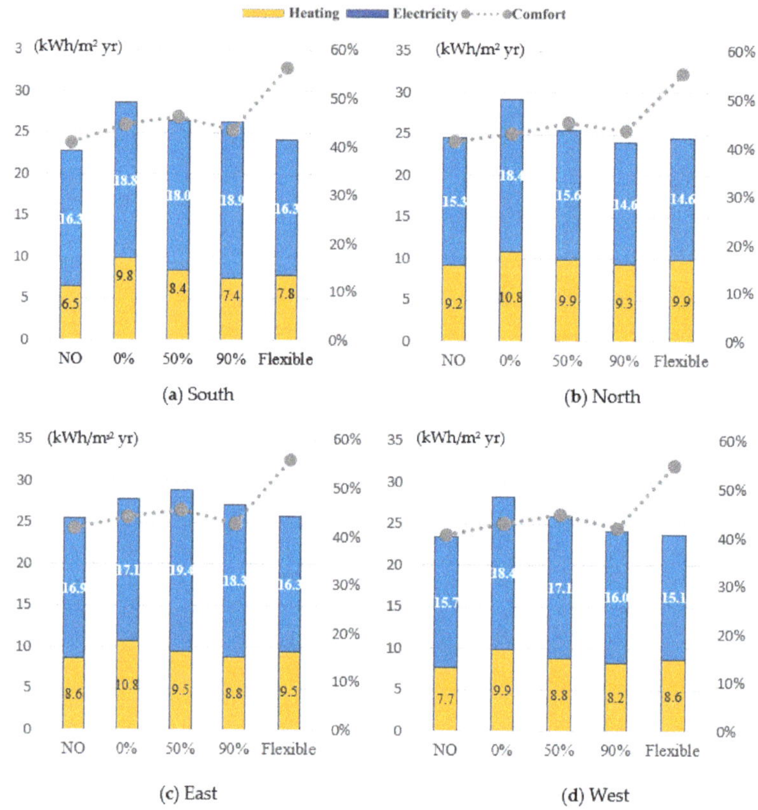

Figure 8. Energy consumption (Heating/electricity kWh/m^2 yr) and outdoor thermal comfort (%) in London.

The flexible canopy is the most effective in all the orientations, as is evident in Figure 9 and Table 7. An obvious fact is the improvement of 17% observed in the case of the south-facing classroom without a canopy. In addition, the canopy with 0% transparency shows low performance in the north- and west-facing classroom with performance decreases of 18% and 13%, respectively.

The canopy performance in Manchester is presented in Figure 10; it is slightly different as Manchester is located in the middle part of UK with relatively more cloudy days than London. The flexible canopy performed effectively in the north, east, and west orientations while the canopy with 0% transparency increases the energy consumption and discomfort in outdoor areas. East-oriented fixed canopies with different transparencies did not show satisfactory performance. A 7% improvement was observed in the west orientation with canopy, while decrements of 13, 7, and almost 0% enhancement were observed with canopies having transparencies of 0%, 50%, and 90%, respectively, as observed in Table 8. The south-oriented canopy with 50% transparency showed the worst performance. Moreover, the flexible type canopy is suitable for north, east, and west orientations.

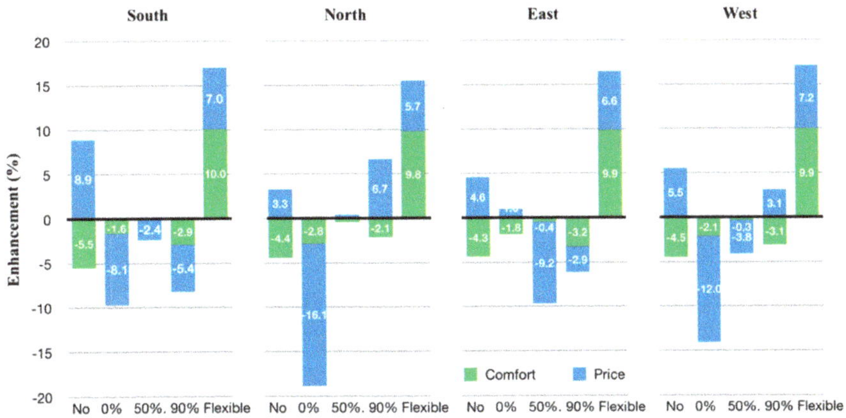

Figure 9. Enhancement (%) in thermal comfort and energy price (bill) in London.

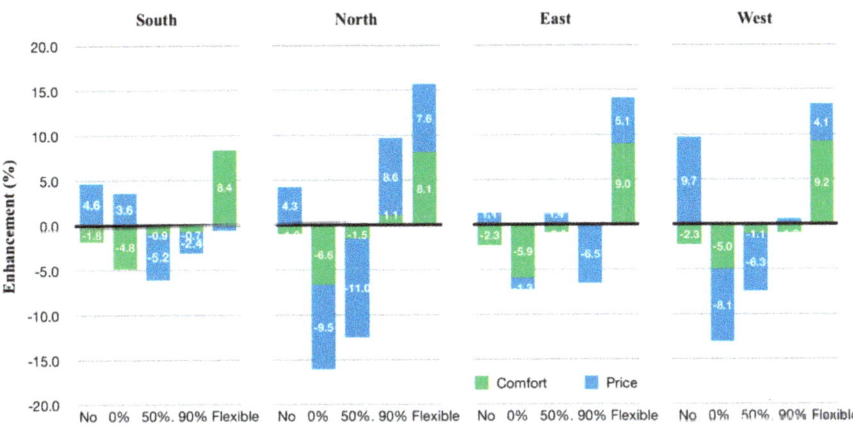

Figure 10. Enhancement (%) in thermal comfort and energy price (bill) in Manchester.

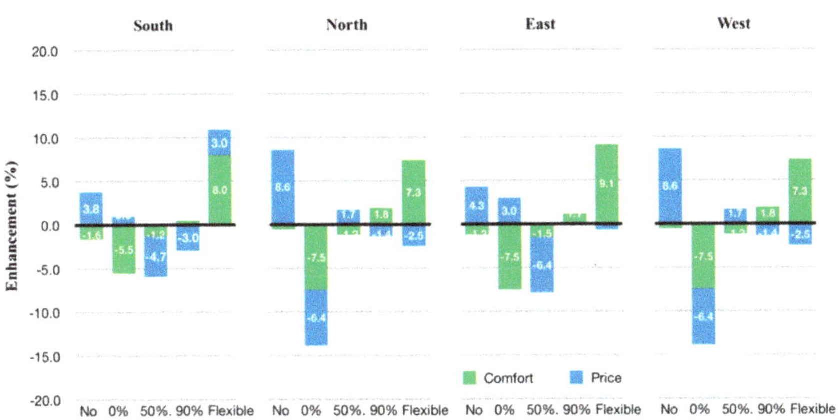

Figure 11. Enhancement (%) in thermal comfort and energy price (bill) in Glasgow.

Table 7. Enhancement (%) in thermal comfort and energy price from the average.

		NO	0%	50%	90%	Flexible
South	Comfort	−5.51	−1.64	0.00	−2.89	10.04
	Price	8.88	−8.08	−2.39	−5.37	6.96
North	Comfort	−4.42	−2.82	−0.42	−2.12	9.78
	Price	3.28	−16.05	0.40	6.66	5.72
East	Comfort	−4.34	−1.84	−0.44	−3.24	9.86
	Price	4.58	0.98	−9.23	−2.93	6.60
West	Comfort	−4.48	−2.08	−0.28	−3.08	9.92
	Price	5.50	−11.99	−3.80	3.14	7.16

Table 8. Enhancement (%) in thermal comfort and energy price from the average in Manchester.

		NO	0%	50%	90%	Flexible
South	Comfort	−1.84	4.84	−0.94	−0.74	8.36
	Price	4.60	3.57	−5.16	−2.41	−0.61
North	Comfort	−1.02	−6.62	−1.52	1.08	8.08
	Price	4.26	−9.47	−10.99	8.62	7.58
East	Comfort	−2.32	−5.92	−0.92	0.18	8.98
	Price	1.36	−1.29	1.36	−6.53	5.10
West	Comfort	−2.26	−4.96	−1.06	−0.96	9.24
	Price	9.67	−8.07	−6.34	0.60	4.14

Glasgow is infamous for its shorter summers and longer winters compared to other places in the UK. Therefore, the canopy with 0% transparency (opaque) showed the worst performance in the north and west orientations, as observed in Figure 11. There is no significant overall enhancement in canopy performance, compared to London and Manchester. Improvements of 10% and 9% improvement in the south and east orientations, respectively, were observed when using the flexible canopy tabulated, as presented in Table 9. One of the crucial functions of the canopy is to provide shelter from rain and strong winds; therefore, the appropriate type of canopy should be selected using an approach that considers both energy consumption and outdoor thermal comfort.

Table 9. Enhancement (%) in thermal comfort and energy price from the average in Glasgow.

		NO	0%	50%	90%	Flexible
South	Comfort	−1.64	−5.54	−1.24	0.46	7.96
	Price	3.77	0.94	−4.71	−2.95	2.95
North	Comfort	−0.56	−7.46	−1.16	1.84	7.34
	Price	8.59	−6.41	1.66	−1.38	−2.46
East	Comfort	−1.20	−7.50	−1.50	1.10	9.10
	Price	4.25	2.96	−6.39	−0.14	−0.69
West	Comfort	−0.56	−7.46	−1.16	1.84	7.34
	Price	8.59	−6.41	1.66	−1.38	−2.46

4. Findings and Discussion

Throughout the study, the interconnected effects of using a flexible canopy as a microclimate modifier on outdoor thermal comfort, as well as the daylight levels in indoor environments and the subsequent energy demands were revealed. The intention behind carrying out this work was to provide a better understanding of the microclimatic characteristics of outdoor canopy spaces in schools through examining common building elements and to understand the implications this would have on the comfort level of students and the energy demands of the building. The simulation results showed that using a flexible canopy can enhance the microclimate and also reduce the energy demand of the classroom by a greater degree compared to using fixed canopies with different transparencies.

The idea of utilizing a flexible canopy is based on the findings of simulations involving fixed canopies with different transparencies; a canopy with 90% transparency performs well in winter, and one with 0% transparency performs well in summer. This can save energy while maximizing outdoor comfort, leading to a potential reduction of 20% in the lighting energy demand in London and an increase of 12%–17% in outdoor comfort compared to that when using a fixed canopy (opaque).

In this research, we demonstrated that having sheltered outdoor canopy spaces in schools could be beneficial to the local microclimate of the area, thereby contributing to the creation of a comfortable outdoor environment while minimizing energy demands. A canopy used as the main architectural feature of a transitional space should serve as an extended classroom throughout the year. The simulations showed that a flexible canopy is more effective than a fixed canopy in improving outdoor thermal comfort and indoor daylighting. Most of the methods for calculating the environmental performance of a flexible canopy are similar to those used for a fixed canopy. Its flexible nature that makes this canopy adaptable for use in most areas; however, its design details and operations are extremely critical. Considerations need to be made on whether it should be controlled manually or automatically although automatic control proved effective in the simulations. Users would need to learn how to operate the canopy and understand when it may be necessary to adjust the canopy depending on the environmental conditions and purpose for using the transitional space or indoor classroom. An unsophisticated operation method may worsen the outdoor or indoor environment. The initial costs also need to be considered in terms of the value of the buildings and the length of the payback period.

The flexible configuration of a canopy can improve its adaptability to unstable environments depending on the needs of the occupant. However, the mechanical details their operation, as well as the cost and feasibility of manufacturing these canopies need to be considered. In summer, the canopy can help improve outdoor thermal comfort by preventing direct solar radiation from passing through, while at the same time increasing the wind speed either by being partly or fully open. In winter, the wind speed should be minimized and solar radiation maximized. However, as maximizing solar radiation to increase indoor daylight can cause increased glare, the operation of the canopy should be manually controllable so that it can be adjusted whenever necessary.

As a shading device, a flexible canopy can help minimize the glare problem, the main reason for the increased use of electric lights in schools. An automatically operated canopy is recommended; however, it should also be possible to operate the canopy manually should the occupants need to do so or if power supply is cut.

One vital feature of an environmentally improved transitional outdoor space is its flexibility in dealing with the changing environmental needs of the occupants. The canopy space in schools should be flexible enough to cope with the students' needs, depending on whether they are using an indoor classroom or its adjacent outdoor space. For example, if students and teachers assemble in the canopy space to listen to stories or sing songs, the canopy can be adjusted to improve the microclimate; it can either allow more solar radiation in winter block direct sunlight altogether in summer according to the environmental conditions. However, designing an operation mechanism for a flexible canopy could ultimately be difficult because of the complexities of the operating system, as it would need to be robust and secure for teachers/students to handle. Table 10 summarizes the performance of both fixed and flexible canopies according to the environmental criteria discussed in this study. This is a summary of the theoretical performance considering ideal occupant behavior and operational conditions.

Moreover, it is recommended to use flexible canopies with a combined system facilitating both manual and automatic operations so teachers can operate the canopy whenever required. Thus, it is possible to prove the hypothesis that outdoor spaces under canopies linked with indoor spaces can be environmentally enhanced in terms of daylighting levels and outdoor thermal comfort using flexible building elements.

Table 10. Comparison of the environmental performance of fixed and flexible canopies.

	Fixed Canopy	Flexible Canopy
Outdoor thermal comfort	▪ There is a limit for maximizing outdoor thermal comfort achieved mainly by protecting against solar radiation in summer and allowing it to enter in winter. ▪ Transparency of the canopy should be carefully selected.Relatively easy to install.	▪ It is the best way to provide outdoor thermal comfort by controlling the amount of direct solar radiation. ▪ More effective for south-oriented spaces than for other orientations. ▪ Automatic operation is recommended, but options for manual operation should be included.
Daylighting	▪ Transparency of the canopy does not really affect the use of electric lights in the parent building. ▪ How blinds are used in the classroom (either manual or automatic) plays an essential role in determining whether or not electric lights are used. ▪ South-facing classrooms have greater daylight autonomy than classrooms with other orientations, thus showing more potential to manipulate daylight depending on the design of the blinds provided.	▪ Theoretically, automatic control of the canopy can give the best outcome in preventing glare when operated like external blinds. ▪ Performs better in south-oriented spaces than in spaces with east or west orientations. ▪ As a shading device, a flexible canopy can minimize the use of blinds in a classroom. ▪ If an automatic blind is used in the classroom, the role of the flexible canopy in controlling daylight is lessened.
Heating energy	▪ The lower the transparency, the higher the heating energy demand in any orientation. ▪ However, heating energy demand can vary according to the size of the glazed area and materials.	▪ The operation and transparency of the canopy's front panel affect energy performance. ▪ The performance in terms of reducing heating energy is better than that of a 50% transparency canopy but worse than of a 90% transparency of fixed canopy. ▪ There is no significant improvement in performance when compared to other types canopies.

5. Conclusions

This study demonstrated flexible operation of architectural elements allowing the building to adapt in response to seasonal or daily climatic variations. A canopy was selected as the architectural element; it can help overcome any limitations resulting from building orientation and materials that can deteriorate both the indoor and outdoor environments. To analyze the performance of canopies, simulations using different transparencies and orientations were performed for flexible and fixed canopies in London, Manchester, and Glasgow, having different climates. Poor environmental performance was observed in some cases (in terms of the daylighting level, heating demand, and outdoor comfort), worse than that observed without a canopy. However, the south-facing canopy with a highly transparent pane showed better performance compared to fixed and opaque canopies in all locations. In particular, the flexible canopy significantly reduced thermal discomfort in outdoor spaces during summer by blocking direct solar radiation; moreover, it increased outdoor comfort in spring and autumn by allowing sunlight to pass through.

This study also developed a method for selecting an appropriate canopy based on the performance enhancement (%) combined with outdoor thermal comfort and indoor energy consumption including electricity used for both lighting and heating. A building element should be selected using a holistic approach than merely focusing on specific environmental factors (such as daylighting and thermal comfort). As architecture is a complex combination of design and engineering performance, building elements must be selected using combined approaches to meet the performance requirements. Moreover, the adaptability of flexible building elements can be beneficial under rapidly changing climatic conditions. Adapting to changing environmental conditions is one of the most crucial properties of building elements considering climatic changes that can occur in the future. Flexible building elements are often used in traditional architecture in different countries according to climatic conditions. Flexible elements may not be required in buildings under constant climatic conditions throughout the year as opposed to those subject to extreme conditions; still, architects and engineers need to draw from vernacular architecture for designing buildings in different regions.

Further, this study has some limitations. As there is no single software package that can simulate the effects of using canopies in both outdoor and indoor spaces at a time, coupling different software packages to observe these effects requires a skilled and experienced engineer. The methods used in this study can be changed if advanced software packages become available in the future for analyzing the environmental performance of building elements. The criteria of thermal comfort and daylighting levels can be varied using different functions depending on the types of spaces and user activities. For example, daylighting levels of 300–500 lux are needed in spaces involving visual activities such as classrooms and libraries and 200 lux for circulation areas. Perceptions of thermal comfort in different countries would vary, which can be explored in further studies. Climate data for rural and urban areas should be considered carefully as the microclimate is sometimes considerably different in different environments.

Environmental assessment (e.g., BREEAM, LEED, etc.) with flexible operation of building elements can be beneficial to prove their enhanced performance in specific cases. Therefore, the results may vary depending on the building elements and locations. Nevertheless, with these simulation techniques and methods, designers and engineers can establish more detailed metrics for using flexible building elements with the combined analysis both indoor and outdoor spaces to maximize human comfort and minimize building energy demands.

Author Contributions: Conceptualization, K.J.L. and C.W.K.; methodology, C.W.K. and S.C.; software, C.W.K.; validation, S.C. and K.J.L.; formal analysis, C.W.K.; investigation, C.W.K.; resources, S.C.; data curation, S.C.; writing—original draft preparation, C.W.K.; writing—review and editing, K.J.L.; visualization, C.W.K.; supervision, K.J.L. and S.C.; project administration, K.J.L.; funding acquisition, K.J.L.

Funding: This research was funded by the research fund of Hanyang University, South Korea (HY-2013).

Acknowledgments: This study is thankfully supported by the Industry-University Cooperation Foundation, Erica campus, Hanyang University.

Conflicts of Interest: The authors declare no conflict of interest.

References

1. Givoni, B. *Passive and Low Energy Cooling of Buildings*; Van Nostrand Reinhold: New York, NY, USA, 1994; p. vii. 263p.
2. Thomas, R. *Environmental Design: An Introduction for Architects and Engineers*, 3rd ed.; Taylor & Francis: London, UK, 2006; 272p.
3. Hootman, T. *Net Zero Energy Design: A Guide for Commercial Architecture*; John Wiley & Sons: Hoboken, NJ, USA, 2013; 441p.
4. Yannas, S. *Design of Educational Buildings; Environment & Energy Studies Programme*; Architectural Association Graduate School on behalf of the Commission of the European Communities: London, UK, 1994; Volumes 1–3, 95p.
5. Webb, A.L. Energy retrofits in historic and traditional buildings: A review of problems and methods. *Renew. Sustain. Energy Rev.* **2017**, *77*, 748–759. [CrossRef]
6. Steemers, K.; Baker, N.; Fanchiotti, A. *Daylighting in Architecture: A European Reference Book*; Published for the Commission by James & James: London, UK, 1993.
7. Loonen, R.C.G.M.; Favoino, F.; Hensen, J.L.M.; Overend, M. Review of current status, requirements and opportunities for building performance simulation of adaptive facades. *J. Build. Perform. Simul.* **2017**, *10*, 205–223. [CrossRef]
8. Grobman, Y.J.; Capeluto, I.G.; Austern, G. External shading in buildings: Comparative analysis of daylighting performance in static and kinetic operation scenarios. *Archit. Sci. Rev.* **2017**, *60*, 126–136. [CrossRef]
9. Gelfand, L.; Freed, E.C. *Sustainable School Architecture: Design for Primary and Secondary Schools*; John Wiley: Hoboken, NJ, USA; Chichester, UK, 2010.
10. Brook Lapping Productions. Learning outside the classroom primary history. In *Education in Video*; Teachers TV/UK Department of Education: London, UK, 2010.
11. Constable, K. *The Outdoor Classroom Ages 3–7: Using Ideas from Forest Schools to Enrich Learning*; Routledge: Milton Park, Abingdon, UK; New York, NY, USA, 2012.
12. Chen, L.; Ng, E. Outdoor thermal comfort and outdoor activities: A review of research in the past decade. *Cities* **2012**, *29*, 118–125. [CrossRef]
13. Aljawabra, F. Thermal Comfort in Outdoor Urban Spaces: The Hot Arid Climate. Ph.D. Thesis, University of Bath, Bath, UK, 2014.
14. Schiano-Phan, R.; Weber, F.; Santamouris, M. The Mitigative Potential of Urban Environments and Their Microclimates. *Buildings* **2015**, *5*, 783–801. [CrossRef]
15. Santamouris, M. *Energy and Climate in the Urban Built Environment*; James & James: London, UK, 2001; 402p.

16. Taleghani, M.; Kleerekoper, L.; Tenpierik, M.; van den Dobbelsteen, A. Outdoor thermal comfort within five different urban forms in The Netherlands. *Build. Environ.* **2015**, *83*, 65–78. [CrossRef]
17. Schinasi, L.H.; Benmarhnia, T.; De Roos, A.J. Modification of the association between high ambient temperature and health by urban microclimate indicators: A systematic review and meta-analysis. *Environ. Res.* **2018**, *161*, 168–180. [CrossRef]
18. Hawkes, D. *The Environmental Tradition: Studies in the Architecture of Environment*; E. & F.N. Spon: London, UK, 1996; p. 212.
19. Merlier, L.; Frayssinet, L.; Johannes, K.; Kuznik, F. On the impact of local microclimate on building performance simulation. Part II: Effect of external conditions on the dynamic thermal behavior of buildings. *Build. Simul.* **2019**, *12*, 747–757. [CrossRef]
20. Lu, S.; Xia, H.; Wei, S.; Fang, K.; Qi, Y. Analysis of the differences in thermal comfort between locals and tourists and genders in semi-open spaces under natural ventilation on a tropical island. *Energy Build.* **2016**, *129*, 264–273. [CrossRef]
21. Yang, B.; Olofsson, T.; Nair, G.; Kabanshi, A. Outdoor thermal comfort under subarctic climate of north Sweden—A pilot study in Umeå. *Sustain. Cities Soc.* **2017**, *28*, 387–397. [CrossRef]
22. Szokolay, S.V.; Auliciems, A. *PLEA Notes 3: Thermal Comfort PLEA: Passive and Low Energy Architecture International*; revised 2007; University of Queensland: Brisbane, Australia, 1997.
23. CIBSE. *Comfort*; CIBSE: London, UK, 2006.
24. Liu, J.; Niu, J.; Xia, Q. Combining measured thermal parameters and simulated wind velocity to predict outdoor thermal comfort. *Build. Environ.* **2016**, *105*, 185–197. [CrossRef]
25. He, Y.; Jones, P.J.; Rayment, M. A simple parameterisation of windbreak effects on wind speed reduction and resulting thermal benefits to sheep. *Agric. For. Meteorol.* **2017**, *239*, 96–107. [CrossRef]
26. Pomponi, F.; Barbosa, S.; Piroozfar, P.A.E. On The Intrinsic Flexibility of the Double Skin Façade: A Comparative Thermal Comfort Investigation in Tropical and Temperate Climates. *Energy Procedia* **2017**, *111*, 530–539. [CrossRef]
27. Nazarian, N.; Sin, T.; Norford, L. Numerical modeling of outdoor thermal comfort in 3D. *Urban Clim.* **2018**, *26*, 212–230. [CrossRef]
28. Gulyas, A.; Unger, J.; Matzarakis, A. Assessment of the microclimatic and human comfort conditions in a complex urban environment: Modelling and measurements. *Build. Environ.* **2006**, *41*, 1713–1722. [CrossRef]
29. Höppe, P. The physiological equivalent temperature—A universal index for the biometeorological assessment of the thermal environment. *Int. J. Biometeorol.* **1999**, *43*, 71–75. [CrossRef]
30. Spagnolo, J.; de Dear, R. A field study of thermal comfort in outdoor and semi-outdoor environments in subtropical Sydney Australia. *Building and Environment* **2003**, *38*, 721–738. [CrossRef]
31. Hoppe, P. Different aspects of assessing indoor and outdoor thermal comfort. *Energy Build.* **2002**, *34*, 661–665. [CrossRef]
32. CIBSE. *Design for Improved Solar Shading Control*; DTI & Chartered Institution of Building Services Engineers: London, UK, 2006; 29p.
33. Littlefair, P.J. *Summertime Solar Performance of Windows with Shading Devices*; BRE Bookshop for BRE Trust: Garston, UK, 2005; 28p.
34. Van Esch, M.M.E.; Looman, R.H.J.; de Bruin-Hordijk, G.J. The effects of urban and building design parameters on solar access to the urban canyon and the potential for direct passive solar heating strategies. *Energy Build.* **2012**, *47*, 189–200. [CrossRef]
35. DfES. *Schools for the future: Designing School Grounds*; TSO (The Stationery Office): London, UK, 2006; 117p.
36. Dudek, M. *Architecture of Schools: The New Learning Environment*; Architectural Press: Boston, MA, USA, 2000.
37. Tucker, R.; Izadpanahi, P. Live green, think green: Sustainable school architecture and children's environmental attitudes and behaviors. *J. Environ. Psychol.* **2017**, *51*, 209–216. [CrossRef]
38. EDSL Tas v9.1.4 Software. Environmental Design Solutions Limited. Available online: www.edsl.net (accessed on 23 November 2019).

39. Obaidat, M.S. Simulation and Modeling Methodologies, Technologies and Applications. In Proceedings of the International Conference, SIMULTECH 2013, Reykjavík, Iceland, 29–31 July 2013. revised selected papers.
40. Kim, H.; Lee, K.; Kim, T. Investigation of Pedestrian Comfort with Wind Chill during Winter. *Sustainability* **2018**, *10*, 274. [CrossRef]
41. Great Britain. Department for Education and Employment. *BB87 Guidelines for Environmental Design in Schools: (Revision of Design Note 17)*; revised 2003; Stationery Office: London, UK, 1997.

© 2019 by the authors. Licensee MDPI, Basel, Switzerland. This article is an open access article distributed under the terms and conditions of the Creative Commons Attribution (CC BY) license (http://creativecommons.org/licenses/by/4.0/).

sustainability

MDPI

Article

Experimental Confirmation of the Reliability of Fanger's Thermal Comfort Model—Case Study of a Near-Zero Energy Building (NZEB) Office Building

Michał Piasecki [1], Małgorzata Fedorczak-Cisak [2,*], Marcin Furtak [2] and Jacek Biskupski [3]

[1] Building Research Institute, 00-611 Warsaw, Poland; m.piasecki@itb.pl
[2] Faculty of Civil Engineering, Cracow University of Technology, 31-155 Cracow, Poland; mfurtak@pk.edu.pl
[3] Faculty of Energy and Fuels, AGH University of Science and Technology in Cracow, 30-059 Cracow, Poland; biskupski@agh.edu.pl
* Correspondence: mfedorczak-cisak@pk.edu.pl

Received: 27 February 2019; Accepted: 22 April 2019; Published: 26 April 2019

Abstract: Designing and constructing near zero energy buildings (NZEBs) is a challenge not only from a structural point of view, but also from the point of view of ensuring appropriate climate comfort for users. The standards describing how to ensure comfort were created in times when the challenges of building ZEB/NZEB were not yet explored and energy issues were not as important as they are today. Therefore, the assessment of the thermal and climatic comfort of people living and working in such buildings requires a new or revised approach to the methodology of thermal comfort assessment. In this article, the authors present the results of a thermal comfort study based on measurements and thermal sensory tests. Testing was carried out in an experimental office building (passive standard). The main goal of the experiment was to compare the thermal comfort measurement method based on the ISO-Fanger model with the actual comfort results obtained by the panellists in the model office condition. The tests allowed the lowest operating temperature providing thermal comfort (predicted mean vote (PMV) = 0 and −0.5) to be determined. Sensory tests were conducted using three types of questions. The results were compared to the other researchers' findings. It was noted that the panellists showed better thermal comfort sensation at lower temperatures than would result from the traditional Fanger distribution, so the authors proposed the experimental function of percentage of dissatisfied (PPD) = f(PMV). The authors hope that it contributed to the actual state of knowledge as a "small and specific scale" validation of the existing thermal comfort model. The results also revealed that the method of heating has an influence on the subjective thermal sensation.

Keywords: thermal comfort; thermal comfort model; panel tests; NZEB; indoor environmental quality; PMV; PPD

1. Introduction

The issue of human thermal comfort in buildings already has over 50 years of history [1] and was practically standardized by ISO, CEN, and ASHRAE in both hemispheres [2,3]. However, in the actual opinion of scientists [4–19], the subject of comfort, including the thermal satisfaction of building users in relation to technical building systems, still creates a potential field for research and several new questions have arisen in recent years. First, because the construction methods that focus on building energy efficiency and users' wellbeing have changed significantly since the time Fanger published his research and the first international standards in this area were introduced. Changes in the perception of comfort in buildings are partially responsible for ASHRAE revising its standard on thermal environmental conditions for human occupancy [3] in 2017, while CEN currently applies standard prEN 16798-1 to indoor environmental input parameters for the design and assessment of energy performance of buildings, which also addresses thermal comfort [20] and replaces standard EN 15251 [21]. Changes in construction have led to a large difference between comfort in Naturally-Ventilated (NV) buildings [22] and those where mechanical ventilation (HVAC) is used [23]. Studies conducted in recent decades indicate new parameters determining important factors in user comfort, with more parameters having a significant impact on the comfort, wellbeing and even health of people living and spending time in enclosed spaces [24]. The number of parameters that affect comfort increased from the four basic parameters indicated by Fanger to as many as ten that are cited nowadays, in particular for NZEB buildings promoted by the European Energy Directive [25]. As a result, the simple definition drafted 50 years ago, which states that thermal comfort "is that condition of mind, which expresses satisfaction with the thermal environment" [1], no longer meets the current expectations of investors and building residents. A large population of researchers agrees that "indoor comfort" should be considered in a much broader context [26]. Some scientists have gone ahead with defining the building users comfort indicator based on users' predicted satisfaction. Over the last few years, there has been wide discussion in the literature [27–30] on whether the linear model was designed for assessing the Indoor Environmental Quality (IEQ_{index}). This index was designed taking into consideration standard EN15251:2007 [21], mainly for calculating the percentage of people satisfied with indoor environmental quality as a function of the four parameters: air quality and user perception of thermal, acoustic and visual comfort. In fact, the characteristics of the IEQ model should also contain the synergy effect of environmental parameters included in sub-components and their sensory perception, as suggested by ASHRAE Guideline 10 [31] and the effect of measurement accuracy on the IEQ sub-component parameters [32]. The overall assessment index values of perceived IEQ vary on a scale from 0 to 100%, and may provide the basis for classification of the indoor environmental quality of a building. IEQ sub-models use the more or less accepted empirical dependencies that determine the impact of individual aspects on the user's satisfaction, as is the case in the case of thermal impacts and their impact on thermal influence on user satisfaction. However, these models evolve over time as more and more data are collected by scientists in the various analysed cases of modern buildings [33–35]. As part of their research work on IEQ [32,36], the authors have identified the need to also focus and review the thermal comfort model used for IEQ_{index} calculation and validate the thermal model in practice on the energy-efficient building. In the case of modern buildings with integrated HVACR systems, as in our research case study, there is more scope for influencing additional comfort parameters (apart from temperature and humidity, which are still widely used worldwide to assess thermal comfort). It is important that by using Building Management Systems (BMS) it is not only possible to measure the comfort-related factors but also to adjust them. Therefore, it should be possible to discuss the climatic comfort as IEQ and the actual number of satisfied users instead of physical parameters only or thermal comfort alone. And also taking into consideration the justified questions arising from the correlation of the Predicted Percentage of Dissatisfied (PPD) and Predicted Mean Vote (PMV) based on the standards in comparison to the actual satisfaction of users and whether the current sensory responses of users coincide with values based on the thermal model. This was one of the research challenges that has been discussed in the article. In the authors' opinion, thermal measurements based on the ISO 7730 [2] standard alone are not perfect for reflecting the wide spectrum of climate sensations. It seems that the best solution is to compare the measurement results and user responses within strictly defined experimental conditions. In practice, modern HVACR systems offer new possibilities not seen in the times of Fanger, such as individually adjusted climate comfort parameters or on-demand ventilation (VOD) controlled, e.g., by the actual concentration of CO_2 in the air. Control of these parameters allows climatic comfort to be designed and even corrected during the use of the rooms. Other authors also point to the possibility of using a modern heating system to control climate comfort while increasing energy efficiency [37]. The authors therefore believe that PMV and PPD parameters can nowadays not only be measured but also corrected in real time. Another interesting subject is the issue of summer comfort (mainly related to the need for room cooling) vs. winter comfort, as well as linking the differentiation of comfort results with the geographical location, even within Europe [24]. For countries located in "colder parts" of the Earth (e.g., northern or central European countries), to which our country belongs and where these studies were conducted, ensuring full climatic comfort in winter poses a greater challenge than in summer, i.e., in a completely different way than in Australia [38]. The literature also increasingly distinguishes important comfort parameters for residential buildings from buildings such as schools or hospitals [39–41]. In practice, due to the size and the most important parameters determining the comfort of these rooms, one can actually speak of completely different standards of climatic comfort for a hospital [42] and for a single-family house. Which is why it seems so important to define both the region and the type of building in which the research is carried out, even for research using Fanger's methodology. It is worth remembering that the Fanger model was based on results obtained from surveys [7,43] and supported by Nevins questionnaire results [44], but the raw study was based on sixty-four student answers for only 13 temperatures (18.9 °C–32.2 °C) and only one humidity (50%), where it was assumed that the radiation temperature is equal to the measured actual temperature. Despite numerous listed tests to which the ISO 7730 standard refers [45] carried out on various types of objects, in various climatic states and on differently aged populations, the thermal model has not changed to this day. Basically speaking, neutral thermal comfort is established when the heat released by the human body is in equilibrium/balance with its heat production. There are major recognized parameters that affect thermal sensation results. These parameters can be grouped into two categories. One related to the building itself, the so-called "technical parameters of the indoor environment", including: air temperature, the temperature of the surrounding surfaces (so-called radiant temperature), air speed and turbulence, relative air humidity and the parameters of the space tested and parameter gradients. There is a wide range of combinations of these factors where the level of comfort may be satisfactory, which may be called the "comfort zone". The comfort zone is mainly determined by provisions provided in standard ISO 7730. According to a normative approach, it was essential for the authors to include them in the presented research, taking into consideration that: the sultriness limit in relation to the air humidity should not be exceeded; the air speed has to be within closely defined limits (for speeds under 0.1 m/s, the number of dissatisfied occupants due to draughts is less than 6%); the difference between radiant temperature and air temperature should remain small during the test; the difference in the radiant temperature in various directions should remain negligible (less than 5 °C, known as the "radiation temperature asymmetry"); the indoor air temperature stratification has to be less than 2 °C between the head and ankles of a seated person; and the perceived temperatures in the laboratory room should change by no more than 1 °C at different spheres. Bearing these requirements in mind, the authors introduced them to the laboratory office under examination. It is commonly accepted knowledge that the more irregular the thermal field in a room, the greater the expected number of dissatisfied people. As the result of the sensitivity analysis of the effects of physical parameters on the measurable results of PMV and PPD (pre-test activity) the authors knew that the biggest influence on thermal sensation would be the actual temperature, then the temperature of radiation and the humidity followed by the air flow associated with the installation and the number of air exchanges. The second group of parameters affecting the assessment of thermal comfort is the so-called "human factors" group and this group includes parameters related to the panellists, such as: population size; population quality; ethnicity; age; gender; weight; height; body surface area; body mass index; activities; clothing; adaptability to temperature changes; daily and monthly cycle; metabolism; and current nutrition. Authors made the justified assumptions and simplifications for these factors in the provided research and present details of these assumptions in the Method section of this paper. For the authors, academics, the main target group for panel research is students. Students spend more than 40% of their time in classrooms, and climate comfort has a particular impact on the effects of their work and learning – and even on their health. It is also a very open and flexible group, and willing in practice to take part in thermal tests. However, it is also a group which, as shown by [46], may have a specific greater adaptability, achieving subjective comfort earlier than the average adult group. Some authors explicitly indicate that adolescents may feel thermal comfort in a greater range of temperatures than adults [47], which may have a significant impact on the PMV determination. According to Fanger, the sensation of thermal comfort for older people and students should not show statistically representative significant differences. The research presented in our paper was conducted on a group of 50 students similar to Fanger's panellists and the results refer to the whole group without distinction by gender. Numerous authors, including Fanger himself, have shown that the difference between the results obtained by women and men is not statistically significant (confidence level of 5%). Our opinion is rather similar; currently, it is recognized that the differences in thermal sensation between a woman and a man don't depend on gender itself but on the anthropometric parameters like body mass index (BMI), speed of metabolism, the length of hair and the way of dressing it, the thermal resistance of clothing used and muscle mass.

Another research aspect discussed in the paper is that almost all energy-efficient buildings operate at lower heating temperatures. In our opinion, these temperatures might not usually provide high classes of thermal comfort in accordance with [48]. The authors believe that it is important to validate whether the proposed or designed room temperatures provide comfort to users who have to perform certain activities in the rooms, taking into account the various systems' inertia, temperature changes, thermal inertia of partitions, thermal radiation and temperature gradients. The hypothesis is that the use of simple thermal comfort measurements may not fully reflect the actual thermal sensation of the users. Achieving a high level of energy efficiency while providing adequate climate comfort for each type of building is becoming a challenge for the future in which scientists and the authors have already started work. In this context, the definition of the limits of sensory comfort takes on a new meaning. The research currently underway in the experimental office enables PMV tests to be carried out over a shorter period of time on a much larger number of participants, while at the same time observing the impact of the way this comfort is achieved, depending on the extensive set of parameters of the HVACR system. Lowering the temperature of the heating medium, for example, may affect the feeling of climate comfort, especially in rooms with underfloor heating. The experimental office rooms used for such tests were prepared in the energy efficient building of the Małopolska Laboratory of Energy Saving (MLBE) of the Cracow University of Technology. This building was designed and erected as an experimental laboratory building, where both educational classes and advanced climate comfort tests were conducted (see detailed description of room tested in the methodology section). The research conducted made it possible to apply several methods of space heating at the same time, e.g., underfloor and air heating, which in turn enabled quick regulation of climate comfort parameters over a range of temperature and humidity, as well as CO_2 and TVOC concentration (a separate material and article). The study in MLBE also allowed a significant number of people (up to 30) to be examined, assessing comfort in a "right-here-right-now" questionnaire, within a designated period of time. They were tested regardless of the prevailing outdoor temperatures, as opposed to studies conducted throughout the year [48], where the influence of the outdoor temperature on the examined people may prove to be important. One of them turns out to be the differences (e.g., vertical gradient) of floor temperature (usually changing its temperature relatively slowly), in relation to the temperature of the air (supplied), which may cause the additional discomfort noted and described in the literature, e.g., in the standard [21]. The tests in the MLBE laboratory gave the authors the possibility of preparing a spectrum of climatic conditions, and the data to compare with the standardized thermal model and the assessment methodology.

During the thermal sensory output assessment, the authors also discovered that the way of asking a question about actual sensory thermal comfort in the office may affect the final results. The conducted research shows that the question about actual thermal comfort for office work and general thermal comfort gives slightly different results. The results coincide with Fanger's results and constitute a basis for discussing the issue of thermal comfort in almost zero energy buildings.

2. Materials and Methods

2.1. Goals of the Experiment

The main goal of the experiment was to compare the thermal comfort measurement method based on the Fanger/ISO 7730 model with the sensory comfort results obtained by the panellists in the experimental model office. Other aims of the study are:

- conducting an experiment to determine the lowest operating temperature giving thermal comfort (PMV = 0 and −0.5) in the experimental space of the MLBE building by way of physical measurements and sensory questionnaire surveys of a statistically representative panellist group (the size of the test group is 50),
- comparing the results of sensory tests obtained by means of three types of questionnaire questions with environmental measurement results and resulting from the Fanger-ISO 7730 comfort thermal model,
- determining the impact of three question types on the results (seven-scale question, 0–100% scale question, yes/no question),
- comparison and discussion of results obtained by measuring method and results of surveys (PPD = f(PMV),
- comparing the raw results of surveys (PPD) obtained by other researchers with the authors' own survey results,
- proposing an equation for the experimental thermal comfort curve for given boundary conditions on the basis of the obtained results—a "limited scale" validation of the existing thermal comfort model

A graph showing the assumptions and steps of the experiments is presented in Figure 1.

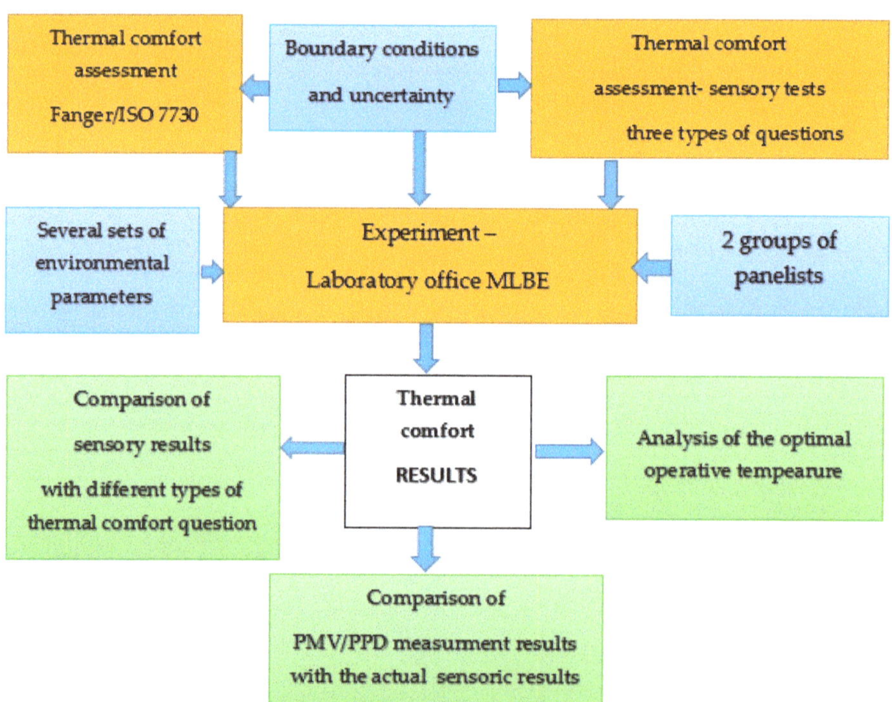

Figure 1. Scheme of the thermal comfort assessment experiment.

One of the intentions of the experiments was to establish a comparative scale for previously conducted studies (American and Danish) in the context of the practical use of the Fanger equation (and ISO 7730) for nZEB buildings and for the provided boundary conditions. The results obtained by the authors are of an illustrative nature for a specific nZEB building case study —not generalizing the issue of thermal comfort for other building user populations. The authors' intention was not to validate the thermal comfort model for different populations that Fanger did and was not to create a compendium of knowledge about the whole issue of thermal comfort assessment taking into consideration all challenges related to specific human reaction aspects (grouping results due to the parameters of the people being tested). Experiments of this type are very labour-intensive, which is why authors conducted them using the limits and simplifications provided as boundary conditions. In the most recent studies, usually one person or a very small group was tested in the research chamber. The intention of the authors was not to directly copy experiments. It was decided to carry out an experiment involving number of panellists at the same time in the laboratory office. The number of panellists corresponded to the assumed number of employees in the office, so the result obtained at one time reflects the average thermal comfort of employees. This approach, in our opinion, may be practical within the context of determining operational temperatures for the NZEB offices where a thermal sensory survey can be more appropriate than widely accepted tests using "the measurement" operating on the ISO 7730 algorithm.

2.2. Thermal Comfort Model

The model for assessing the thermal comfort of an indoor office area is based on the assessment of the indoor environment's physical parameters. For the case-study building equipped with heating and cooling systems, the indicator predicted mean vote (PMV) is determined in accordance to ISO 7730 [2]. PMV is a reference parameter for thermal environmental assessment as provided in the standards EN 15251 [21], draft of FprEN 16798-1 and [20]. PMV is a seven-point scale of thermal sensation in a function of measured physical parameters as presented:

$$PMV = f\ (t_a, t_{mr}, v_a, p_v, M, I_{cl}) \quad (1)$$

where t_a is the air temperature [°C], t_{mr} is the mean radiant temperature [°C], v_a is the relative air velocity [m/s], p_v is the water vapour partial pressure [P_a], M is the metabolic rate [W/m^2] and I_{cl} is the clothing insulation [m^{2}K/W].

The measurement methodology is based on standard methodology [49]. PMV is required to determine the predicted percentage of dissatisfaction which is calculated by the following formula:

$$PPD = 100 - 95 \cdot \exp\ (-0.03353 \cdot PMV^4 - 0.2179 \cdot PMV^2) \quad (2)$$

In practice, the values of PMV and PPD are determined by measurement equipment or can be calculated by the web-tool located at http://comfort.cbe.berkeley.edu/EN of the Center for the Built Environment, University of California, Berkeley.

2.3. Case Study Object and Boundary Conditions

The test on PMV/PPD and the sensory thermal comfort of the NZEB building users was carried out in the experimental building of the Małopolska Laboratory of Energy Efficient Building (MLBE) (Figure 2a), in the lecture/office hall on the third floor. The national plan equates to a "nearly zero energy building" with a "low-energy building" and gives its definition: "A building with low energy consumption" should be understood as a building that meets the requirements related to energy saving and thermal insulation included in the technical regulations-construction, referred to in Article 7 of the Act of 7 July 1994—Construction Law, in particular, Section X and Annex 2 to the Regulation of the Minister of Infrastructure of 12 April 2002 on the conditions technical requirements that should be met by buildings and their location, effective from 1 January 2021, and for buildings occupied by public authorities and owned by them—from 1 January 2019. MLBE building has a high level energy supply systems for the building including: tri-generation system (simultaneous generation of thermal and cooling energy and electric)—powered by natural gas—consists of a co-generation unit (CHP) and an absorption refrigerating device, compressor reversible heat pump type glycol/hot water from the ground—3 vertical probes, 99 m deep each, ground heat exchanger heating in the winter and cooling in the summer part of the ventilation, ventilation heat recovery units with moisture recovery with efficiency greater than 80%, photovoltaic cells placed on sunblinds, and on the roof of the building; the links on the blinds have a system of automatic orientation towards the sun, flat and vacuum solar collectors supporting the hot water preparation system. All energy supply systems for the building are combined into one an adaptive and smart control system with a purpose function related to the minimization of primary energy consumption (USAD system). Internal installations are: supply and exhaust mechanical ventilation system taking into account the periodicity of operation and the regulation of the fresh air stream with the use of CO_2 sensors, exhaust ventilation system operating periodically connected with automatically tilting windows that allows using the building's accumulation capacity for cooling ("night storage cooling" combined with building materials containing variable-phase components), underfloor heating and air heating system depending on the thermal zones of the building. U-values of building elements are: the external walls-ventilated 0.1 W/(m^{2}K), roofs and floor on the ground 0.1 W/(m^{2}K), glass facades and doors 0.7 W/(m^{2}K). The end use energy is at passive house standard EK = 11.6 kWh/(m^2 year). The tests were performed on 19 and 20 December 2018. Figure 2b shows the view of the places on which the respondents were asked about thermal comfort. MLBE is located in Cracow, designed to conduct building physics research under "in situ" conditions.

(a) (b)

Figure 2. Małopolska Laboratory of Energy Efficient Building—view of the building (**a**). The office room in which the tests were carried out (**b**).

Figure 3a,b show the floor section and location of testing devices, as well as the panellists' location (numbers) in the two day study (M1, M2 are main measurement devices for indoor parameters, Po1-Po7 secondary, see section on measurement devices).

(a)

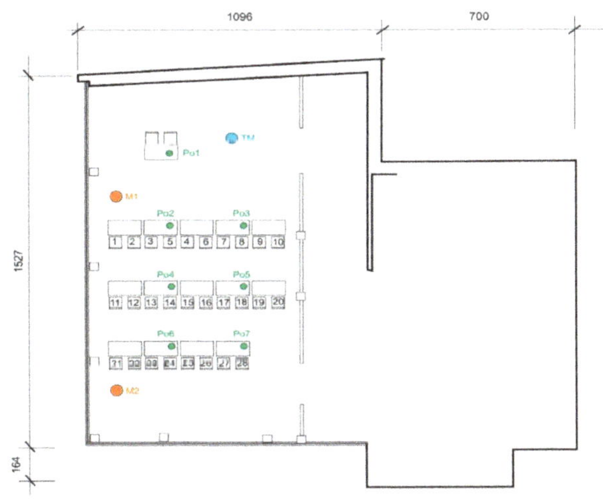

(b)

Figure 3. Location (**a**) of the people participating in the research and the research equipment on 19 December 2018 (28 panellists), location (**b**) of the people participating in the research and the research equipment on 20 December 2018 (22 panellists).

During the tests blinds on the glass facades were lowered, the respondents answered questions in artificial light (range 450–500 lux). The test room was cooled to +17.0 °C as to be the initial temperature for the first thermal comfort test. The initial conditions were maintained for 24 hours before the beginning of the task. The humidity level (RH) was set at the range of 25–35%. The ventilation system used the standard exchange settings of 1.2 changes per hour. The locations of the air vents are shown in Figure 4a. After the respondents took their positions, the heating function was activated to slowly and continuously warm up to +24 °C where the thermal neutral sensation (PMV = 0) was expected. The change in temperature from very cold to neutral level was planned over 2 hours. Two full temperature sessions were done (first and second day). Fan coil units (air heating) located in accordance with Figure 4b were used to heat the room.

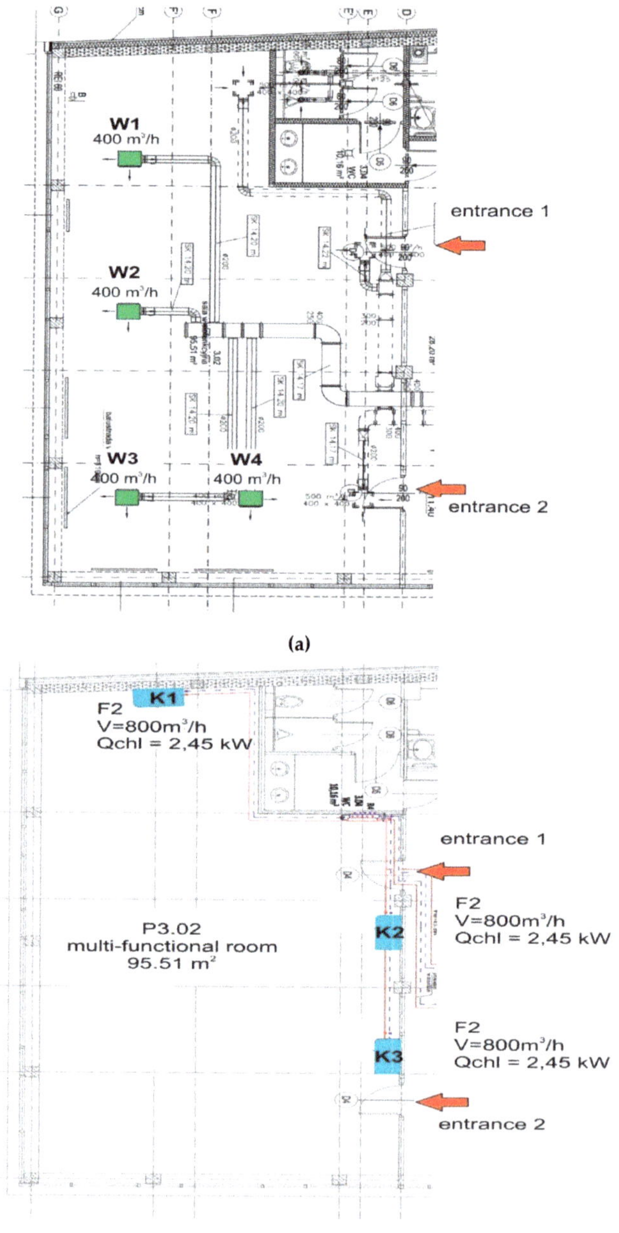

(a)

(b)

Figure 4. (**a**) Location of vents (W1–W4) and fan coil units (FCUs) (K1–K3). Marking in (**b**): F2—fan coil with brushless EC (Energy Efficient Variable Speed EC Motor Fan Coil Unit Solutions) motor FCU, two-pipe.

The measured CO_2 level did not exceed 1000 ppm during the 2 days test, which authors assume did not affect the thermal index results.

2.4. Panel Group

The survey involved: 22 students of AGH University of Science and Technology in Krakow on 19 December 2018 and 28 students of Cracow University of Technology on 20 December 2018. The panellist group was ethnically homogenous—a white human variation, type Caucasian. Participants declared a healthy state before. All necessary anthropometric data characterizing the panel group is provided in Table 1. Table 1 also compares the panel group with the group used in Fanger's research (including standard deviations).

Table 1. Anthropometric data of the tested panel groups (Authors' data and Fanger's data) with expanded uncertainty at the confidence level of 1—α = 0.95.

Group	Gender	Group Size	Age [Years]	Height [cm]	Body Weight [kg]	Skin Surface "DuBois" [m²]	Body Mass Index	Clo [m² K/W]
Academic youth-Author's test	Man	12	23 ± 2.4	175 ± 8.0	74 ± 13.0	1.8 ± 0.25	24.2 ± 3.0	0.7 ± 0.05
	Woman	38	22 ± 2.0	162 ± 6.0	58 ± 15.0	1.6 ± 0.22	22.1 ± 2.4	0.7 ± 0.05
	Mean	50	22 ± 2.2	165 ± 16.0	62 ± 18.4	1.6 ± 0.23	21.2 ± 2.4	0.7 ± 0.05
Academic youth-Fanger's test	Man	64	24 ± 4.6	180 ± 12.0	71 ± 13.0	1.9 ± 0.24	22.2 ± 3.0	0.6 ± 0.05
	Woman	64	23 ± 2.4	168 ± 7.0	57 ± 14.8	1.6 ± 0.24	20.2 ± 2.2	0.6 ± 0.05
	Mean	128	23 ± 4.4	174 ± 16.0	64 ± 21.0	1.8 ± 0.34	21.2 ± 2.6	0.6 ± 0.05

The group of panellists admitted for research meets the parameters given for the group accepted for research by Fanger so the authors' focus is on a parametrically comparable group. According to Fanger's research, it was assumed that the sensation of thermal comfort for older people and academic youth panellists should not show statistically representative significant differences, so the authors decided not to assess any other age group under this research. The group of panellists reflects the potential employees of the academic nZEB office building, where young people are the dominant faction. The average age of panellists was twenty-two, and this value is only 2.5% lower than for Fanger's panellists. The average height was 5% lower, and the body mass index (BMI) was higher by 6%. BMI was developed almost 200 years ago by Adolf Quetelet and is currently used by research centres dealing with health, including thermal comfort. All panellists surveyed had an average BMI of 24.16 at the limit of normal body weight, i.e., 18.5 < BMI < 24.9, similar to that of Fanger where his panellists' index was 22.16. This value is 7% higher. The value of BMI for the group may slightly affect the result. It is common knowledge that people with a higher fat content in body weight may be more tolerant of lower temperatures. The authors do not take into consideration that the slightly higher BMI index from Fanger's panellists could affect the obtained results, because the BMI difference from Fanger is statistically low. The authors have the prodigiousness of other various human factors that may affect the result of thermal comfort test results, including: psycho-physical condition, physiological circadian (day rhythm), ethnicity and nutrition before tests. Fanger stated that the difference between the results obtained by women and men is not statistically significant (confidence level of 5%). Taking this simplification into account, the authors did not focus on the differences in results obtained by men and women, instead analysing the results averaged for the studied group as a whole. The authors acquired the results with a gender distinction so they may be used as part of another publication. The value of clothes' thermal resistance (clo) between women and men was also averaged and calculated despite the fact that some women have long hair and wear extra underwear (e.g., bra). Panellists were wearing long trousers, short-sleeved shirts and shoes, which corresponds to the insulation of I_{clo} clothes at 0.7 [clo] (calculated by the authors) and performed a physical activity at the level of 1.1 [met] (semi-active sitting/working in a sitting position; typing, reading, task solving (cooperation), conversation). The menstrual cycle (as this is a question not recommended ethically for technical assessment studies) was not taken into account in the studies. Currently, it is assumed in the known studies of thermal comfort that the influence of menstruation is statistically insignificant.

The group didn't consume meals up to two hours before the study or during the study. During the test, students were allowed to drink water to supplement the possible needs related to the secretion of sweat.

Physiological circadian rhythm was not included in the research. Research was carried out during the daytime around the time between breakfast and lunch. In our opinion, this factor did not affect the results by disturbing the general results and conclusions.

During the test, there were no effects of sudden change in temperature and other parameters. Both groups remained air-conditioned for 30 minutes in neutral conditions before the tests (at PMV = −0,2; ta = 23.5 °C, RH = 35%).

2.5. Thermal Sensory Tests—Votes

The respondents evaluated their sensory thermal comfort in writing (three types of questions). The students taking part in the survey answered questionnaire questions at intervals of 5 or 10 minutes when new thermal conditions were established (17 °C –25 °C). The temperature increased on average by 0.3 degrees in the interval of 5 minutes. Experiment schedule was two hours of testing with change of temperature and summary 3 hours considering the neutral thermal buffer before main stage of testing. In total, temperature increased by 7 degrees in about two hours.

The surveys were divided into three types of questions as presented;

The first of them is: Determine the feeling of thermal sensation on a 7-degree scale, where the value −3 was marked as very cold, −2 as cold., −1 as quite cold, 0—neutral (comfortable), +1 quite warm, +2 warm, +3 hot (Fanger approach-based).

The second of them is: Determine using a two-degree scale whether the prevailing conditions are comfortable for work (yes/no).

The third of them is: Determine in what percentage the conditions are suitable for work (from 0% [absolutely not suitable] to 100% [the conditions are comfortable/neutral]).

2.6. The Measuring Equipment

Non-dependent on surveys, the following equipment was used for independent testing of indoor air parameters. Thermal comfort meters, marked in Figure 2a,b as M1 and M2. The meters were located at the front and the end of the test area. Figure 5 shows the sensors in the device for measuring thermal comfort (M1, M2). The measured parameters were:

- t_a—actual air temperature measurement,
- t_g—temperature of blackened sphere (heat radiation meter), 15 cm in diameter,
- t_{nw}—natural wet-bulb temperature measurement,
- RH—measurement of relative air humidity,
- v_a—measurement of air flow speed.

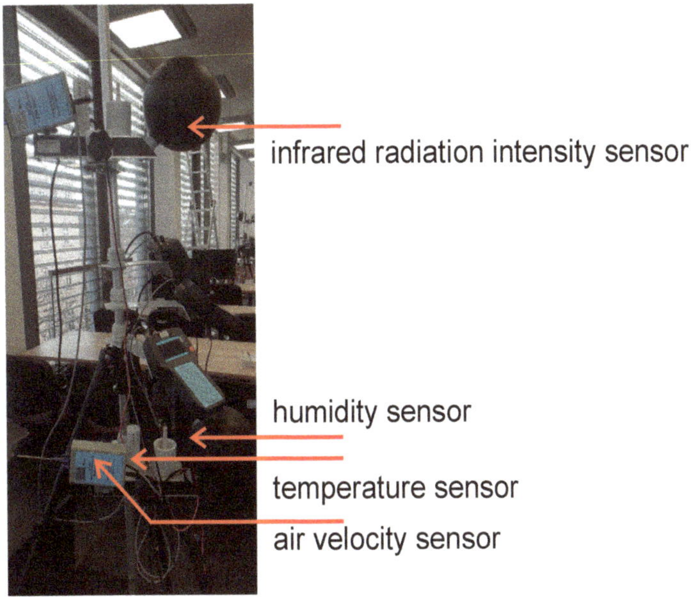

Figure 5. The sensors connected to the testing device EHA-MM101 dedicated to thermal comfort measurements (M1).

Measurements of the physical parameters necessary to determine PMV and PPD were carried out at three heights: 5 cm, 100 cm and 160 cm above floor level in parallel, but only measurements at the panellists' chest level (100 cm) of sitting participants were taken for further analysis. Measurement at three heights allows a possible negative gradient of vertical temperature to be determined, which would disturb the sensory responses of the panellists. The frequency of data collection was every 10 min. The technical data and sensor resolution are presented in Table 2.

Table 2. Sensors' technical data.

Type of Sensor	Measurement Range	Scale	Producer Accuracy
Temperature sensors	−20 °C –50 °C	0.01 °C	0.5 °C
Humidity sensors	0–100%	0.1% RH	1%
Air speed	0.01–10 m/s	0.01 m/s	2%
Radiant temp. measurement	0–50 °C	0.01 °C	2%

In addition (for additional verification purposes), temperature and humidity sensors were also placed on each panellist's measurement table (6 places) in order to verify if the temperatures are within the standard deviation (uncertainty) of the basic/main devices (M1 and M2) used for the tests. An example of this device is shown in Figure 6. and the location of Po2–Po7 devices is presented in Figure 3a,b. The accuracy of these measurements was:

- CO_2 measurement range 0–5000 ppm CO_2 accuracy 50 ppm +3% of the measured value
- Measurement RH [%] range 0–100 %RH accuracy 3% (30–70%RH) /5%(70–90%RH)
- Measurement t [°C] range −5–55 °C accuracy 0.3 °C (−5–20 °C)/0.4 °C (20–55 °C)

Figure 6. Additional test equipment for t [°C], humidity level (RH) [%] and CO_2 [ppm] measurements.

The other assumptions for assessment methodology for determining thermal comfort were based on EN ISO 7730 [2].

On the basis of the measurements obtained, thermal comfort parameters were calculated in accordance with ISO 7730 and Fanger's thermal model. The designated parameters were: PMV—average thermal comfort rating [-] and PPD—percentage of dissatisfied people [%].

On the basis of the thermal sensation answers (vote), the thermal comfort parameters were calculated. The designated parameters were: PMV—average thermal comfort rating [-] and PPD—percentage of dissatisfied people [%].

2.7. Measurement Uncertainty

The "overall systematic uncertainty" (or combined) is calculated by adding (in quadrature) all the calculated measurement uncertainties (A-class) and other like panel test uncertainty (B-class). The uncertainty approach is based on the international reference document JCGM 100:2008 Evaluation of measurement data—Guide to the expression of uncertainty in measurement published by BIPM, Sevres 2010. The overall uncertainty is the experimenter's best estimate of how far an experimental thermal result might be from the "true value". The realistic uncertainty were calculated according to EN-ISO 7730 with the PMV measurement uncertainty taking into account the uncertainty designated by Alfano A, Palella BI., Riccio in a paper titled The role of measurement accuracy on the thermal environment assessment by means of PMV index presented in Building and Environment no46:1361-1369., providing the measurement uncertainty of the parameters: t_a, t_{mr} and v_a, p_a. The estimate does not include the effect of two hard-to-measure PMV parameters known as the thermal resistance of clothing Icl in units clo, and the level of metabolic activity of humans Met. The combined standard uncertainty is taking into account the readings from PMV tables compiled according to ISO 7730, with readability U = 0.1 PMV and the standard deviation of the PPD(PMV) model. Author's estimation was confirmed using the uncertainty budget of PMV calculated for a similar example by Ekici Can in Measurement Uncertainty Budget of the PMV Thermal Comfort Equation. As a part of the test, the uncertainty of measurement for all measuring elements was determined. On this basis, the uncertainty of the determination of the measured PMV coefficient was determined and then the PPD coefficient was adjusted as presented in Table 3. PPD_POM is the dissatisfaction percentage resulting from the measurement of physical values based on the Fanger model and the ISO 7730 standard. The uncertainty of measuring the temperature in the indoor environments is recommended by a measurement producer as 0.5 °C and is consistent with the literature.

Table 3. Realistic measurement uncertainty of PPD_POM assessment ($\pm SD_{real(PPD)}$) taking into consideration the provisions of ISO 7726.

Parameter	Standard Deviation%	Range
Air temperature t_a °C	0.5 °C ⇒ 0.08 PMV ⇒ 0.6% PPD	−20 °C–50 °C
Radiant temperature t_{mr} °C	2 °C ⇒ 0.28 PMV ⇒ 3% PPD	0–50 °C
Relative humidity RH %	5% RH ⇒ 0.07 PMV ⇒ 0.5% PPD	0–90%
Relative air velocity va m/s	|0.01 + 0.01va|m/s ⇒ 0.03 PMV ⇒0.2%PPD	0.01–10 m/s
PPD- table error	0.1 PMV ⇒ 0.73% PPD	
$SD_{real(PPD)} = (0.36+9+0.25+0.04+0.54)^{0.5} = 3.2\%$		

The realistic measurement uncertainty of PPD determination using the measurement method was assessed as $SD_{real(PPD)} = 3.2\%$ (Table 3).

The uncertainty of thermal sensory vote was determined for each PMV calculated for each of the temperature conditions, i.e., measurements. Thermal Sensation Index calculated as a weighted average response of sensory results from a seven-point scale is PMV_ANK. Standard deviations for PMV_ANK ranged from 0.5 to 0.8 PMV. The calculated standard deviation of panel 'votes' was calculated as 11.9%. In comparison, 'vote' standard deviation $SD_{votePPD}$ values for corresponding tests known from the literature [32,50,51] are 4–20%. The specified extended uncertainty of PPD assessment for the provided survey is 24.6%. Table 4 schows overall uncertainty Uoverall of PPD based on realistic measurement uncertainties $SD_{realPPD}$ and vote standard deviation (k = 2, a level of confidence 95 %).

Table 4. Overall uncertainty $U_{overall}$ of PPD based on realistic measurement uncertainties $SD_{realPPD}$ and vote standard deviation (k = 2, a level of confidence 95 %).

Parameters	$SD_{realPPD}$%	$SD_{votePPD}$%	$U_{overall}$%
$PPD_{(PMV)}$	3.2	11.9	$2\cdot(10.24 + 141.6)^{0.5}=24.6$

2.8. Other Assumptions and Boundary Conditions

Special boundary conditions regarding temperature measurements are required to test the field of temperature in a case study room. The purpose of using several temperature sensors was to check what the temperature distribution was in the room heated by floor and Split AC unit in heating mode. Measurements of temperature at points M1 (front of panellists) and M2 (rear of panellists) at a height of 1 metre showed similar values within the range of standard deviation, i.e., not more than 0.5 °C for the entire measurement series. Therefore, it was assumed that the analysis will take readings from the meters located to the front of the panellists, i.e., M1. Additional measurements on stations at a height of 1 metre also did not show significant anomalies in the horizontal temperature distribution at 1 m height. Simultaneously, the authors assume the existence of homogeneous thermal conditions at a height of 1 m in the area occupied by the panellists.

While conducting the experiment, the authors kept in mind the level of energy of users related to metabolism, i.e., the demand for food (so authors decided on 2 hours). A longer experiment could cause disturbances in the concentration of young people, and they were supposed to maintain activity at the same level for 2 hours (office work at sitting position).

The authors are aware that an additional factor of uncertainty which may affect the results of the users' thermal comfort is the use of an upper (just below the ceiling) fan-coil heating system, as this causes additional air recirculation. Some authors claim that the mere fact of using such a means of heat distribution may disrupt climate comfort. The authors did not consider this fact as an additional factor in the boundary conditions for the experiment.

The reason we tested from cold to neutral was our focus on looking for the lowest operative temperature at which neutral comfort (PMV = 0) will appear (not the lowest one from the warm side at which comfort will disappear). This approach was in our understanding natural for our climatic zone. In most cases when people enter the office, the room warms up to the reference temperature and does not cool down from being too warm.

3. Results

As part of the experiment, the thermal physical parameters of the laboratory office room were measured to determine PMV (later measured PMV is named PMV_POM). For each measurement of PMV_POM, the thermal satisfaction/dissatisfaction of the panellists (PPD) was determined simultaneously by means of three surveys;

- PPD_ANK, a seven-point thermal sensation scale (−3 to 3),
- PPD_WAR, a two-point thermal sensation scale (yes/no), a conditional question on whether you are satisfied with the thermal conditions,
- PPD_ZAD_ANK, percentage scale (0–100%), a conditional question on what is the percentage of your satisfaction with the thermal conditions.

The results of key measured parameters and the results of surveys are presented in Table 5.

Table 5. Thermal assessment of office space based on three questionnaires (−3 to 3 scale as PPD_ANK, yes/no scale as PPD_WAR, 0–100% scale as PPD_ZAD_ANK).

t_a	t_{mr}	RH	v_a	met	clo	PMV_POM	PPD_POM	PPD_ZAD_ANK	PPD-WAR	PPD_ANK
[°C]	[°C]	[%]	[m/s]	[met]	[clo]	[-]	[%]	[%]	[%]	[%]
17.2	16.8	30.8	0.01	1.2	0.7	−1.8	65	57	55	55
17.4	16.6	32.1	0.02	1.2	0.7	−1.7	64	55	64	55
18.0	17.2	32,3	0.02	1.2	0.7	−1.6	56	44	27	43
19.5	17.9	33.7	0.08	1.2	0.7	−1.3	40	26	23	18
20.4	19.1	33.7	0.01	1.2	0.7	−1.0	27	25	18	14
21.4	19.6	31.2	0.01	1.2	0.7	−0.8	19	23	14	5
22.2	20.2	32.1	0.01	1.2	0.7	−0.6	13	21	9	5
22.7	20.4	31.6	0.01	1.2	0.7	−0.5	11	19	5	0.0
23.2	21.1	31.5	0.01	1.2	0.7	−0.4	8	18	9	0.0
23.7	20.9	31.3	0.02	1.2	0.7	−0.3	7	17	9	0.0
24.1	21.2	30.9	0.02	1.2	0.7	−0.2	6	19	13	0.0
24.6	21.7	31.1	0.01	1.2	0.7	−0.1	5	23	16	4

The percentage distribution of thermal sensation answers for each thermal condition is presented in Table 6. The obtained results were used for further analysis of thermal comfort (Figures 1–5).

Table 6. Results of the thermal sensation answers of panellists for exemplary thermal conditions.

T_A	PMV_POM	Thermal Sensation Answers-Sensory Result [%]							PPD_ANK
[°C]	[-]	−3	−2	−1	0	1	2	3	[%]
17.2	−1.8	9.1	45.5	45.5	0	0	0	0	55
17.4	−1.7	18.2	36.4	45.5	0	0	0	0	55
17.9	−1.6	4.5	38.2	43.6	13.6	0	0	0	43
19.5	−1.3	4.5	13.6	22.7	59.1	0	0	0	18
20.4	−1.0	4.5	9.1	22.7	63.6	0	0	0	14
21.4	−0.8	0	4.5	31.8	63.6	0	0	0	5
22.2	−0.6	0	4.5	40.9	50	4.5	0	0	5
22.7	−0.5	0	0	31.8	63.6	4.5	0	0	0
23.2	−0.4	0	0	31.8	59.1	9.1	0	0	0
23.7	−0.3	0	0	31.8	54.5	13.6	0	0	0
24.1	−0.3	0	0	31.8	36.4	31.8	0	0	0
24.6	−0.1	0	0	22.7	31.8	40.9	4.5	0	5

Figure 7 shows the distribution of dissatisfaction (PPD) resulting from the measurement of physical air parameters based on the Fanger model and the ISO 7730 standard (PPD_POM) as well as dissatisfaction resulting from the surveys (PPD_ANK) for PMV_POM. Similar to Fanger's study, the number of dissatisfied was counted, including those who answered −3 or −2 as unsatisfied in the survey. This condition was also used by Fanger in 1973 [1].

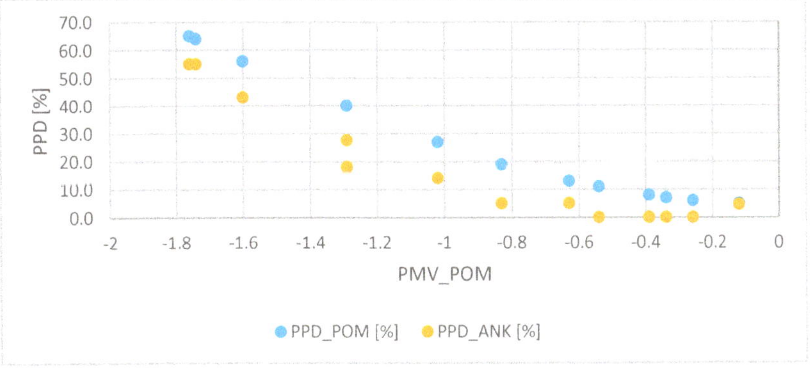

Figure 7. Distribution of dissatisfaction on the basis of measurements ISO 7730) and panellists' responses on a seven-point scale (where PPD_POM is dissatisfaction percentage resulting from the measurement of physical values based on the Fanger model and the ISO 7730 standard and PPD_ANK is dissatisfaction percentage resulting from experimental thermal sensory surveys for a seven-point scale, the number of dissatisfied was counted, including those who answered −3 or −2 as dissatisfied in the survey).

Figure 8 shows the distribution of dissatisfaction (PPD) resulting from the measurement of physical values based on the Fanger model and the ISO 7730 standard (PPD_POM) as well as dissatisfaction based on "yes/no" surveys.

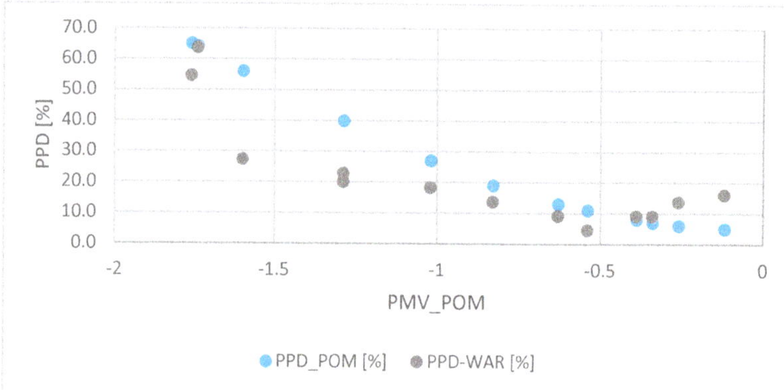

Figure 8. Distribution of dissatisfaction resulting from thermal measurements and resulting from experimental surveys as a yes/no question on actual thermal comfort (where PPD_POM is dissatisfaction percentage resulting from the measurement of physical values based on the Fanger model and the ISO 7730 standard and PPD_WAR- dissatisfied percentage resulting from the surveys with a question "Are you satisfied with thermal comfort yes/no").

Figure 9 shows the distribution of dissatisfaction resulting from the measurement of thermal physical values as well as dissatisfaction resulting from the surveys carried out—"what is the percentage of your satisfaction with the thermal conditions, scale 0–100%", (PPD_ZAD_ANK).

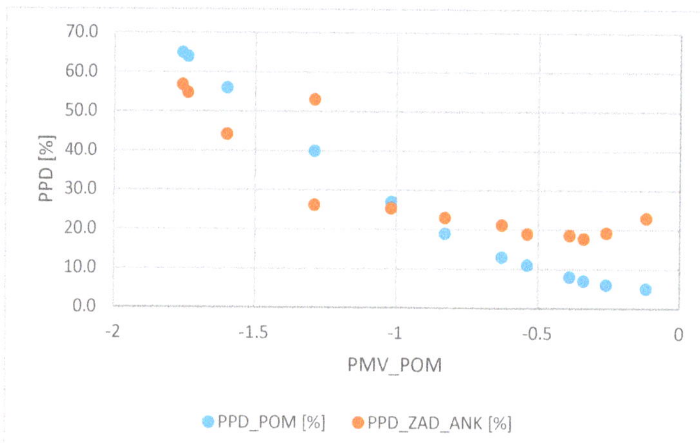

Figure 9. Distribution of dissatisfaction resulting from thermal measurements and resulting from experimental surveys (what is the percentage of your satisfaction with the thermal conditions, scale 0–100%). (where PPD_POM is dissatisfaction percentage resulting from the measurement of physical values based on the Fanger model and the ISO 7730 standard and PPD_ZAD_ANK dissatisfaction PPD resulting from experiment surveys (How would you rate the level of your thermal comfort on a scale of 0–100%).

The PMV was also calculated (additionally to measured PMV) from the survey results as the weighted average response of panellists on a seven-degree scale (later as PMV_ANK). Figure 10 presents the distribution of dissatisfied measurements and questionnaires for the PMV_ANK based on the questionnaires, similar to the calculation PMV in the previous figures. Thermal comfort is achieved slightly earlier than in the result of the theoretical Fanger model.

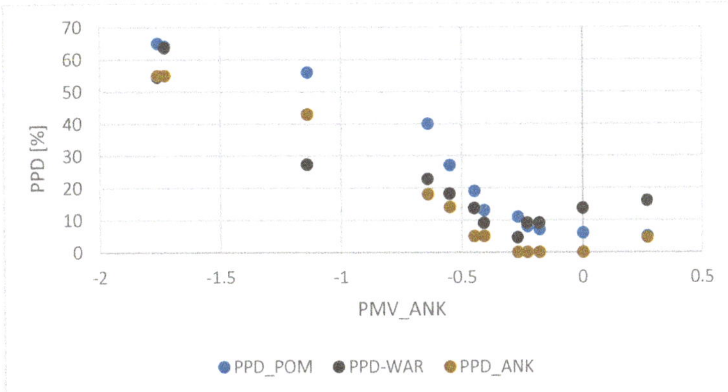

Figure 10. Distribution of dissatisfaction resulting from three experimental surveys for PMV_ANK resulting/calculated from a seven-point survey results scale. (where PMV_ANK is Thermal Sensation Index calculated as a weighted average response of sensory sensations from a seven-point scale, PPD_ANK is dissatisfaction percentage resulting from experimental thermal sensory surveys for a seven-point scale, the number of dissatisfied was counted, including those who answered −3 or −2 as dissatisfied in the survey, PPD_WAR- dissatisfied percentage resulting from the surveys with a question "Are you satisfied with thermal comfort yes/no").

4. Discussion

4.1. Comparison of Results with the Fanger Model

A comparison of the thermal comfort measurement results based on physical parameters in accordance with ISO 7730 [2] to the results obtained by the survey method using a seven-degree comfort scale indicates that: thermal comfort for our respondents has been achieved slightly better (on the colder side) than as presented by the ISO 7730 model. This is an important observation due to the potential space needed to reduce the energy demand for the comfort of the office rooms. For PMV = −0.5 to 0, i.e., in the area of thermal comfort, satisfaction results ranged from 0% to 20% satisfied. The indication of "slightly better adaptation" of the surveyed panellists to the measurement conditions of PMV may be due to their natural adaptive ability resulting from living in a "cooler climate" or actual winter time. The authors of the publication noted [52] a similar phenomenon, but in relation to secondary school students in a warm and tropical climate, observing that they were "better than would result from calculations".

As part of the PMV experiment, a seven-level questionnaire was also calculated as the mean of the number of satisfied for a given scale response. This is PMV_POM, which was taken from measurements and also PMV_ANK from questionnaires. It was shown that these results are similar and the trend line of the comfort model is preserved.

Bearing in mind the wide acceptability of the Fanger model, the results obtained from the conducted surveys were compared with the results of the Fanger and Nevins questionnaires [1,44]. Fanger's theoretical model was created over 50 years ago in a study of 64 students for 13 temperatures (18.9–32.2) and 50% humidity, assuming clo = 0.6, met = 1 and ta = tmr). The results of the comparison of our raw results with the Fanger results are presented in Figure 11. Where F_PPD_ANK is the predicted percentage of dissatisfied - raw results taken from Fanger experiment and PPD_ANK_EXP is the predicted percentage of dissatisfied as a new model based on experimental results.

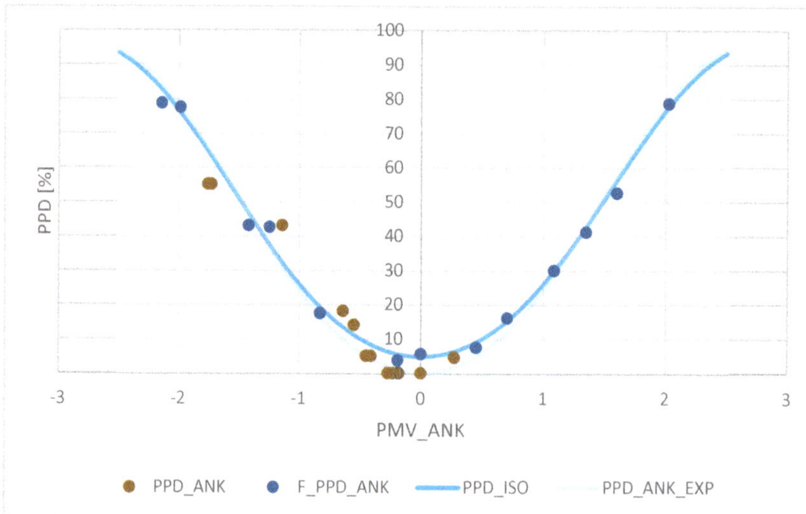

Figure 11. Comparison of the results of surveys reported by Fanger and the results of the surveys.

After analysing the distribution of thermal sensation results obtained from the surveys, Fanger proposed his famous curve presented in Figure 11 (F_PPD_ANK&PPD_ISO). The results presented in our paper obtained from this study's questionnaires are correlated with the Fanger comfort curve with one difference. In the vicinity of thermal comfort, i.e., PMV = 0, as a part of the experiment, there were a larger number of satisfied panellists, i.e., people filling the seven-level questionnaire as −1, 0 or 1. In the context of the obtained results, an experimental curve is proposed. Curve weights were also slightly corrected by the least-squares method and regression gave the experimental dependence expressed by the formula (PPD_ANK_EXP in Figure 11):

$$PPD = 100 - 99.9 \cdot \exp\left(-0.0355 \cdot PMV^4 - 0.242 \cdot PMV^2\right) [\%] \tag{3}$$

As provided, the obtained function does not deviate from Fanger's chart by more than 5% for $-1 <$ PMV < 0. Differences that appear in relation to the Fanger model appear for the results of surveys and sensory tests. In our opinion, the differences (though insignificant) result from a few reasons. The first one is that the Fanger curve is the result of statistical processing and simplification. The significant weakness was the counting of sensory results "−1 PMV to 1" as satisfied. Fanger in the vicinity of PMV = zero due to the research of single panellists had a larger spread of results (this was roughly 2.5 PMV, the width of the entire distribution of results). In our case, due to the compact group, this dispersion was less than 2 PMV. This means that using Fanger's counting method for PMV = −1 to 0 authors got 100% result counted as satisfied. One would have to admit that the whole group was satisfied; with Fanger, several people fell out, which gave him 95% satisfied for PMV = 0. Another issue is that Fanger did not attempt to show a statistical analysis of his results, but this is partly understandable because the test group was relatively small. Fanger also did not specify the uncertainty of his results, which was probably at a level comparable to ours, i.e., 15–25%. Research proves that "the truth" is somewhere near the Fanger equation—with some probability (uncertainty).

If, for the assumed boundary conditions and the parameters tested, significant compliance with the universal Fanger equation was proven, then this should not be automatically expected for other parameters; with another population or other gender distribution or other mass/BMI of panellists, this compliance will not be observed. This requires additional research.

4.2. Obtained Results in the Context of NZEB Buildings

In the context of seeking an optimal and cost-justified operating temperature ensuring thermal comfort in the office area, the measurement results indicate that at 30% RH humidity, PMV = −0.5 and PPD = 10% thermal comfort was obtained for approx. 22.8 °C and full comfort PMV = 0 at 24.7 °C. In accordance with the thermal standard of ASHRAE 55-2017 [3], this is the minimal operative temperature allowing for users' comfort (red dot in Figure 12). These temperatures are rather economically unjustified for any nearly zero-energy buildings and in almost every known case study the BMS system works on lower temperatures.

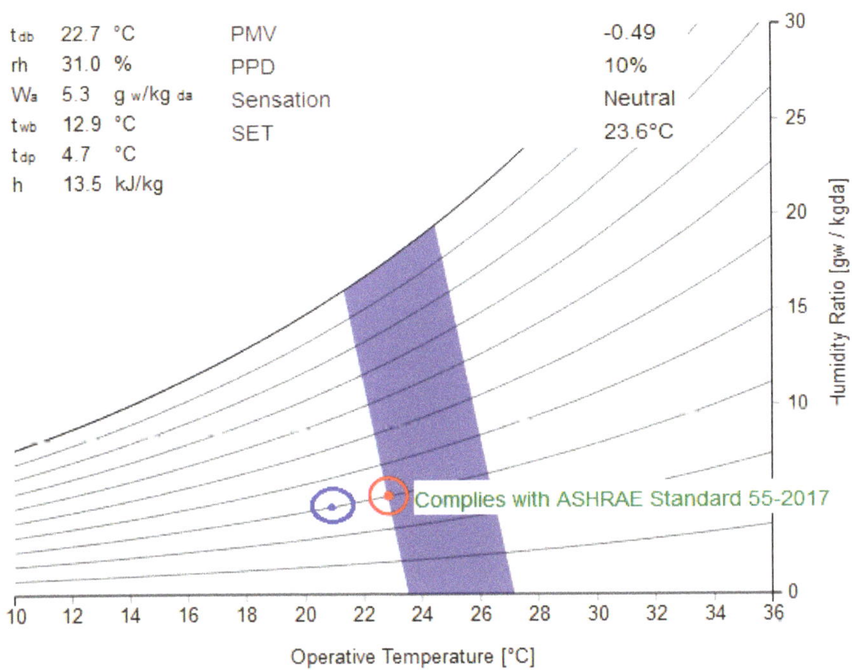

Figure 12. Thermal comfort chart on the function of operative temperature and humidity. Minimum operative temperature for this case-study office for thermal comfort by ASHRAE 55-2017 (PPD = 10%, red-dot) and minimum operative temperature for case-study office for thermal comfort based on sensory answers of users (blue-dot).

It is promising, therefore, that actual thermal comfort as proven on sensory tests can be satisfied by users at slightly lower temperatures, i.e., only 10% dissatisfied at 20.5 °C degrees and, for example, 5% at 21.3 °C degrees (blue dot in the Figure 12). These temperatures don't comply with ASHRAE 55 and EN 15251. In this context, one would consider whether it wouldn't be justified to carry out similar verifications to the office facilities in the context of building energy saving management plan development. Perhaps higher operative temperatures should be used for design purposes in accordance with the standards and then verified for the actual thermal comfort of users in order to obtain maximum energy savings. Table 7 shows the thermal comfort classes for offices based on the recommended ISO 7730 values and shows the operative temperatures at which this class was achieved in the MLBE case-study case for the ISO 7730 measurements (t_{a_pom}) and resulting from the seven-level questionnaire (t_{a_ank}). ta_ANK is taken from the real measured air temperature (°C) and it relates to the thermal comfort class. Index ANK relates to the temperature at which thermal comfort is achieved by panel group based on the results of votes on a 7-point scale. From the table of test panel results, authors read the first measurement temperature ta for which PPD is at minimum 6% (A class) or minimum 15% (C class building).

Table 7. Room categories depending on the PMV indicator (cold side of sensation, PMV < 0).

Room Category	Coefficients: PMV [-]	PPD [%]	t_{a_pom}	t_{a_ank}
Best (A)	−0.2 < PMV	<6	24.2 °C	22.2 °C
Min. (C)	−0.7 < PMV	<15	21.6 °C	20.4 °C

In MLBE building there are large possibilities to control environmental conditions. The experiment showed that the building is able to bring the building to a state of comfort felt by people in a short time without disturbing the temperature gradients (air+floor heating). It seems that rooms with wider environmental condition control, such as our laboratory office (NZEB), may be more comfortable then offices with lower controllability. Our results showed a higher level of satisfaction in cooler PMV levels. There are a few other studies on students showing that students in rooms with full air conditioning (HVAC) showed a faster feeling of climate comfort than in the case of gravity ventilation (NV). This may also lead to the conclusion that the BMS system is not so much a desirable as it is a necessary requirement of low energy lecture halls/offices.

4.3. Results in the Context of the Impact of the Question Type on PMV

An interesting result of the conducted experiment is the clear influence of the type of questionnaire question about thermal comfort on the results. As part of the experiment, three questions were asked by the interviewers. One of them concerned the determination of thermal comfort on a seven-degree scale; the second was a two-degree one using a yes/no answer as to whether there was thermal comfort; while the third was an assessment using the question of whether these conditions are good enough (thermal comfort wise) to perform office work (on a scale of 0–100%). The most similar results to those obtained by other researchers, [53], of course, came from the seven-degree scale (PPD_ANK). The "yes/no survey" gave the shape of a somewhat mangled comfort curve. In our opinion, this is due to the fact that the panellists were better able (quicker) to say that it was already warm enough as such. Both surveys, however, gave shapes similar to the PPD_ISO / PPD_POM model curve slightly shifted to the left (lower temperatures). The survey on thermal comfort using a scale of user satisfaction from 1 to 100%, gave different results from the previous two. A higher number of dissatisfied users was found in the area of thermal comfort as were in Fanger's [1]. There are even 20–25% dissatisfied around PMV = 0. In our opinion this should be explained by the fact that the respondents had the chance to claim that they were satisfied in only 80% or 90% of the comfort conditions, leaving a psychological buffer for the expected "better" conditions with the next experiment temperature set, which usually doesn't give 100%. The influence of the scale affects the quantification of results. For example, in the case of the measured PMV = −1.25 we can obtained the following survey/panel test results: on a "2-point scale yes/no" (YES, i.e., PMV from −1 to 1), on a scale of 100% (50%, i.e., so PMV = −1.5), on a 7-degree scale (PMV = −2, or −1). The problem concerns a subjective capture to a certain scale of thermal sensation. Another example is the area of the measured PMV = 0: on the scale yes/no (NO, so PMV from −2 to −3), on a scale of 100% (80%, i.e., so PMV = −1), on a 7-degree scale (PMV = 0). Authors take the position that it is best to use a 7-point scale (in practice it is a 5-point scale) partly just because other researchers follow the same way (so all can compare each other). We do not recommend using a two-step scale or, in particular, a 100% scale.

4.4. Other

The authors believe that the influence of panellist age on the results requires comment. This question was also raised in other scientists' research [47,54]: whether Fanger's model is an equally suitable model for adults as regards adolescents or youngsters. Current comfort standards, such as ISO 7730, and ASHRAE's Standard 55 determine indoor design values for operative temperatures indoors based on the heat-balance and adaptive thermal comfort models. There is no assurance that results obtained from field studies in offices or universities, or experiments conducted in climate chambers, will accurately reflect the thermal sensations and preferences of office users (in our case students). Furthermore, researchers [55] revealed that young people's thermal preferences were cooler than those predicted by the adaptive standard in EN 15251 [14,56,57]. However, this indicates a certain need to take age into account in the evaluation of room comfort keeping in mind the intended use of the rooms (further research).

All results of thermal comfort tests on test panels enrich our basic knowledge but still, there are open more detailed questions. The next research focus authors consider as necessary is the impact of the combination of existing installations and smart BMS actions on the thermal comfort of users including in details: location of the exhaust mechanical ventilation system (gradients), ventilation operating connected with automatically tilting windows that allows using the building's accumulation capacity for cooling ("night storage cooling" combined with building materials containing phase-changing components), variables of sets of underfloor heating and air heating system modes (time reactions).

Validating the model on the warm side is also interesting. There are voices that the model is not symmetrical on the warm side. For example [57] noted that the thermal comfort of healthcare occupants in a tropical region is slightly warmer than the neutral temperature.

5. Conclusions

The authors believe, as was shown in the article, that comfort measurement/assessment in new NEZB buildings should also be supported by a survey of users and employees. The actual thermal sensory test is much more reliable because the real function of the building is to achieve the users' thermal comfort, not the set of parameters theoretically related to comfort. In cold climatic conditions, effective heating of NZEBs, especially lecture halls in winter, is a great challenge. Therefore, the rooms are often heated periodically, for example, by means of heat pumps operating by a warm air supply—exactly as in the experiment. In our experiments, from the energy side, the authors found that this kind of heating may not affect the perceived thermal comfort. Thermal comfort tests carried out in an experimental office area of a low-energy building MLBE on a sample of 50 students gave some interesting results that offer some new views on the subjective preferences of the thermal comfort of office users. The results presented based on sensory assessment positively validate the existing Fanger model and ISO 7730 approach for NZEB building. In our case, the overall results' deviation from the Fanger function was no significant. A comparison of the thermal comfort measurement results based on physical parameters in accordance with ISO 7730 to the results obtained by the survey method using a seven-degree comfort scale indicates that: thermal comfort was obtained slightly faster (on the cold side) than appears from the ISO model. This is important information due to the worldwide trend to reduce building energy demand. The experimental dependence expression of PPD = f(PMV) for a presented case study was proposed and it can be a good reference for future studies. The test results create the basis for further analyses concerning determination of optimal overall comfort IEQ of low-energy buildings where thermal comfort satisfaction TC_{index} is one of the main sub-components.

Author Contributions: The conceptualization. M.P., methodology M.P., J.B., M.F.-C.; formal analysis. M.P.; investigation. J.B., M.F.-C.; resources. M.P., M.F.-C.; data curation. M.F.-C., M.P.; writing—original draft preparation. M.P.; writing—review and editing. J.B., M.F.-C., M.F.; visualization. M.F.-C., M.P.; project administration. M.F.; funding acquisition. M.P., M.F.

Funding: No external funding for this research.

Acknowledgments: We would like to thank Krystyna Kostyrko for an inspiration and consultancy.

Conflicts of Interest: The authors declare no conflict of interest.

References

1. Fanger, P.O. Assessment of man's thermal comfort in practice. *Occup. Environ. Med.* **1973**, *30*, 313–324. [CrossRef]
2. BSI. *ISO 7730: Ergonomics of the Thermal Environment Analytical Determination and Interpretation of Thermal Comfort Using Calculation of the PMV and PPD Indices and Local Thermal Comfort Criteria*; ISO: Geneva, Switzerland, 2005.
3. *ANSI/ASHRAE Standard 55-2017, Thermal Environmental Conditions for Human Occupancy*; ANSI: Piscataway, NJ, USA, 2017.
4. Nagashima, K.; Tokizawa, K.; Marui, S. Thermal comfort. In *Handbook of Clinical Neurology*; Elsevier: Amsterdam, The Netherlands, 2018.
5. Djongyang, N.; Tchinda, R.; Njomo, D. Thermal comfort: A review paper. *Renew. Sustain. Energy Rev.* **2010**, *14*, 2626–2640. [CrossRef]
6. Taleghani, M.; Tenpierik, M.; Kurvers, S.; van den Dobbelsteen, A. A review into thermal comfort in buildings. *Renew. Sustain. Energy Rev.* **2013**, *26*, 201–215. [CrossRef]
7. van Hoof, J. Forty years of Fanger's model of thermal comfort: Comfort for all? *Indoor Air* **2008**, *18*, 182–201. [CrossRef] [PubMed]
8. Rupp, R.F.; Vásquez, N.G.; Lamberts, R. A review of human thermal comfort in the built environment. *Energy Build.* **2015**, *105*, 178–205. [CrossRef]
9. Peeters, L.; de Dear, R.; Hensen, J.; D'haeseleer, W. Thermal comfort in residential buildings: Comfort values and scales for building energy simulation. *Appl. Energy* **2009**, *86*, 772–780. [CrossRef]
10. Frontczak, M.; Wargocki, P. Literature survey on how different factors influence human comfort in indoor environments. *Build. Environ.* **2011**, *46*, 922–937. [CrossRef]
11. van Hoof, J. Thermal comfort: Research and practice. *Front. Biosci.* **2010**, *15*, 765–788. [CrossRef]
12. Cheng, Y.; Niu, J.; Gao, N. Thermal comfort models: A review and numerical investigation. *Build. Environ.* **2012**, *47*, 13–22. [CrossRef]
13. de Dear, R.J.; Akimoto, T.; Arens, E.A.; Brager, G.; Candido, C.; Cheong, K.W.D.; Li, B.; Nishihara, N.; Sekhar, S.C.; Tanabe, S.; et al. Progress in thermal comfort research over the last twenty years. *Indoor Air* **2013**, *23*, 442–461. [CrossRef]
14. Mors, S.T.; Hensen, J.L.M.; Loomans, M.G.L.C.; Boerstra, A.C. Adaptive thermal comfort in primary school classrooms: Creating and validating PMV-based comfort charts. *Build. Environ.* **2011**, *46*, 2454–2461. [CrossRef]
15. Halawa, E.; van Hoof, J. The adaptive approach to thermal comfort: A critical overview. *Energy Build.* **2012**, *51*, 101–110. [CrossRef]
16. Enescu, D. A review of thermal comfort models and indicators for indoor environments. *Renew. Sustain. Energy Rev.* **2017**, *79*, 1353–1379. [CrossRef]
17. Nicol, J.F.; Roaf, S. Rethinking thermal comfort. *Build. Res. Inf.* **2017**, *45*, 711–716. [CrossRef]
18. Yao, R.; Li, B.; Liu, J. A theoretical adaptive model of thermal comfort—Adaptive Predicted Mean Vote (aPMV). *Build. Environ.* **2009**, *44*, 2089–2096. [CrossRef]
19. Mora, R.; Bean, R. Thermal comfort: Designing for people. *Ashrae J.* **2018**, *60*, 2.
20. Final draft EN 16798-1, Energy performance of buildings—Ventilation of buildings—Part 1: Indoor environmental input parameters for design and assessment of energy performance of buildings addressing indoor air quality, thermal environment, lighting and acoustics (Module M1–6). 2018. Available online: https://www.cibse.org/getmedia/bb1c34b1-19fe-4199-a44f-acb981077ab0/Draft-BS-EN-16798-1-Energy-performance-of-buildings.pdf.aspx (accessed on 24 April 2019).
21. *ISO 15251, Indoor Environmental Input Parameters for Design and Assessment of Energy Performance of Buildings Addressing Indoor Air Quality, Thermal Environment, Lighting and Acoustics*; ISO: Geneva, Switzerland, 2007.
22. de Dear, R. Thermal comfort in practice. *Indoor Air.* **2004**, (Suppl. 7), 32–39. [CrossRef]
23. Charles, K.E. *Fanger's Thermal Comfort and Draught Models*; National Research Council Canada: Halifax, NS, Canada, 2003.
24. Ormandy, D.; Ezratty, V. Health and thermal comfort: From WHO guidance to housing strategies. *Energy Policy* **2012**, *49*, 116–121. [CrossRef]

25. *Directive 2010/31/EU of the European Parliament and of the Council of 19 May 2010 on the energy performance of buildings*; Official Journal of the European Union: Brussels, Belgum, 2010.
26. Fedorczak-Cisak, M.; Furtak, M.; Gintowt, J.; Kowalska-Koczwara, A.; Pachla, F.; Stypuła, K.; Tatara, T. Thermal and vibration comfort analysis of a nearly zero-energy building in Poland. *Sustainability* **2018**, *10*, 3774. [CrossRef]
27. Fassio, F.; Fanchiotti, A.; Vollaro, R.D. Linear, non-linear and alternative algorithms in the correlation of IEQ factors with global comfort: A case study. *Sustainability* **2014**, *6*, 8113–8127. [CrossRef]
28. Ncube, M. *The Development of a Methodology for a Tool for Rapid Assessment of Indoor Environment Quality in Office Buildings in the UK*; University of Nottingham: Nottingham, UK, 2012.
29. Heinzerling, D. *Commercial Building Indoor Environmental Quality Evaluation: Methods and Tools*; UC Berkeley: Berkeley, CA, USA, 2012.
30. Piasecki, M.; Kozicki, M. The Approach of Including TVOCs Concentration in the Indoor Environmental Quality Model (IEQ)—Case Studies of BREEAM Certified Office Buildings. *Sustainability* **2018**, *10*, 3902. [CrossRef]
31. *Interactions Affecting the Achievement of Acceptable Indoor Environments*; ASHRAE Standard: Atlanta, GA, USA, 2011.
32. Piasecki, M.; Kostyrko, K.B. Indoor environmental quality assessment, part 2: Model reliability analysis. *J. Build. Phys.* **2018**, *5*, 288–315. [CrossRef]
33. Liang, H.H.; Chen, C.P.; Hwang, R.L.; Shih, W.M.; Lo, S.C.; Liao, H.Y. Satisfaction of occupants toward indoor environment quality of certified green office buildings in Taiwan. *Build. Environ.* **2014**, *72*, 232–242. [CrossRef]
34. Mihai, T.; Iordache, V. Determining the Indoor Environment Quality for an Educational Building. *Energy Procedia* **2016**, *85*, 566–574. [CrossRef]
35. Pei, Z.; Lin, B.; Liu, Y.; Zhu, Y. Comparative study on the indoor environment quality of green office buildings in China with a long-term field measurement and investigation. *Build. Environ.* **2015**, *84*, 80–88. [CrossRef]
36. Piasecki, M.; Kostyrko, K.; Pykacz, S. Indoor environmental quality assessment: Part 1: Choice of the indoor environmental quality sub-component models. *J. Build. Phys.* **2017**, *41*, 264–289. [CrossRef]
37. Freire, R.Z.; Oliveira, G.H.C.; Mendes, N. Predictive controllers for thermal comfort optimization and energy savings. *Energy Build.* **2008**, *40*, 1353–1365. [CrossRef]
38. Spagnolo, J.; de Dear, R. A field study of thermal comfort in outdoor and semi-outdoor environments in subtropical Sydney Australia. *Build. Environ.* **2003**, *38*, 721–738. [CrossRef]
39. Zomorodian, Z.S.; Tahsildoost, M.; Hafezi, M. Thermal comfort in educational buildings: A review article. *Renew. Sustain. Energy Rev.* **2016**, *59*, 895–906. [CrossRef]
40. Trebilcock, M.; Soto-Muñoz, J.; Yañez, M.; Martin, R.F. The right to comfort: A field study on adaptive thermal comfort in free-running primary schools in Chile. *Build. Environ.* **2017**, *114*, 455–469. [CrossRef]
41. Zeiler, W.; Boxem, G. Effects of thermal activated building systems in schools on thermal comfort in winter. *Build. Environ.* **2009**, *44*, 2308–2317. [CrossRef]
42. Khodakarami, J.; Nasrollahi, N. Thermal comfort in hospitals—A literature review. *Renew. Sustain. Energy Rev.* **2012**, *16*, 4071–4077. [CrossRef]
43. Fanger, P.O. Assessment of thermal comfort practice. *Br. J. Ind. Med.* **1973**, *30*, 313–324.
44. Rohles, F.; Nevins, G.R. Nature of thermal comfort for sedentary man. *ASHRAE Trans.* **1971**, *77*, 239–246.
45. Alfano, F.R.D.; Ianniello, E.; Palella, B.I.; Riccio, G. Thermal comfort design in indoor environments: A comparison between EU and USA approaches. In Proceedings of the HB 2006—Healthy Buildings: Creating a Healthy Indoor Environment for People, Lisboa, Portugal, 4–8 June 2006.
46. de Dear, R.J.; Brager, G.S. Thermal comfort in naturally ventilated buildings: Revisions to ASHRAE Standard 55. *Energy Build.* **2002**, *34*, 549–561. [CrossRef]
47. de Dear, R.; Kim, J.; Candido, C.; Deuble, M. Adaptive thermal comfort in australian school classrooms. *Build. Res. Inf.* **2015**, *43*, 383–398. [CrossRef]
48. Olesen, B.W. Revision of EN 15251: Indoor Environmental Criteria. *Rehva J.* **2012**, *49*, 6–12.
49. BSI. *Ergonomics of the Thermal Environment—Instruments for Measuring Physical Quantities*; Bs En Iso 77262001; ISO: Geneva, Switzerland, 2002.
50. Alfano, F.R.D.A.; Palella, B.I.; Riccio, G. Notes on the Calculation of the PMV Index by Means of Apps. *Energy Procedia* **2016**, *101*, 249–256. [CrossRef]

51. Wang, J.; Wang, Z.; de Dear, R.; Luo, M.; Ghahramani, A.; Lin, B. The uncertainty of subjective thermal comfort measurement. *Energy Build.* **2018**, *181*, 38–49. [CrossRef]
52. Kim, J.; de Dear, R. Thermal comfort expectations and adaptive behavioural characteristics of primary and secondary school students. *Build. Environ.* **2018**, *127*, 13–22. [CrossRef]
53. Buratti, C.; Ricciardi, P. Adaptive analysis of thermal comfort in university classrooms: Correlation between experimental data and mathematical models. *Build. Environ.* **2009**, *44*, 674–687. [CrossRef]
54. Carlucci, S.; Bai, L.; de Dear, R.; Yang, L. Review of adaptive thermal comfort models in built environmental regulatory documents. *Build. Environ.* **2018**, *137*, 73–89. [CrossRef]
55. Teli, D.; Jentsch, M.F.; James, P.A.B. Naturally ventilated classrooms: An assessment of existing comfort models for predicting the thermal sensation and preference of primary school children. *Energy Build.* **2012**, *53*, 166–182. [CrossRef]
56. Haddad, S.; King, S.; Osmond, P. Enhancing thermal comfort in school buildings. *Proc. Heal. Build.* **2012**. Available online: https://www.isiaq.org/docs/PDF%20Docs%20for%20Proceedings/7F.2.pdf (accessed on 24 April 2019).
57. Sattayakorn, S.; Ichinose, M.; Sasaki, R. Clarifying thermal comfort of healthcare occupants in tropical region: A case of indoor environment in Thai hospitals. *Energy Build.* **2017**, *149*, 45–57. [CrossRef]

© 2019 by the authors. Licensee MDPI, Basel, Switzerland. This article is an open access article distributed under the terms and conditions of the Creative Commons Attribution (CC BY) license (http://creativecommons.org/licenses/by/4.0/).

sustainability

MDPI

Article

The Optimization of Visual Comfort and Energy Consumption Induced by Natural Light Based on PSO

Yonggang Zhang [1,2], Yongwei Zhong [1,2,3,*], Yingda Gong [1] and Lirong Zheng [1,2]

1 School of Information Science and Technology, Fudan University, Shanghai 200433, China; zyg0521@126.com (Y.Z.); 17210720123@fudan.edu.cn (Y.G.); lrzheng@fudan.edu.cn (L.Z.)
2 Shanghai Institute of Intelligent Electronics & Systems, Shanghai 200433, China
3 OnePlus Technology (Shenzhen) Co., Ltd., Shenzhen 518040, China
* Correspondence: yongweizhong@fudan.edu.cn; Tel.: +86-21-65643533-8011

Received: 9 December 2018; Accepted: 19 December 2018; Published: 21 December 2018

Abstract: This paper presents the "model construction method", an optimization method and industrial internet of things (IIoT) technology that is proposed for nearly zero energy buildings (nZEB), providing a comfortable visual environment by only utilizing natural light while improving its induced indoor air conditioner energy consumption (ACEC). The incident light is sampled by light sensors, and this data is sent to the cloud server. The visual comfort and indoor ACEC, both of which are induced by incident light, are combined as the optimization objective, and the area of windows covered by curtains is used as the optimal parameter in the particle swarm optimization (PSO). The visual comfort and indoor ACEC induced by incident light are modeled, and the construction method is independent of the geographical location. Five modes are defined for applications with different purposes, the performance of which are investigated and compared carefully. The result shows that natural light could provide comfortable visual comfort, while the ACEC induced by it could be reduced effectively.

Keywords: visual comfort; air conditioner energy consumption; natural light; particle swarm optimization; Industrial Internet of Things

1. Introduction

Nearly zero energy buildings (nZEB) have attracted lots of interests and work over the last decade, as building energy consumption plays an import role in total energy consumption [1]; about 40% of total energy in developed countries [2], and about 25% in China [3]. As a result, developing nZEB technology could improve total energy consumption effectively, which is valuable for global sustainability.

Building energy efficiency improvement and natural energy, like solar radiation and wind, are two important technologies for realizing nZEB. Building energy consumption forecast methods have been developed to improve building design and equipment setting [4]. In [5], an approach used in designing building envelopes is developed, which ensures conditions of thermal, visual, and acoustic comfort with a minimum of energy consumption. In a building that is well designed thermally, the electricity consumption of the lighting system is the dominant source of consumption [6,7]. In [8], the required visual comfort for every single user by utilizing visual comfort zones is provided, while more than 60% energy savings could be achieved. In [9], a smart energy efficient lighting system is developed based on the presence of occupancy and occupancy localization information.

Natural light could also be utilized in nZEB. In [10], energy saving (<25%) by using natural light and artificial light is realized, which is controlled by a self-adaptive integrated system. Another study [11] finds that creating a luminance contrast of between 11:1 and 12:1 on the window wall with a 45% window-to-exterior-wall ratio using a supplementary LED system could improve energy consumption, because it will diminish the likelihood of users turning on the ceiling lights. However, the incident solar radiation will increase the energy consumption of the air conditioner. In [12], shading devices installed on windows are designed to reduce energy consumption, while artificial light is used.

Furthermore, smart controlling systems plays a key role in nZEBs [13–18]. In [13], an efficient home energy management controller (EHEMC) based on genetic harmony search algorithm (GHSA) is proposed to reduce electricity expense, peak to average ratio (PAR), and maximize user comfort. For multiple homes, modes of operation for the appliances, according to their energy consumption, with varying operation time slots are classified, and the constrained optimization problem is solved using heuristic algorithms. In [14], the hybrid gray wolf differential evolution (HGWDE) is proposed for jointly optimizing electricity costs and user preferences. When the electricity tariff is in peak periods, the load is shifted into the slots with low price. Similar works are reported in [15] and [16], which employ the binary multiple knapsack problem formulation technique and finite state machine, respectively. In [17,18], energy saving is achieved based on a Multi-Agent architecture, which is the representative architecture in this field. In these works, the visual comfort is not taken into account.

In this paper, an IIoT-based smart controlling system and the corresponding models, model construction method and optimization method, are proposed for providing a comfortable visual environment by utilizing only natural light, while improving its induced ACEC effectively. In the proposed system, the incident light is sampled by light sensors, and the cloud-based service analyzes this data by employing the proposed model and PSO, then the corresponding order is sent to drive automatic curtains. A framework of the smart controlling system is proposed for conducting experiments, and the cloud-based IIoT technology makes the system low cost, easy to deploy, and usable for large-scale deployment [19]. The curtains could stop the incident light, and influence the occupants' visual comfort and ACEC significantly. Therefore, the area of windows covered by curtains is used as the optimal parameter in the PSO [20]. A model combining visual comfort with indoor ACEC is proposed, in which there are several windows in different situations. Five modes are defined for different applications by setting the model coefficients a and b with different values: (a) mode 1 (a = 1 and b = 0), only energy consumption is concerned; (b) mode 2 (a = 0 and b = 1), only visual comfort is concerned; (c) mode 3 (a = 2 and b = 1), energy consumption has priority; (d) mode 4 (a = 1 and b = 2), visual comfort has priority; and (e) mode 5 (a = 1 and b = 1), the moderate policy. The performance of 5 modes is investigated carefully. Compared with mode 2, which has the best visual comfort, mode 3 reduces air conditioner power about 117 w (24.5%), while the visual comfort is nearly the same. These results show that it is practicable to provide good visual comfort by utilizing only natural light, while the ACEC caused by natural light could be reduced effectively.

2. Framework of IIoT-based Smart Controlling System

The framework of the proposed IIoT-based smart controlling system is shown in Figure 1. The proposed IIoT-based smart controlling system is composed of cloud-based services, smart gateways, and automatic curtains with light sensors. Automatic curtains are connected to gateways wirelessly (In Figure 1, it is via Zigbee wireless network), and gateways submit the data to cloud-based services via the Internet.

Light sensors sample the incident light and periodically submit the information to services on the cloud through gateways, which implement protocol conversions between Zigbee and Internet. Though one curtain includes one or more light sensors in Figure 1, it is obvious that curtains could share these sensors.

The service on the cloud periodically analyzes this data from gateways by employing the proposed model and PSO, and sends the corresponding order to drive automatic curtains through gateways. The model depends on the building, and five modes are defined for different applications, by setting the model coefficients a and b with different values, which will be discussed in detailed in Section 3.

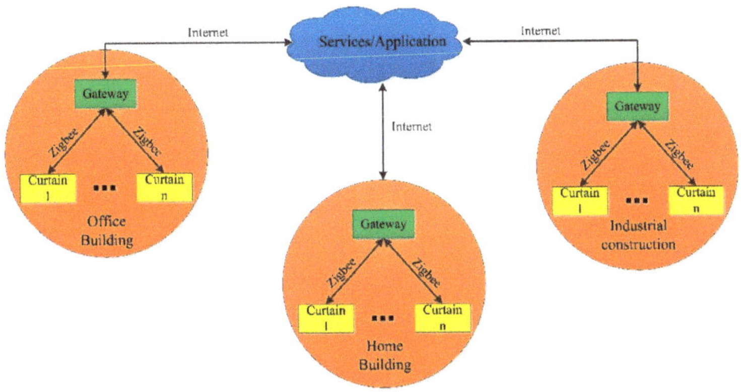

Figure 1. The framework of the proposed IIoT-based smart controlling system.

The proposed framework could be deployed in home buildings, office buildings, and industrial constructions. The cloud-based IIoT technology makes it low cost, easy to deploy, and usable for large-scale deployment. Furthermore, PSO is suitable for multi-variable optimization, therefore the proposed system will have superiority in complex architecture.

In our work, the comfortable visual environment of the home building in Shanghai is provided by natural light, and the ACEC that is induced by natural light is improved.

3. Model and Particle Swarm Optimization

This section introduces the model, the model construction method, and function and optimization method that was researched and used in this paper, including the model of energy consumption induced by incident light, the fit function of natural light levels induced by incident light, the fit function of indoor visual comfort, and the particle swarm optimization. Strong sunshine is required to make this model work effectively, and that's the reason why 12:00 on July 15 in Shanghai is chosen in this work. The model construction method is independent of the geographical location, though the house in this work is in Shanghai, China.

3.1. Model of Energy Consumption Induced by Incident Light

As shown in Figure 2, the dimensions of the house are $6 \times 4 \times 2.8$ m^3, and there are two windows on the south wall and east wall, the sizes of which are 6×2 m^3 and 4×2 m^3, respectively. The vertical distance between the ground and the center of the window is 1.6 m. There is another house with the same dimensions and structure on the east of the research area, and the distance between the two houses is 4 m.

Curtains on the south wall and east wall are used to stop incident light, as shown in Figure 2b,c. The visible transmittance of windows (single glass) is 0.737, and it is assumed that the curtain could stop incident light completely (100%). Then the incident light power could be expressed as:

$$\begin{aligned} P &= P_{u1} \times W_2 \times H_2 + P_{u2} \times W_3 \times H_4 \\ &= P_{u1} \times L_1 \times H_1 \times x_1 + P_{u2} \times W_1 \times H_1 \times x_2 \end{aligned} \quad (1)$$

where x_1 is the ratio of none curtain-covered area of the south window to the whole area of the south wall, and x_2 is the ratio of none curtain-covered area of the east window to the whole area of the east wall. The range of x_1 and x_2 are both from 0 to 0.7. P_{u1} and P_{u2} are the incident light power per square meter on the south and east wall, respectively, which could be obtained by light sensors or employing simulation tools such as Autodesk Ecotect Analysis.

The incident light provides indoor visual comfort, and also increases the room temperature, which should be removed by air conditioner to keep the room temperature constant. The energy efficiency ratio (EER) of the air conditioner is supposed to be 3.4, namely it is a product of China energy efficiency index (EEI) 2. Then the air conditioner power induced by incident light is:

$$P_{ac} = P/3.4 \quad (2)$$

The P_{ac} at 12:00 am on 15 July is listed in Table 1.

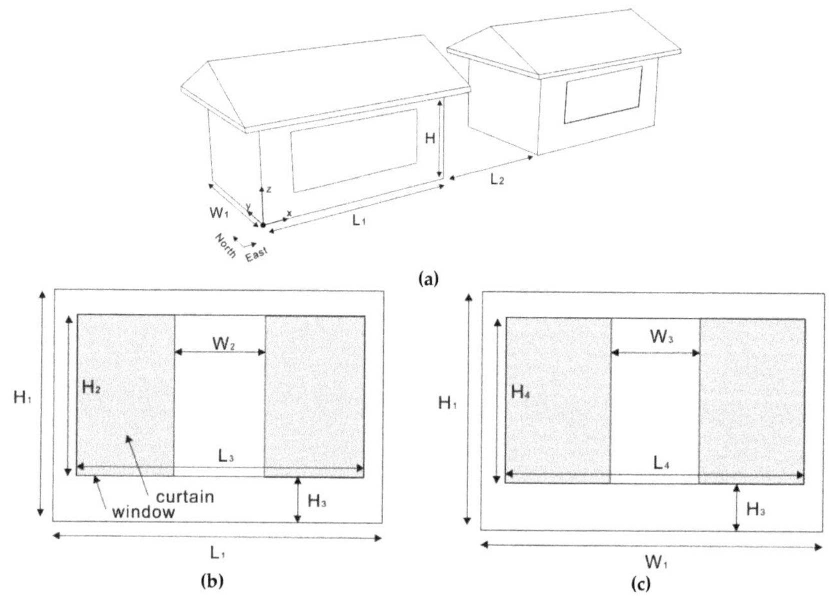

Figure 2. The dimension and structure of the proposed model. (**a**) $L_1 = 6$, $W_1 = 4$, $H = 2.8$, $L_2 = 4$. (**b**) The dimension of the south wall (m): $L_1 = 6$, $H_1 = 2.8$, $H_3 = 0.6$, $L_3 = 5.88$, $H_2 = 2$, $0 \leq W_3 \leq 5.88$. (**c**) The dimension of the east wall (m): $W_1 = 4$, $H_1 = 2.8$, $H_3 = 0.6$, $L_4 = 3.92$, $H_4 = 2$, $0 \leq W_3 \leq 3.92$.

Table 1. P_{ac} at 12:00 am on 15 July.

Pac (W)	x_1	x_2	**Pac (W)**	x_1	x_2
70.8	0.1	0	37.0	0	0.1
141.5	0.2	0	74.0	0	0.2
218.0	0.3	0	111.0	0	0.3
283.0	0.4	0	153.0	0	0.4
372.9	0.5	0	184.8	0	0.5
413.0	0.6	0	221.8	0	0.6
441.7	0.7	0	249.9	0	0.7

The fit function of P_{ac} is:

$$\begin{aligned} P_{ac} &= P_{ac1} + P_{ac2} \\ &= 646.7 \times x_1 + 18.6 + 360.1 \times x_2 + 3.3 \\ &= 646.7 \times x_1 + 360.1 \times x_2 + 21.9 \end{aligned} \quad (3)$$

where their root mean squared errors (RMSE) are 20.7 and 4.0, respectively.

3.2. Model of Energy Consumption Induced by Incident Light

The indoor natural light level (LL) provided by incident light is sampled at position A at 12:00 am on July 15. The coordinate of A is (3, 2, 0.8), and the origin of the coordinate is shown in Figure 2a. The light level at position A is used to calculate the indoor visual comfort at 12:00 am on July 15, and the data is obtained by employing Autodesk Ecotect Analysis, as shown in Table 2.

Table 2. Light level at position A at 12:00 am on 15 July.

LL (Lux)	x_1	x_2	LL (Lux)	x_1	x_2	LL (Lux)	x_1	x_2
201.3	0.1	0	282.9	0	0.6	531.4	0.1	0.7
311.3	0.2	0	321.9	0	0.7	655.3	0.2	0.7
413.1	0.3	0	261.6	0.1	0.1	894.8	0.5	0.7
474.1	0.4	0	426.5	0.2	0.2	490.2	0.3	0.1
536.7	0.5	0	582.3	0.3	0.3	680.8	0.3	0.5
582.5	0.6	0	709.0	0.4	0.4	762.1	0.3	0.7
629.1	0.7	0	814	0.5	0.5			
52.2	0	0.1	916.6	0.6	0.6			
91.4	0	0.2	1003.2	0.7	0.7			
140.7	0	0.3	728.5	0.7	0.1			
198.9	0	0.4	777.1	0.7	0.2			
240.8	0	0.5	824.5	0.7	0.5			

The fit function of indoor light level is:

$$LL = -1206 \times x_1^2 + 1719 \times x_1 + 460.3 \times x_2 + 63.93 \times x_1 \times x_2 + 17.23 \quad (4)$$

where the RMSE is 27.0.

3.3. Fit Function of Indoor Visual Comfort

The curtain could change the incident light and light level in the room dramatically, therefore the occupants' visual comfort [21,22] is significantly influenced by it. In [22], the acceptable light levels are examined in an office environment by interviewing 293 occupants, the data of which is used for fitting visual comfort function, as listed in Table 3. The occupants' acceptability of light level is defined as the ratio of the number of acceptance to the sample number (293), and is used to represent visual comfort (VC).

Table 3. Occupants' responses towards local visual environment [22].

VC	LL (Lux)	VC	LL (Lux)
0.59	250	0.95	750
0.67	350	0.92	850
0.85	450	0.85	950
0.9	550	0.78	1050
0.97	650		

The fit function of visual comfort is:

$$VC = -1.73 \times 10^{-6} \times x^2 + 2.5 \times 10^{-3} \times x + 0.0531 \quad (5)$$

where the RMSE is 0.03. Substituting (4) into (5), then we have:

$$\begin{aligned} VC = & -1.73 \times 10^{-6} \times (-1206 \times x_1^2 + 1719 \times x_1 + 460.3 \times x_2 + 63.93 \times x_1 \times x_2 + 17.23)^2 \\ & + 2.5 \times 10^{-3} \times (-1206 \times x_1^2 + 1719 \times x_1 + 460.3 \times x_2 + 63.93 \times x_1 \times x_2 + 17.23) + 0.0531 \end{aligned} \quad (6)$$

3.4. Particle Swarm Optimization

PSO consists of 6 steps [20,23], and the pseudocode is shown in Appendix A:

1. Initiate the positions p_n and velocities v_n of n particles.
2. Evaluate the value of the PSO fitness function for each particle.
3. Compare and update each particle's own best position p_{bestn}.
4. Compare and update n particles' global best position g_{best}.
5. Update the positions and velocities:

$$v_n = w \times v_n + c_1 \times \text{random} \times (p_{bestn} - p_n) + c_2 \times \text{random} \times (g_{best} - p_n) \quad (7)$$

$$p_n = p_n + v_n \quad (8)$$

6. After N iterations, stop the program.

Both c_1 and c_2 are chosen to be 2 [20], inertia weight W is 0.8, and the fitness function is:

$$F = a \times 0.2 \times \lg(646.7 \times x_1 + 360.1 \times x_2 + 21.9) - b \times [-1.73 \times 10^{-6} \times (-1206 \times x_1^2 + 1719 \times x_1 + 460.3 \times x_2 + 63.93 \times x_1 \times x_2 + 17.23)^2 + 2.5 \times 10^{-3} \times (-1206 \times x_1^2 + 1719 \times x_1 + 460.3 \times x_2 + 63.93 \times x_1 \times x_2 + 17.23) + 0.0531] \quad (9)$$

The coefficients a and b, used to define different modes, represent the weight of air conditioner energy consumption and visual comfort, respectively.

4. Analysis and Results

The incident light increases the room temperature. Therefore, the air conditioner should work to remove it from the room to keep the room temperature unchanged. The incident light could be reduced by the curtain, which also influences occupants' visual comfort. PSO is used to optimize the curtain-covered area of windows, and five modes are investigated for different applications, which are defined as below:

(a) mode 1: a = 1 and b = 0, indicates that only energy consumption is concerned.
(b) mode 2: a = 0 and b = 1, indicates that only visual comfort is concerned.
(c) mode 3: a = 2 and b = 1, indicates that energy consumption has priority.
(d) mode 4: a = 1 and b = 2, indicates that visual comfort has priority.
(e) mode 5: a = 1 and b = 1, indicates the moderate policy.

The optimal solutions of 5 modes are listed in Table 4.

Table 4. Optimal solutions of 5 modes.

	x_1	x_2	ACEC (W)
mode 1	0	0	0
mode 2	0.7	0.18	477
mode 3	0.36	0.29	360
mode 4	0.36	0.45	417
mode 5	0.36	0.4	319

Figure 3a,b shows the complexity of PSO applied in this work. It is found that the optimal solution is obtained after 300 iterations for most cases, as shown in Figure 3, therefore N is set to be 300.

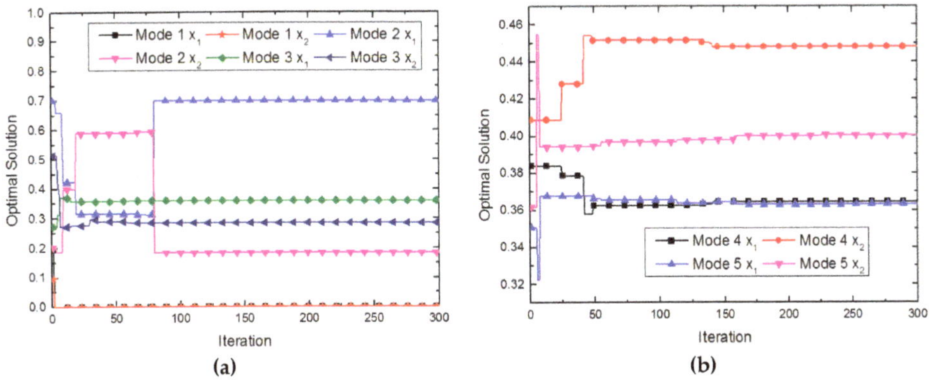

Figure 3. The optimal solution after 300 iterations. (**a**) Mode 1, 2, and 3. (**b**) Mode 4 and 5.

4.1. Mode 1 ($a = 1$, $b = 0$)

As shown in Table 4, the optimal solution is $x_1 = 0$ and $x_2 = 0$, which indicates the entire windows are covered by curtains, and no solar radiation could enter the room. Therefore the air conditioner energy consumption induced by the incident light is 0, while occupants' visual comfort is the worst.

4.2. Mode 2 ($a = 0$, $b = 1$)

The optimal solution is $x_1 = 0.7$ and $x_2 = 0.18$, as shown in Table 4. Substituting this solution into (6), the corresponding visual comfort (occupants' acceptability) is 0.96, which indicates that the natural light level is the best. The corresponding air conditioner power is 477 W.

4.3. Mode 3 ($a = 2$, $b = 1$)

The optimal solution is $x_1 = 0.36$ and $x_2 = 0.29$, as shown in Table 4. The corresponding visual comfort (occupants' acceptability) is 0.94, and the corresponding air conditioner power is 360 W. Compared with mode 2, mode 3 achieves nearly the same visual comfort, while the air conditioner energy consumption is reduced by 24.5%.

4.4. Mode 4 ($a = 1$, $b = 2$)

The optimal solution is $x_1 = 0.36$ and $x_2 = 0.45$, as shown in Table 4. The corresponding visual comfort (occupants' acceptability) is 0.96, and the corresponding air conditioner power is 417 W. Compared with mode 3, mode 4 has nearly the same visual comfort, but the air conditioner energy consumption increases dramatically. It is observed that it is the east window not the south window that is adjusted to improve the visual comfort, which could explain the stronger incident light on the south window.

4.5. Mode 5 ($a = 1$, $b = 1$)

The optimal solution is $x_1 = 0.36$ and $x_2 = 0.4$, as shown in Table 4. The corresponding visual comfort (occupants' acceptability) is 0.95, and the corresponding air conditioner power is 399 W. Compared with mode 3, mode 5 has nearly the same visual comfort and larger air conditioner energy consumption.

By investigating the performance of all modes, it is found that mode 3 has the best performance. Compared with mode 2 which has the best visual comfort, mode 3 could reduce the air conditioner power about 117 w (25%), while the visual comfort is nearly the same.

5. Conclusions

The visual comfort and indoor AECE, both of which are induced by natural light at noon during summer in Shanghai, are optimized by employing the proposed model and optimization method. A fitness function measuring the visual comfort and the corresponding ACEC in the building was constructed, and the area of windows covered by curtains was used as the optimal parameter in the PSO. Five modes were defined and investigated for different applications, and mode 3 achieved nearly the best visual comfort with energy consumption reduced by 24.5%. The results verify that the proposed model and optimization method could work effectively to provide a good visual environment by utilizing natural light while reducing the ACEC of buildings.

Author Contributions: Conceptualization, Y.Z. (Yonggang Zhang) and Y.Z. (Yongwei Zhong); Methodology, Y.Z. (Yonggang Zhang); Writing—Original Draft Preparation, Y.Z. (Yonggang Zhang); Writing—Review & Editing, Y.Z. (Yongwei Zhong) and Y.G.; Supervision, L.Z.; Project Administration, Y.Z. (Yongwei Zhong).

Funding: This research was funded by the National Natural Science Foundation of China under grant number [61571135].

Acknowledgments: This work is supported by the National Natural Science Foundation of China under Grant No.61571135. The authors would like to thank the anonymous reviewers for their constructive comments.

Conflicts of Interest: The authors declare no conflict of interest.

Appendix A Pseudocode of the Proposed Optimization

```
%Main function
%%%%%%%%%%%%%%%%%%%%
[xm,fv,C]=PSO(@fitness,20,2,2,0.8,300,2,0,0.7);% call PSO function
xxx2(:,1)=C(1,:);
xxx2(:,2)=C(2,:);
plot(xxx2) %plot optimal solution
xlabel('Iteration');
ylabel('Optimal solution')

%Fitness function
%%%%%%%%%%%%%%%%%%%%
function f=fitness(a,b)
f=A*0.2*log10(646.7*a+18.6+360.1*b+3.3)-B*(-1.73e-6*(17.23+1719*a+460.3*b-1206*a^2+63.93*a*b)^2
    +2.5e^-3*(17.23+1719*a+460.3*b-1206*a^2+63.93*a*b)+0.0531);
end

%PSO function
%%%%%%%%%%%%%%%%%%%%
function[xm,fv,C]=PSO(fitness,N,c1,c2,w,M,D,lb,ub)
% N-particle number, c1, c2-acceleration constants, w-inertia weight, M-iterations, D-particle dimension

%initialization:
x(i,j)=rand()*(ub-lb)+lb; %initialize position
```

```
v(i,j)=rand(); %initialize speed

%initialize fitness
p(i)=fitness(x(i,1),x(i,2));
y(i,:)=x(i,:);

%initialize local and global extremum
gbest=[x(1,1),x(1,2)];
for i=2:N
    if fitness(x(i,1),x(i,2))<fitness(gbest(1),gbest(2))
        gbest=x(i,:);
    end
end

% iterations
C = zeros(2,M);
for t=1:M
    for i=1:N
        v(i,1)=w*v(i,1)+c1*rand*(y(i,1)-x(i,1))+c2*rand*(gbest(1)-x(i,1));
        v(i,2)=w*v(i,2)+c1*rand*(y(i,2)-x(i,2))+c2*rand*(gbest(2)-x(i,2));
        x(i,:)=x(i,:)+v(i,:);
        for j = 1:D
            if x(i,j)<lb
                x(i,j) =lb;
            end
             if x(i,j)>ub
                x(i,j) =ub;
             end
        end
        if fitness(x(i,1),x(i,2))<p(i)
           p(i)=fitness(x(i,1),x(i,2));
           y(i,:)=x(i,:);
        end
        if p(i)<fitness(gbest(1),gbest(2))
            gbest=y(i,:);
        end
    end
```

```
        C(1,t) = gbest(1);
        C(2,t) = gbest(2);
end
xm=gbest';%optimal solution
fv=fitness(gbest(1),gbest(2));
```

References

1. Xu, Z.; Jia, Q.; Guan, X.; Xie, X. A new method to solve large-scale building energy management for energy saving. In Proceedings of the 2014 IEEE International Conference on Automation Science and Engineering (CASE), Taipei, Taiwan, 18–22 August 2014; pp. 940–945.
2. Gouda, M.M.; Danaher, S.; Underwood, C.P. Building thermal model reduction using nonlinear constrained optimization. *Build. Environ.* **2002**, *37*, 1255–1265. [CrossRef]
3. Jiang, P.; Tovey, K. Overcoming barriers to implementation of carbon reduction strategies in large commercial buildings in China. *Build. Environ.* **2010**, *45*, 856–864. [CrossRef]
4. Deb, C.; Zhang, F.; Yang, J.; Lee, L.S.; Shah, K.W. A review on time series forecasting techniques for building energy. *Renew. Sustain. Energy Rev.* **2017**, *74*, 902–924. [CrossRef]
5. Oral, G.K.; Yener, A.K.; Bayazit, N.T. Building envelope design with the objective to ensure thermal, visual and acoustic comfort conditions. *Build. Environ.* **2004**, *39*, 281–287. [CrossRef]
6. Muhamad, W.N.W.; Zain, M.Y.M.; Wahab, N.; Aziz, N.H.A.; Kadir, R.A. Energy efficient lighting system design for building. In Proceedings of the International Conference on Intelligent Systems, Modelling and Simulation, Liverpool, UK, 27–29 January 2010; pp. 282–286.
7. Meredith, S. *Beyond Light Bulbs: Lighting the Way to Smarter Energy Management*; Emerald Book: Austin, TX, USA, 2008; ISBN 9781934572078.
8. Lee, W.; Immadi, P.; Song, K.; Cho, H. A light control technique for energy-efficient buildings in visual comfort constrained environment. In Proceedings of the 2017 2nd International Multidisciplinary Conference on Computer and Energy Science (SpliTech), Split, Croatia, 12–14 July 2017; pp. 1–5.
9. Basnayake, B.A.; Amarasinghe, Y.W.; Attalage, R.A.; Jayasekara, A.G. Occupancy identification based energy efficient Illuminance controller with improved visual comfort in buildings. In Proceedings of the 2017 Moratuwa Engineering Research Conference (MERCon), Moratuwa, Sri Lanka, 30–31 May 2017; pp. 304–309.
10. Guillemin, A.; Morel, N. An innovative lighting controller integrated in a self-adaptive building control system. *Energy Build.* **2001**, *33*, 477–487. [CrossRef]
11. Amirkhani, M.; Garcia-Hansen, V.; Isoardi, G.; Allan, A. An energy efficient lighting design strategy to enhance visual comfort in offices with windows. *Energies* **2017**, *10*, 1126. [CrossRef]
12. Touma, A.A.; Ouahrani, D. Shading and day-lighting controls energy savings in offices with fully-Glazed facades in hot climates. *Energy Build.* **2017**, *151*, 263–274. [CrossRef]
13. Hussain, H.M.; Javaid, N.; Iqbal, S.; Hasan, Q.U.; Aurangzeb, K.; Alhussein, M. An efficient demand side management system with a new optimized home energy management controller in smart grid. *Energies* **2018**, *11*, 190. [CrossRef]
14. Naz, M.; Iqbal, Z.; Javaid, N.; Khan, Z.A.; Abdul, W.; Almogren, A.; Alamri, A. Efficient power scheduling in smart homes using hybrid grey wolf differential evolution optimization technique with real time and critical peak pricing schemes. *Energies* **2018**, *11*, 3155. [CrossRef]
15. Rasheed, M.B.; Javaid, N.; Ahmad, A.; Jamil, M.; Khan, Z.A.; Qasim, U.; Alrajeh, N. Energy optimization in smart homes using customer preference and dynamic pricing. *Energies* **2016**, *9*, 503. [CrossRef]
16. Piccolo, A.P.S.; Graditi, G.; Atrigna, M. Designing and testing decision support and energy management systems for smart homes. *J. Ambient Intell. Hum. Comput.* **2013**, *4*, 651–661.
17. González-Briones, A.; Chamoso, P.; De La Prieta, F.; Demazeau, Y.; Corchado, J.M. Agreement technologies for energy optimization at home. *Sensors* **2018**, *18*, 1633. [CrossRef] [PubMed]
18. González-Briones, A.; Prieto, J.; De La Prieta, F.; Herrera-Viedma, E.; Corchado, J.M. Energy optimization using a case-based reasoning strategy. *Sensors* **2018**, *18*, 365. [CrossRef] [PubMed]

19. Al-Ali, A.R.; Zualkernan, I.A.; Rashid, M.; Gupta, R.; AliKarar, M. A smart home energy management system using IoT and big data analytics approach. *IEEE Trans. Consum. Electron.* **2017**, *63*, 426–434. [CrossRef]
20. Kennedy, J.; Eberhart, R. Particle swarm optimization. In Proceedings of the IEEE International Conference on Neural Networks 6, Perth, WA, Australia, 27 November–1 December 1995; pp. 1942–1948.
21. Wymelenberg, K.G. Visual comfort, discomfort glare, and occupant fenestration control: Developing a research agenda. *LEUKOS* **2014**, *10*, 207–221. [CrossRef]
22. Mui, K.W.; Wong, L.T. Acceptable illumination levels for office occupants. *Archit. Sci. Rev.* **2006**, *49*, 116–119. [CrossRef]
23. Si, W.; Ogai, H.; Hirai, K.; Takahashi, H.; Ogawa, M. An improved PSO method for energy saving system of office lighting. In Proceedings of the Sice Annual Conference (SICE), Tokyo, Japan, 13–18 September 2011; pp. 1533–1536.

© 2018 by the authors. Licensee MDPI, Basel, Switzerland. This article is an open access article distributed under the terms and conditions of the Creative Commons Attribution (CC BY) license (http://creativecommons.org/licenses/by/4.0/).

sustainability

MDPI

Article

Energy and Hydraulic Performance of a Vegetated Roof in Sub-Mediterranean Climate

Patrizia Piro [1], Marco Carbone [1], Marilena De Simone [2], Mario Maiolo [3], Piero Bevilacqua [2] and Natale Arcuri [2,*]

[1] Department of Civil Engineering, University of Calabria, Arcavacata di Rende, 87036 Cosenza, Italy; patrizia.piro@unical.it (P.P.); marco.carbone@unical.it (M.C.)

[2] Department of Mechanical, Energy and Management Engineering, University of Calabria, Arcavacata di Rende, 87036 Cosenza, Italy; marilena.desimone@unical.it (M.D.S.); piero.bevilacqua@unical.it (P.B.)

[3] Department of Environmental and Chemical Engineering, University of Calabria, Arcavacata di Rende, 87036 Cosenza, Italy; mario.maiolo@unical.it

* Correspondence: natale.arcuri@unical.it; Tel.: +39-0984-49-4851

Received: 24 August 2018; Accepted: 26 September 2018; Published: 28 September 2018

Abstract: Widespread overbuilding, the prevalence of asphalt surfaces on green areas, and the use of building materials with low heat dissipation abilities are among the main causes of Urban Heat Islands. Within urban areas, evapotranspiration and shade from plants can significantly reduce the UHI phenomenon, help in stormwater management, and reduce building energy consumption. The goal of this work is to analyze the hydraulics and energy performances of an experimental extensive green roof at the University of Calabria (Italy) in Mediterranean area. This study confirmed that green roofs significantly mitigate storm water runoff generation in terms of runoff volume reduction and peak attenuation, and improve the thermal performance of buildings and the internal comfort of indoor spaces.

Keywords: green roof; energy-hydraulic performance; rainwater management

1. Introduction

Green roofs have been widely investigated in many cities around the world as a tool to solve many problems in urban environments. They can reduce the energy demands of air-conditioned spaces, and hence greenhouse gas emissions, through direct shading of the roof, evapotranspiration, and improved insulation values [1–4]. If widely adopted, green roofs could reduce the urban heat island [5,6] (an elevation of temperature relative to the surrounding rural or natural areas due to the high concentration of heat-absorbing dark surfaces such as rooftops and pavements) which would further lower energy consumption in urban areas. Moreover, based on the assumption that traditional approaches in urban stormwater drainage may be insufficient to manage water cycle changes associated with urbanization [7,8], green roofs, used as part of the storm water management strategy, represent sustainable solutions for mitigating runoff volumes and hydrograph peaks. Part of the rain is stored in the growing medium temporarily, to be subsequently taken up by the plants and returned to the atmosphere through evapotranspiration [9].

Integrated water and energy management in the urban context is one major route towards environmental sustainability and the reduction of carbon dioxide emissions. These objectives are also linked to the life cycle of infrastructures for the integrated management of these resources [10,11]. The understanding of mass (water) and energy fluxes between the building envelope and the surrounding environment is indeed a topical issue whenever sustainability and comfortable living conditions are addressed for urban areas [12].

From a hydraulic point of view, green roofs delay runoff into the sewage system, thus helping to reduce the frequency of combined sewage overflow (CSO) events, which is a significant environmental problem for many major cities in Europe [13]. Given the beneficial effects that green roofs produce on storm water retention, in the last years, several studies have been carried out. Some of them focus on the effects of the hydrological and physical parameters on green roof hydraulic efficiency; a review of those studies can be found in Garofalo et al. [14]. A study carried out in Portland has proven how ecoroofs can be considered an effective means for urban storm water management [15]. While Moran et al. [16], by monitoring two different green roofs in North Carolina, analyzed green roof performance on water quality improvement and runoff reduction. Carter and Rasmussen [17] have shown how green roof precipitation retention decreases with precipitation depth. In New York City, Carson et al. [18] analyzed three full-scale green roofs, obtaining retention values of 36%, 47%, and 61% respectively, and developed a characteristic runoff equation (CRE) for each green roof. Another study was focused on the evaluation of the survival, cover, roof cooling, and stormwater retention proprieties of 15 native species in coastal regions of Atlantic Canada in extensive green roof monoculture [19]. DeNardo et al. [20] and Spolek [21] have analyzed the effects of green roofs both on stormwater mitigation and thermal benefits.

The plants and the growing medium can also remove airborne pollutants picked up by the rain, thus improving the quality of the runoff. In addition, green roofs can improve air quality, provide additional green space in urban areas, and increase property value [4].

Vegetated roofs on the bare roofs of buildings can offer a good thermal protection that reduces the high thermal load from which the construction suffers during the summer period [22]. This is an acceptable ecological solution, which contributes not only to the reduction of the thermal loads of the building's shell, but also to the improvement of densely-built urban centers with little natural environment. Plants have an effect on the climate; foliage protects buildings from solar radiation and controls the temperature and humidity of the indoor environment. In closed spaces with planted roofs, the air temperature beneath the plants is lower than that of the air above. The difference between a vegetated roof and bare roof of a single building is both qualitative and quantitative [23]. The process of heat transfer into the planted roof is very different. Solar radiation, external temperature, and relative humidity are reduced as they pass through the foliage that covers the roof. Plants, owing to their biological functions like photosynthesis, respiration, transpiration, and evaporation, absorb a portion of solar radiation. The remaining solar radiation influences the internal climate as it passes through the building elements of the roof.

In general, researchers suggest that green roofs contribute to reducing temperature fluctuations on external and internal surfaces [24], lowering energy consumption for the maintenance of indoor comfort conditions [25]. Furthermore, vegetated roofs increase the thermal inertia of the building envelope, generating a time delay and attenuation of the external solicitation on the roof surface [26]. From the point of view of storm water management, a green roof system is able to significantly reduce storm water runoff generation in terms of volume, peak attenuation, and increase of concentration time [27].

This study presents an analysis of the energy and hydraulic performance of an experimental green roof in Cosenza, a Southern Italian city with a temperate climate. As part of the Research Project PON "Integrated and Sustainable Management Services for the Water-Energy Cycle in Urban Drainage Systems", a real scale model of a green roof was installed at the University of Calabria in order to evaluate the influence of green roofs on rainwater management and energy consumption for air conditioning in the Mediterranean area. The roof sections were instrumented to monitor temperature profiles within the roofing systems, solar radiation, relative humidity, soil moisture content, and storm water runoff. This set-up allows direct comparison of the performance and benefits of the green roof and the conventional roof.

The measurements for evaluating the energy performance of the green roof were conducted for each sector on a seasonal basis in order to assess the thermal behavior of the extensive green roof throughout a whole year.

The hydraulic performances were evaluated during the monitoring campaign in September–October 2015.

2. Methodology

2.1. Experimental Site

To fulfill the aims of the research project "Integrated and Sustainable Management Services for the Water-Energy Cycle in Urban Drainage Systems", a scale model of the green roof was installed at the University of Calabria. A typical green roof stratigraphy has different layers of construction. A protection layer directly above the structural roof construction serves to protect the roof from moisture. This layer includes some combination of waterproof membranes and root barriers. A drainage layer that may include small moisture-retention reservoirs and voids resides directly above the protection layer to allow excess moisture to flow out of the green roof, either into waterspout or to some form of storage, and provide aeration to the vegetation roots. The growing medium (soil) layer resides above the drainage layer, which is usually separated by a geotextile filter layer. This medium layer is typically a light-weight combination of sand, aggregate, and organic matter. Finally, vegetation, chosen in accordance to the climate of the location, completes the installation.

In order to facilitate the comparison of different design solutions, the experimental site was organized into four sectors (Figure 1). Sectors 1 and 2 were vegetated with the same native Mediterranean species, but with different materials used in the stratigraphy. Sectors 1 and 3 had the same stratigraphy, but sector 3 was not initially vegetated. Sector 4 was left as a conventional roof, equipped with the classic dark waterproof membrane; for this reason, it was considered as the reference roof. The four sectors have a slope of 1% for correct water management, and are hydraulically independent.

Figure 1. Experimental green roof [28] at the Urban Hydraulics and Hydrology Lab at the University of Calabria, Italy.

The vegetation species used in the green roof are *Carpobrotus edulis, Dianthus gratianopolitanus,* and *Cerastium tomentosum*. These species were chosen for their particular features: high drought tolerance, low nutritional requirements, good ground coverage, rapid multiplication, short roots, and low maintenance. For their high drought tolerance, the water supply of sectors 1 and 2 is guaranteed by reusing the green roof's outflow, collected in a specific storage tank and distributed through a drip irrigation system during drought periods.

This study was carried out in two main phases.

The first phase was focused on the analysis of the thermal performance of the vegetated roof in Mediterranean climate by comparing experimental data from three sectors.

Finally, in order to evaluate the performance of the green roof in terms of storm water mitigation, the real runoff data from sector 1 were collected and compared with the real outflow from the reference roof. Thus, the hydraulic efficiency was analyzed by evaluating two hydrological indexes (Retained Volume and Peak Flow Attenuation).

2.2. Equipment and Measurement Period

The monitoring system of the experimental site was equipped with a data acquisition system which acquires sensors output at the experimental site every minute, and records the measured data in a computer.

Meteorological data are monitored by a weather station installed on the roof of the same building of the experimental set-up. This provides data of ambient air temperature, relative humidity, precipitations, atmospheric pressure, and wind direction and speed. The temperature sensor with a solar shield has an operative range from −40 °C to +80 °C, with an accuracy of 0.4 °C. The relative humidity sensor works in the range of 0–100%, with an accuracy of 3%. The wind speed sensor has an operative range from 1.5 m/s to 79 m/s, and an accuracy of 5%. The rain gauge sensor has a resolution of 0.2 mm. For the measurement of total horizontal solar radiation, a secondary standard EPLAB pyranometer is used.

In each plot, four probes simultaneously measure the temperature and volumetric water content (WC) in the substrate. The values presented in the following sections are obtained as the average of the four measures. Along the vertical profile, two resistance temperature detectors (RTD) were installed at the interface between the lightweight concrete and the root-barrier waterproof layer, and between the traditional roof surface and the lightweight concrete. A further RTD was placed on the top of the insulation layer in sector 3 (Figure 2). The soil probe has an accuracy of ±2% for the measurement of soil moisture in an operative range from 0% to saturation, and an accuracy of ±1% for measurements of temperature, with an output from −50 °C to +50 °C. The RTD sensors are four PT100 class1/3 wires with an accuracy of ±0.1 °C at 0 °C and ±0.27 °C at 100 °C. The green roof is equipped with a drip irrigation system managed by modules that operate activation and shut-down according to the values provided by adequate water presence sensors installed in the drainage layers. Irrigation is provided during hot months in summer to prevent water stress on the plants.

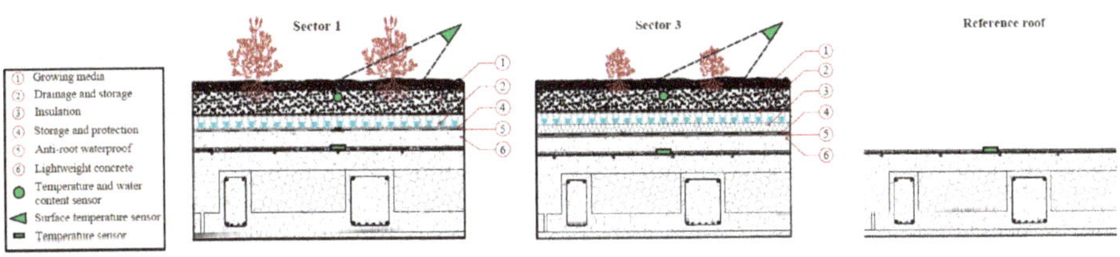

Figure 2. Sector 1 (**left**), Sector 3 (**center**) and Reference roof (**right**) sections with sensors location [23].

In addition, to analyze the hydraulic performance of one specific sector for two continuous months (September 2015 and October 2015), rainfall data was collected every minute by a tipping bucket rain gauge of 0.254 mm resolution, located on the experimental site. Specifically, for hydraulic parameters, the outflow rates from the conventional roof and from the green roof were recorded by flowmeters installed at the outlet of each compartment. The water level in each device was measured by a pressure transducer (Ge Druck PTX1830), and continuously recorded in a SQLITE database system with a resolution of 1 min. To describe the green roof hydraulic performance, two synthetic indexes–were evaluated. The first is *the Retained Volume* (RV)—calculated as the percentage difference between the total volume of rain and the total discharged volume—while the second is *the Peak Flow Reduction* (PFR)—evaluated as the percentage difference between conventional peak flow and the green roof peak flow.

A first analysis was conducted to evaluate the seasonal thermal behaviour of the two different layering solutions installed in the extensive green roof, and to compare the results with those of the traditional reference roof. For this purpose, in each season, two periods were selected according to the climatic conditions, chosen to be representative of the average seasonal conditions, considering one sunny week and one week with the presence of precipitations. In summer, only one sunny period was selected because the analysis of thermal response can be extended to the whole period, considering the scarcity of rainfall that occurred in summer 2015. Data employed for this analysis were collected in year 2014 and 2015. A summary of selected days is reported in Table 1.

Table 1. Periods of analysis.

Dates	Condition	Season
13 October 2014–19 October 2014	Sunny	Autumn
7 November 2014–13 November 2014	Rainy	Autumn
1 January 2015–7 January 2015	Sunny	Winter
24 December 2014–30 December 2014	Rainy	Winter
11 April 2015–17 April 2015	Sunny	Spring
2 April 2015–8 April 2015	Rainy	Spring
16 July 2015–23 July 2015	Sunny	Summer

The variables considered in the following evaluations are:

→ The green plots surface temperature;
→ The substrate temperature;
→ The temperature at the bottom of the of the green layering;
→ The water content of the substrate;
→ The surface temperature of the traditional reference roof.

3. Results and Discussion

In Figures 2 and 3 the trends of the monitored variables for the selected seasonal periods are reported. In particular, the figures report the trends of the plots' surface temperature, substrate temperature, temperature at the bottom of the green layering, water content of the growing medium, and the surface temperature of the reference roof. Since the thermal response of sectors 1 and 2 appeared to be very similar, the results are reported only for sectors 1 and 3, referred to in the following as "green roof GR" and "insulated green roof I-GR".

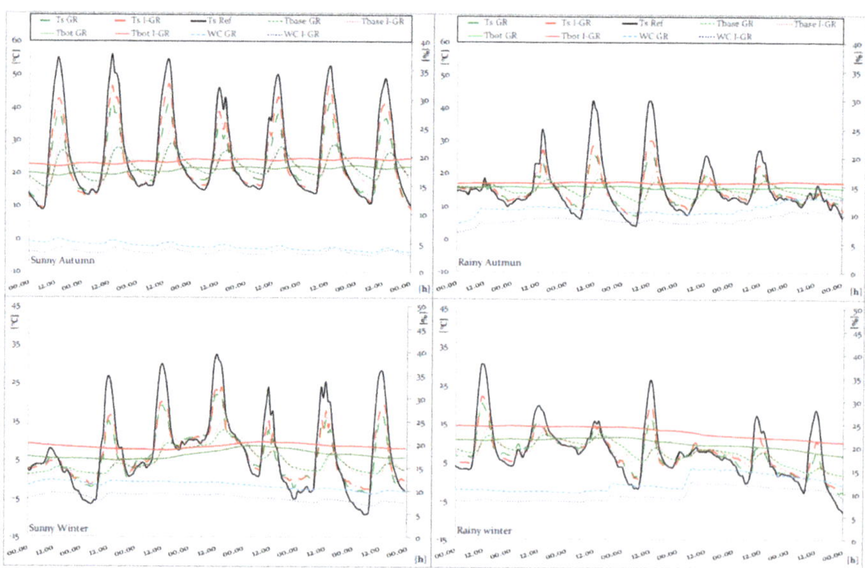

Figure 3. Green plots water content and surface, substrate base, and bottom temperatures, together with reference roof surface temperature in the sunny and rainy autumn and winter periods.

3.1. Autumn Thermal Analysis

In the sunny autumn week, the reference roof reached daily maximum temperatures greater than the green sectors, whereas the daily minimum values attained were very close to an almost-coincident nocturnal pattern. In the hours around noon, a rather significant difference is appreciated between surface temperatures of the green plots: the non-insulated green roof (sector 1) reached a maximum value of 42.4 °C, compared to 47.0 °C of the insulated one (sector 3). The minimum daily values were, conversely, very similar, resulting in a maximum daily surface temperature excursion for the non-insulated plot of 28.3 °C, and of 33.8 °C for the insulated plot. The lower presence of vegetation, the effect of the insulation layer that contained the heat flux towards the indoor environment, and the slight inferior water content of the substrate result in an increased surface temperature in this last plot. The reference roof showed considerably higher daily excursions, with a maximum of 46.2 °C, to be attributed to the superior temperature registered during the day in this plot, since the night values were instead very similar to those of the green plots. In sunny days, the effect of solar radiation is consistently greater because of the lower albedo of the dark bituminous membrane with a rapid and considerable temperature rise. The difference in surface temperature is also reflected in the values measured at the bottom of the substrate layer. In both green plots, a marked attenuation of the daily excursion is visible, and even though the thickness of the growing medium is much reduced, a time shift of the peaks is already appreciable. Moving down the layers, an almost steady trend of temperature was found at the interface with the structural roof. Here the overall effect of the overlying green roof is summarized in an average temperature of 21.0 °C in GR plot, and of 23.6 °C in I-GR plot. Evidently, the presence of the insulation layer in this plot led to a higher temperature level at the base of the green roof. For comparison, the average surface temperature of the reference roof in the week was 26.0 °C.

In the selected rainy period, two conspicuous rain events occurred, during which the trends of surface temperature of the green plots and of the reference roof tended to overlap almost completely. In these days, consequently, the temperature measured at the substrate base also showed almost identical values in the insulated and non-insulated green plots. As soon as solar radiation appeared on clear days, the surface temperatures tended to swiftly rise, reaching a maximum of 26.5 °C in GR plot, 30.1 °C in I-GR, and 42.4 °C in the reference roof. Again, the minimum daily values attained during nighttime were close for all the plots, but in this period, the increased water content of the growing medium allowed for moderate surface temperature in both plots and lower peak temperatures. Compared to the sunny week, the higher level of water in the substrate had the overall effect of strongly attenuating the daily temperature excursion that, in this period, reached a maximum of 19.1 °C in GR and 23.6 °C in I-GR. Again, at the base of the green roof layers, it is possible to appreciate an almost steady trend of temperature with mean values of 15.9 °C in the non-insulated plot and 17.3 in the insulated one. The average temperature of the reference roof was 15.6 °C.

3.2. Winter Thermal Analysis

Regarding the winter behaviour of the green roof, the extremely variable climatic conditions in the investigated years, along with the occurrence of several precipitation events during the winter months, made it difficult to make a marked distinction between sunny and rainy weeks. However, it was possible to select one serene period with the presence of a considerable amount of solar radiation during different days, while in the rainy selected period, abundant rainfall events occurred for a total precipitation of 60 mm (Figure 2). Because of the elevated water content in the substrate of both green roof plots, no significant difference in the daily maximum surface temperatures was detected. The slight higher values of I-GR that are appreciable from the graph were mainly due to the lower water content of this plot compared to the non-insulated one. Furthermore, the reduced solar radiation in winter period did not produce a sharp increase in surface temperature, helping to limit the difference between the two plots. The effect of the insulation layer was instead more visible at the base of the substrate, where I-GR reached a higher peak than GR almost every day, with a maximum difference in the week of 2.4 °C. At the bottom of the stratigraphy, the trend was again very stable around the mean value of 6.4 °C for the non-insulated green roof, and around the value of 8.5 °C for the insulated one. Considering the trend of the surface temperature of the traditional reference roof, it can be seen from the figure that this plot reached considerably higher daily peaks and lower night values compared to the green roof. The night temperature even dropped below zero in some of the selected days, especially in clear sky conditions, because the infrared thermal exchange with the sky intensively cooled the roof surface, causing a dramatic drop at night. The maximum and minimum temperatures were, respectively, 32.4 °C and −9.0 °C, around a mean weekly value of 7.2 °C.

In the rainy winter week, the precipitation events determined a rapid increase of the substrate water content that, however, did not result in significant surface temperature difference compared to the sunny week. In this sense, the limited evapotranspiration effect, because of the low level of solar radiation, low air temperature, and conspicuous level of relative humidity, did not produce considerably lower trends and daily excursions of the surface and substrate temperatures in insulated and non-insulated plots. However, the excursions of the reference roof were instead significantly reduced because of the limited climatic solicitations. It is interesting to note that the daily minimum temperatures attained by the green plots and the reference roof were markedly superior to those of the sunny week. This can be explained by considering that the greater sky covers consistently limited the night infrared cooling of the surfaces, and consequently the drops in temperature. It is interesting to note how, in cloudy and rainy winter conditions, the green plots were able to maintain considerably higher temperature levels at the base of the layering where a mean temperature of 10.8 °C was registered in GR and a mean of 13.0 °C in I-GR. In the reference, roof, in contrast, a weekly mean temperature of 6.5 °C was observed.

3.3. Spring Thermal Analysis

As expected, the overall thermal behavior of the selected spring weeks was quite similar to that of the autumn weeks. In this season, the maximum daily values of the green plots surface temperatures, conversely to those of autumn, were very close. This can be explained by the growth of colonizing species that germinate strongly in spring and early summer, modifying the surface thermal balance of the insulated plot that becomes similar to that of the non-insulated one. The reference roof again showed substantial higher peak temperatures, with a maximum of 61.5 °C registered during the last day of the period, and high daily excursions, with a maximum of 55 °C in the same day. The night surface temperature values were similar for the green plots, where minimum values reached 4.3 °C and 3.7 °C, whereas in the reference roof, a value of 1.7 °C was reached. Again, the effect of the insulation in I-GR is clearly appreciable at the base of substrate, where a marked difference in the daily maximum values is observed with a peak of 35.6 °C in I-GR in comparison with 30.0 °C of GR. The insulated plot was consequently able to provide a slightly higher temperature at the bottom of the green roof, with a mean value of 18.3 °C compared to the value of 16.2 °C in the non-insulated one (Figure 4).

Also in the rainy week, the behavior was similar to the autumn week. When precipitation events occurred, the surface temperatures of the green plots and of the reference roof tended to assume similar values, while at the appearance of solar radiation, the reference roof temperature quickly rose to greater values, reaching a weekly maximum of 50.7 °C. In the green plots, the surface temperature trends were very close, with reduced daily fluctuations that did not exceed 24.3 °C. The base substrate temperature, similar to those of the other seasons, assumed lower peak values in the GR plot, whereas the daily minima were higher than I-GR, with overall lower daily excursions in the non-insulated plot that exhibited a maximum excursion of 11.0 °C compare to 17.0 °C of the insulated one. In the analyzed period, it is possible to appreciate that in the GR plot, the water level was always greater, reaching 14.6%, and being on average 3.6 percentage points higher than I-GR, thereby contributing to lowering the growing medium temperature. The bottom green roof temperature was again higher for the insulated plot, with a weekly mean of 17.9 °C compared to 15.2 °C of the non-insulated one (Figure 4).

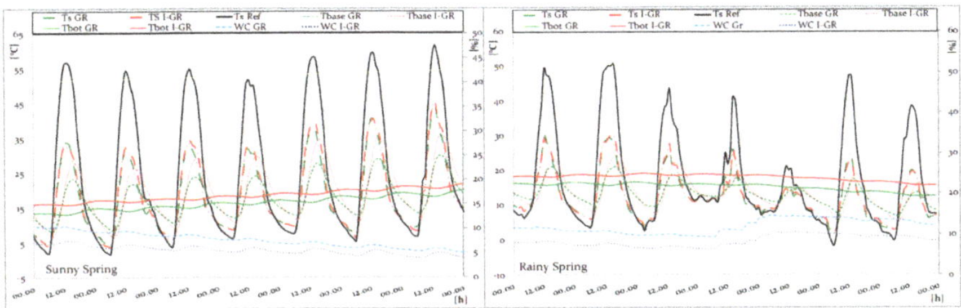

Figure 4. Green plot water content and surface, substrate base, and bottom temperatures, together with reference roof surface temperature in the sunny and rainy spring periods.

3.4. Summer Thermal Analysis

Summer thermal behaviour needs to be discussed separately since the activation of the irrigation system strongly influenced the results in terms of temperature. Figure 3 clearly shows the typical asymmetrical bell-shaped trend of the reference roof and the conspicuously superior temperature value attained by it during the day, with peaks slightly higher than 70 °C. The minimum values were instead close to those of the green plots. In fact, the night cooling of the bituminous membrane was limited compared to the spring season, and did not reach the same minimum temperatures registered in the sunny spring period. In I-GR and GR, although the shape of the surface temperature pattern was quite similar, a substantial difference can be appreciated in the temperatures reached. In this period, in contrast to the others which were analysed, the non-insulted plot showed surface greater temperatures than the insulated one. The daily surface peaks of GR were never lower than 46.0 °C, whereas in I-GR, they were, at most, 35.3 °C. The minimum daily values were instead very close, even though the insulated plot could reach slightly lower values. In fact, to prevent plant stress due to water scarcity, in the critical summer period, the irrigation system is activated, providing water to the plots for one hour from 01:00 to 02:00 in GR, and for two hours from 20:00 to 22:00 in I-GR. For the last plot, in July 2015, the planting of lawn occurred, and consequently, in order to assist the growth of the vegetation, abundant irrigation was provided, resulting in a significant major level of water entering the substrate. As visible in Figure 5, average water content values of 9.0% in GR and 12.6% in I-GR were found. As a result, the excursion of surface temperatures was much more limited in the insulated plot, with an average weekly value of 17.7 °C compared to 28.1 °C of the non-insulated plot. For comparison, the average daily excursion of the traditional roof was 54.5 °C. At the base of the substrate, however, the difference between the two plots' trends is not strongly marked. Thanks to the additional water in the substrate, the insulated plot could generate slight lower temperatures, especially far from midday. However, daily peaks remained quite comparable, with a maximum of 38.0 °C in I-GR and a value of 38.2 °C in GR.

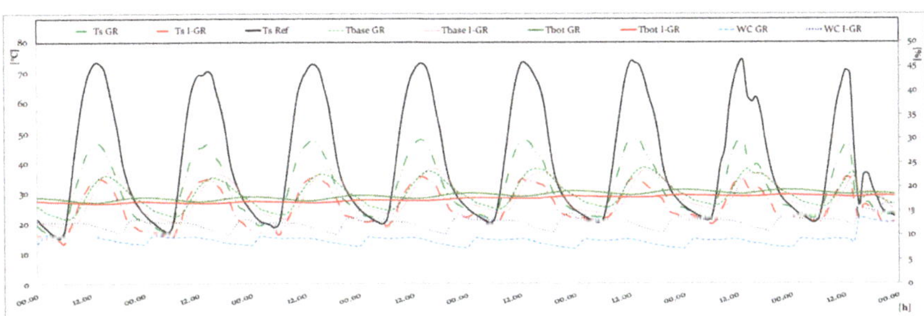

Figure 5. Green plots water content and surface, substrate base, and bottom temperatures, together with reference roof surface temperatures in the summer periods.

Finally, at the bottom of the green roof, temperature showed an almost steady trend in both plots, with average values of 29.2 °C in GR and 28.0 °C in I-GR. This result highlights how a proper watering strategy can be an effective means for temperature regulation in green roofs, enhancing the properties of the insulation into the stratigraphy. As demonstrated by the winter behaviour analysis, the presence of a thermal insulation layer yielded considerably higher temperature levels at the base of the green roof stratigraphy, resulting in consistent smaller heat fluxes toward the outdoor environment, and lowering the energy lost through the building roof cover. In summer, if irrigation is not provided adequately, the surface temperature of the insulated plot can easily reach huge values, consequently generating heat fluxes which are much more dramatic than those of the non-insulated plot [3].

3.5. Hydraulic Results

This study considers part of the monitoring campaign carried out at the experimental site of the University of Calabria, and is still in progress. It can be divided into two phases; the first concerns data collected from the whole Conventional Roof (outflows from the sector 4); the second studies the whole new Vegetated Roof system (outflows from the sector 1). The impervious system (conventional roof) serves as a reference for comparisons with the green roof's hydraulic performances.

Individual events were also defined as being separated by continuous dry periods of at least 6 h [29] and Rain Depth greater than 2 mm [30].

The selected rainfall data included seven events collected in the phase of the monitoring campaign for September and October 2015. The Cumulative Runoff Depth for the Vegetated Roof and for the Conventional one, and the corresponding hyetograph for four rainfall events with different Rain Depth, are illustrated in Figure 6.

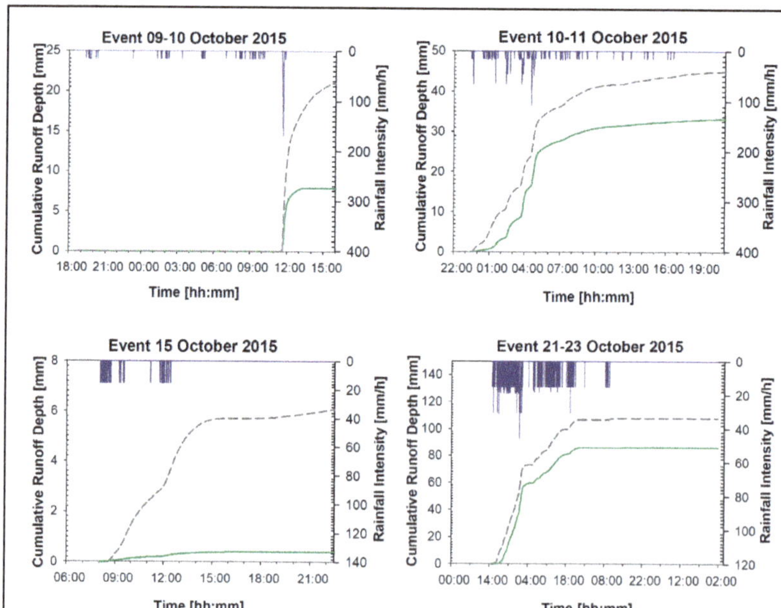

Figure 6. Comparison of Cumulative Runoff Depth for the Vegetated Roof and for the Conventional one for four rainfall events: Rainfall Event 9 October 2015 (Rain Depth = 24.1 mm); Rainfall Event 10–11 October 2015 (Rain Depth = 48.3 mm); Rainfall Event 15 October 2015 (Rain Depth = 6.4 mm); Rainfall Event 21–23 October 2015 (Rain Depth = 120.1 mm).

Table 2 shows the hydraulic performance of the green roof in terms of Retained Volume (RV) and Peak Flow Reduction (PFR) at the event scale.

Table 2. Events observed at the green roof of the University of Calabria.

Event	Rain Depth [mm]	RV [%]	PFR [%]
9–10 September 2015	99.8	15	25
21–23 September 2015	2.8	100	100
7 October 2015	42.2	52.5	65.4
9 October 2015	24.1	55.5	17.9
10–11 October 2015	48.3	35.7	13.3
15 October 2015	6.4	93.9	92
21–23 October 2015	120.1	16.8	28.3
29–31 October 2015	63.3	26.6	52.6
Mean	-	57.5	72.3
Dev. Std.	-	35	43.4

By analysing the results shown in Figure 6, in which the Cumulative Runoff Depth from the Vegetative Roof is compared with the Conventional one, the ability of this specific extensive green roof to attenuate the stormwater runoff volume with respect to the impervious roof is evident.

Furthermore, the findings listed in Table 2 at the event scale show a high variability of the two performance indexes evaluated at the event scale. More in detail, the retained volume, ranging between 15% and 100%, and the peak flow reduction, ranging from 25% to 100%, denote, in agreement with other studies [14,29,31], that the weather conditions prior a rainfall event and the hydrological features of the rainfall events itself significantly affect the hydraulic response of a green roof.

On the other hand, by considering the results obtained in terms of mean values (Table 2), it emerges that the performance of this specific green roof as a device for storm water control in the Mediterranean area appears good. The mean retained volume value of 57.5% is, in fact, really close to that obtained by Palla et al. [32] for a green roof installed at the University of Genoa, another city in Mediterranean climate conditions. Garofalo et al. [14] have evaluated, from the analyses on different studies under various climate conditions and locations, a subsurface coefficient mean value ranging from 0.32 to 0.89. By considering that the subsurface runoff coefficient is the complementary index of the retained volume index, the mean value here obtained falls perfectly within this range of variation. Thus, by comparing these results with others studies, this specific green roof is able to reduce storm water runoff generation in the Mediterranean region.

On the basis of these findings, an extension of these to the spatial scale is needed. In this way, it would be possible to assess the role of green roof installations in preventing flooding phenomena in the urban areas and in limiting the impact of storm water on waste water treatment plants.

4. Conclusions

Observations from the green roof installed at the University of Calabria allowed us to discuss the energy and water related performance of vegetated roofs within the specific climatic context of the Mediterranean region.

The dataset obtained during this study is useful to contribute to a deeper understanding of mass (water) and energy fluxes between the building envelope and the surrounding environment.

This is expected to support sustainable construction in urban areas, where the intrinsic interaction between energy and water plays a major role in the strategies towards sustainability and improved quality of life in cities.

The monitored performance was shown to be very promising in the Mediterranean context, where the characteristics of precipitation and the general climatic conditions are less favorable to the growth of vegetation on top of buildings. This applies to the attenuation of solar radiation through the vegetation layer, as well as to the thermal insulation performance of the green roof structure. The thermal analysis was developed over one year in the four seasons in order to assess the performance of the green roof under the different climatic conditions of a typical Mediterranean locality. The results demonstrated how in winter in sunny conditions, the performance of the reference roof yielded slightly better results than the non-insulated green plot, but was always lower than the insulated green roof. In cloudy conditions, because the absence of solar radiation led to lower levels for the reference roof, this showed the worst performance of all the analyzed roof solutions.

In summer, thanks to the abundant irrigation provided to the insulated green plot, the excursions of surface temperature were much more limited, with an average weekly value of 17.7 °C compared to 28.1 °C of the non-insulated plot, and a value of 54.5 °C of for the traditional roof.

In intermediate seasons, the overall thermal behavior of the green plots was quite similar. In cloudy autumn conditions, at the base of the green roof layers, average weekly temperatures of 15.9 °C in the non-insulated plot and 17.3 in the insulated one were observed, with the reference roof at 15.6 °C. In sunny conditions, the average base temperature resulted respectively in 21.0 °C in the GR plot, 23.6 °C in the I-GR plot, and 26.0 °C in the reference roof. In sunny spring conditions, the insulated plot was able to provide a slight higher temperature at the bottom of the green roof, with a mean value of 18.3 °C compared to the 16.2 °C of the non-insulated one. In cloudy conditions, the bottom green roof temperature was higher for the insulated plot, with a weekly mean of 17.9 °C compared to 15.2 °C of the non-insulated one.

In addition, it was confirmed that green roofs significantly mitigate storm water runoff generation, even in a Mediterranean climate in terms of runoff volume reduction and peak attenuation. Results show a mean value of the Retained Volume for the two months considered of 57.5%. This finding confirms the good performance of the specific experimental green roof, and it is in agreement with other studies carried out in Mediterranean area. On the other hand, by observing the high variability of the same index evaluated at the event scale, it is possible to conclude that the weather conditions prior a rainfall event and the hydrological features of the storm water event itself significantly affect the hydraulic response of a green roof.

Green roofs are a solution to increase the energy and hydraulic performance and the sustainability of the building.

This study confirms that green roofs contribute to reducing thermal fluctuation on the external and internal surfaces, thereby lowering the energy consumption for the maintenance of indoor comfort conditions. From the point of view of storm water management, a green roof system is able to significantly reduce storm water runoff generation in terms of runoff volume reduction and peak attenuation.

Author Contributions: The authors contributions in the realization of the paper has been as follows. Conceptualization: P.P., M.M. and N.A. Data curation: M.C. and P.B. Formal analysis: M.D.S. Principal Investigator: P.P. Supervision: P.P., M.M. and N.A.

Funding: This work was funded within the project PON01_02543 "Integrated and sustainable management service for the water–energy cycle in urban drainage systems" of the National Operational Program-Research and Competitiveness 2007–2013 (PON-REC) and within the project I-BEST "Innovative Building Envelope through Smart Technology", CUP: B28I17000290008.

Acknowledgments: This work was financially supported within the project PON01_02543 "Integrated and sustainable management service for the water–energy cycle in urban drainage systems" of the National Operational Program-Research and Competitiveness 2007–2013 (PON-REC), co-financed by the European Regional Development Fund and National Resources Grant.

Conflicts of Interest: The authors declare no conflict of interest.

References

1. Mentens, J.; Raes, D.; Hermy, M. Green roofs as a tool for solving the rainwater runoff problem in the urbanized 21st century? *Landsc. Urban Plan.* **2006**, *77*, 217–226. [CrossRef]
2. Takebayashi, H.; Moriyama, M. Surface heat budget on green roof and high reflection roof for mitigation of urban heat island. *Build. Environ.* **2007**, *42*, 2971–2979. [CrossRef]

3. Theodosiou, T.G. Summer period analysis of the performance of a planted roof as a passive cooling technique. *Energy Build.* **2003**, *35*, 909–917. [CrossRef]
4. Bevilacqua, P.; Mazzeo, D.; Bruno, R.; Arcuri, N. Experimental investigation of the thermal performances of an extensive green roof in the Mediterranean area. *Energy Build.* **2016**, *122*, 63–79. [CrossRef]
5. Kumar, R.; Kaushik, S.C. Performance evaluation of green roof and shading for thermal protection of buildings. *Build. Environ.* **2005**, *40*, 1505–1511. [CrossRef]
6. Wong, N.H.; Chen, Y. *Tropical Urban Heat Islands: Climate, Buildings and Greenery*; Routledge: London, UK, 2008.
7. Raimondi, A.; Becciu, G. Probabilistic design of multi-use rainwater tanks. *Procedia Eng.* **2014**, *70*, 1391–1400. [CrossRef]
8. Palermo, S.A.; Zischg, J.; Sitzenfrei, R.; Rauch, W.; Piro, P. Parameter Sensitivity of a Microscale Hydrodynamic Model. In *New Trends in Urban Drainage Modelling: UDM 2018. Green Energy and Technology*; Mannina, G., Ed.; Springer: Cham, Switzerland, 2019.
9. Carbone, M.; Garofalo, G.; Nigro, G.; Piro, P. A conceptual model for predicting hydraulic behaviour of a green roof. *Procedia Eng.* **2014**, *70*, 266–274. [CrossRef]
10. Maiolo, M.; Carini, M.; Capano, G.; Nigro, G.; Piro, P. Life cycle assessment of a bitumen antiroot barrier in a green roof in the Mediterranean area. *Int. J. Petrochem. Res.* **2017**. [CrossRef]
11. Maiolo, M.; Carini, M.; Capano, G.; Piro, P. Synthetic sustainability index (SSI) based on life cycle assessment approach of low impact development in the Mediterranean area. *Cogent Eng.* **2017**, *4*, 1410272. [CrossRef]
12. Kassem, M.A.; Kaseb, S.; El-Refaie, M.F. Solar heat gain through vertical cylindrical glass. *Build. Environ.* **1998**, *34*, 253–262. [CrossRef]
13. Piro, P.; Carbone, M.; Garofalo, G.; Sansalone, J. Size distribution of wet weather and dry weather particulate matter entrained in combined flows from an urbanizing Sewershed. *Water Air Soil Pollut.* **2010**, *206*, 83–94. [CrossRef]
14. Garofalo, G.; Palermo, S.; Principato, F.; Theodosiou, T.; Piro, P. The Influence of Hydrologic Parameters on the Hydraulic Efficiency of an Extensive Green Roof in Mediterranean Area. *Water* **2016**, *8*, 44. [CrossRef]
15. Hutchinson, D.; Abrams, P.; Retzlaff, R.; Liptan, T. *Stormwater Monitoring Two Ecoroofs in Portland*; City of Portland Bureau of Environmental Services: Portland, OR, USA, 2003.
16. Moran, A.C.; Hunt, W.F.; Smith, J.T. Green roof hydrologic and water quality performance from two field sites in North Carolina. In *Managing Watersheds for Human and Natural Impacts: Engineering, Ecological, and Economic Challenges*; ASCE: New York, NY, USA, 2005; pp. 1–12.
17. Carter, T.L.; Rasmussen, T.C. Hydrologic behavior of vegetated roofs. *JAWRA J. Am. Water Resour. Assoc.* **2006**, *42*, 1261–1274. [CrossRef]
18. Carson, T.B.; Marasco, D.E.; Culligan, P.J.; McGillis, W.R. Hydrological performance of extensive green roofs in New York City: Observations and multi-year modeling of three full-scale systems. *Environ. Res. Lett.* **2013**, *8*, 024036. [CrossRef]
19. MacIvor, J.S.; Lundholm, J. Performance evaluation of native plants suited to extensive green roof conditions in a maritime climate. *Ecol. Eng.* **2011**, *37*, 407–417. [CrossRef]
20. DeNardo, J.C.; Jarrett, A.R.; Manbeck, H.B.; Beattie, D.J.; Berghage, R.D. Stormwater mitigation and surface temperature reduction by green roofs. *Trans. ASAE* **2005**, *48*, 1491–1496. [CrossRef]
21. Spolek, G. Performance monitoring of three ecoroofs in Portland, Oregon. *Urban Ecosyst.* **2008**, *11*, 349–359. [CrossRef]
22. Bevilacqua, P.; Coma, J.; Pérez, G.; Chocarro, C.; Juárez, A.; Solé, C.; de Simone, M.; Cabeza, L.F. Plant cover and floristic composition effect on thermal behaviour of extensive green roofs. *Build. Environ.* **2015**, *92*, 305–316. [CrossRef]
23. Lazzarin, R.M.; Castellotti, F.; Busato, F. Experimental measurements and numerical modelling of a green roof. *Energy Build.* **2005**, *37*, 1260–1267. [CrossRef]
24. Bevilacqua, P.; Mazzeo, D.; Bruno, R.; Arcuri, N. Surface temperature analysis of an extensive green roof for the mitigation of urban heat island in southern mediterranean climate. *Energy Build.* **2017**, *105*, 318–327. [CrossRef]
25. Teemusk, A.; Mander, Ü. Greenroof potential to reduce temperature fluctuations of a roof membrane: A case study from Estonia. *Build. Environ.* **2009**, *44*, 643–650. [CrossRef]

26. Bevilacqua, P.; Mazzeo, D.; Arcuri, N. Assessment of the thermal inertia of an experimental extensive green roof in summer conditions. *Build. Environ.* **2018**, *131*, 264–276. [CrossRef]
27. Carbone, M.; Turco, M.; Nigro, G.; Piro, P. Modeling of hydraulic behaviour of green roof in catchment scale. In Proceedings of the 14th GeoConference on Water Resources, Forest, Marine and Ocean Ecosystems, Albena, Bulgaria, 19–25 June 2014; Volume 1, pp. 471–478, ISBN 978-619-7105-13-1/ISSN 1314-2704.
28. Patrizia, P. *CICLO ACQUA-ENERGIA–Ricerca, Progettazione, Innovazione*; Edibios: Cosenza, Italy, 2015; ISBN 978-88-97181-38-5.
29. Stovin, V.; Vesuviano, G.; Kasmin, H. The hydrological performance of a green roof test bed under UK climatic conditions. *J. Hydrol.* **2012**, *414*, 148–161. [CrossRef]
30. Voyde, E.; Fassman, E.; Simcock, R. Hydrology of an extensive living roof under sub-tropical climate conditions in Auckland, New Zealand. *J. Hydrol.* **2010**, *394*, 384–395. [CrossRef]
31. Berndtsson, J.C. Green roof performance towards management of runoff water quantity and quality: A review. *Ecol. Eng.* **2010**, *36*, 351–360. [CrossRef]
32. Palla, A.; Gnecco, I.; Lanza, L.G. Hydrologic restoration in the urban environment using green roofs. *Water* **2010**, *2*, 140–154. [CrossRef]

© 2018 by the authors. Licensee MDPI, Basel, Switzerland. This article is an open access article distributed under the terms and conditions of the Creative Commons Attribution (CC BY) license (http://creativecommons.org/licenses/by/4.0/).

sustainability MDPI

Article

A Novel Dynamic Insulation System for Windows

Michael Gruner and Barbara Szybinska Matusiak *

NTNU, Faculty of Architecture and Design, Wellington 6011, New Zealand; michael.gruner@ntnu.no
* Correspondence: Barbara.matusiak@ntnu.no; Tel.: +47-41809981

Received: 11 July 2018; Accepted: 13 August 2018; Published: 16 August 2018

Abstract: One of the measures to reduce energy consumption in buildings in Nordic countries is limiting the window area, as windows contribute to significantly higher heat loss than walls during a long hot season. This conflicts with user needs for daylight and views out, especially in buildings situated in dense urban areas. The purpose of the project was to test if dynamic insulation can be used to reduce heat loss through windows during periods when view out is not needed. The paper presents a new dynamic insulation system for windows in a form of an exterior sliding shutter. The development of the system started from an existing poorly insulating sliding door system that has been equipped with vacuum insulation panels and re-designed according to the new purpose. The new system was both numerically simulated using THERM and tested in full-scale in a Hot-box apparatus at the laboratory of SINTEF Building and Infrastructure. The results are promising and encourage further development towards a commercial product.

Keywords: dynamic insulation; window; shutter; thermal performance; cold climate

1. Introduction

A highly insulated envelope is a prerequisite in cold Nordic climate to reduce energy loss and thereby minimizing heating demand in order to improve the energy performance of buildings. While the insulation thickness of opaque building elements can be easily increased to meet new requirements, thermal performance of windows is relatively limited and windows account for a large portion of heat loss of the building [1].

In order to improve the thermal performance, improvements to the window system comprising glazing, frame and possibly shading, are proposed. Jelle et al. [2] provide an overview of potential strategies for the further development of fenestration. However, proposed measures to increase thermal insulation for insulating glazing units (IGU) and frames such as low emissivity coatings, better cavity fillings, improved spacers and frame constructions are mostly standard in state-of-the-art windows in moderate and cold climates. Other mentioned improvements such as aerogel, PCM (phase change materials), transparent insulation materials, or vacuum glazing may improve one aspect of fenestration, for example insulation but interfere with other basic window functions such as provision of unobstructed view. As Grynning [3] points out, improvements in thermal performance should not impair other beneficial properties. In fact, many advanced solutions are known and discussed since the 1990s [4] but have not succeeded with market penetration. Consequently, best practice of window systems in a Norwegian context has levelled out at triple-glazing with one or two low-emissivity coatings and warm-edge spacers in sufficiently thick that is insulating window frames to avoid excessive material use and weight, avert difficult handling and mounting and ensure robustness and cost-efficiency [3]. The additional layers applied on glass, as electrochromic, photochromic or even low-emissivity coatings have also impact on spectral transmittance of glazing. Perception of colors in interiors, as well as outdoor colors observed through the glazing with such coatings, changes significantly, something that has a negative consequence for the perceived quality of architectural spaces. It may also compromise the color composition applied in interiors, as some colors appear more or less intense or bright than others [5,6].

Another possibility is considering the window system as responsive building element (RBE) as defined by IEA Annex 44. According to van der Aa [7], RBEs *"maintain an appropriate balance between optimum interior conditions and environmental performance by reacting in a controlled and holistic manner to changes in external or internal conditions and to occupant intervention"*. Characteristics of RBEs include dynamic behavior, adaptability, the capability to perform different functions and intelligent control. With respect to the thermal performance of window, this can be done by covering the window with a temporary opaque or semi-transparent, insulating device, often called dynamic, movable or night-time insulation. Advanced marked-available systems provide a range of functionalities. Besides the possibility of dynamic insulation, they can provide solar protection to reduce cooling loads or improve visual comfort, integrate elements for natural ventilation, or even integrate heat storage (PCM) and decentralized mechanical ventilation.

Measures to reduce heat loss through openings/apertures are addressed in architecture through all times by mounting additional elements permanently or temporarily on the outside or inside of the opening. Systematic and scientific efforts have been made since the time of increasing awareness of energy issues in 1970s and 1980s. Langdon [8] and Shurcliff [9] present systematizations of various movable insulation elements. Airtightness to avoid convection between window insulating unit, condensation in the in-between space and the need for good control strategies were found to be important parameters in order to achieve an adequate effect. Exemplarily, Ruyssevelt and Littler [10] developed and thoroughly tested a system of insulating roller blind. In addition to confirming the above-mentioned issues, they encountered more practical problems such as choice of mechanisms and materials, overoptimistic tolerances, robustness of parts and mounting of the system, which resulted in poor performance or failure. More recently, and in the Nordic context, Bülow-Hübe and Lundgren [11] performed a theoretical study with energy simulation software on the performance of insulating shutters. They suggest, that in order to achieve significant improvement of present-day highly insulated windows, thermal resistances of the shutters must be bigger than 1 m^2K/W. Considering limitations regarding the thickness, this may only be achieved with super-insulating materials such as vacuum insulation panels (VIP). The results highlighted the importance of airtightness. Already small gaps around the shutter system minimized the insulating effect of the window-shutter-system and resulted in little reduction of heating and net energy demand for highly insulated windows. However, a less airtight shutter still improved the thermal performance of poorly insulated windows. Internal shutters did not improve insulation significantly. Generally, most effect of the investigated shutter was connected to its ability to provide solar shading and minimize cooling loads. The study was limited to the IGU and did not investigate the influence of the frame nor was a solution proposed for it. Another question posed is the appropriate control of such shutter system and how this can be provided in practice.

This work investigates the development of a simple dynamic insulation system potentially suited for state-of-the-art highly insulating windows in Nordic climate. Firstly, the system is designed based on requirements found in earlier studies. Secondly, numerical simulations are performed to assess the thermal performance of the developed system and to choose the alternative with most potential for further testing. Thirdly, the system is tested in real scale and the performance compared to the numerical results. Lastly, suggestions for further development are given.

The study is limited to thermal performance under steady-state conditions in a laboratory setup without active deployment of the system. No control strategies, interaction with other processes, or additional features were tested.

2. System Development

Based on the previous studies, requirements for a dynamic insulation system can be summarized:

1. High thermal resistance
2. Low thermal bridge values regarding insertion of panel in the frame
3. Low thermal bridge values regarding fitting the system in the wall construction
4. Airtightness
5. Little total thickness/integration in facade construction
6. Exposure to wind, rain, drainage, sun
7. Handling wear and tear due to frequent moving
8. Robustness of construction/installation
9. Simplicity of construction/costs
10. Additional functionalities

These issues must be addressed during the design by the system components: panel, frame, joints, actuators and control. However, since verifying thermal performance was priority in the experiments, actuators for automatic deployment and development of control strategies based on sensor inputs were not included at this stage.

2.1. Components

Figure 1 shows the principal details of the developed system as built.

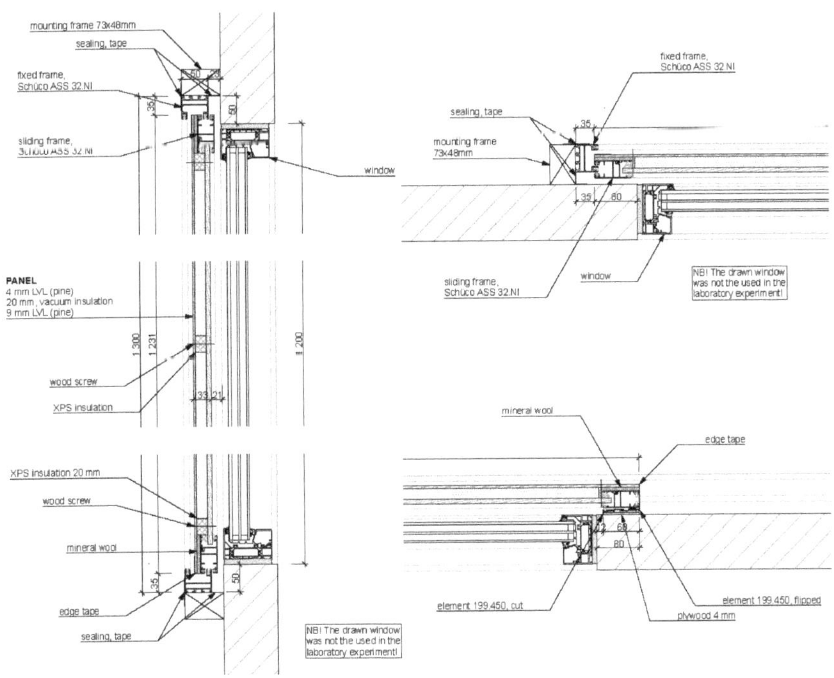

Figure 1. Vertical and horizontal sections through the system as built for the laboratory test.

2.1.1. Frame

Several moving systems were considered. Hinged shutters, roller shutters, folding shutters and so forth, were not deemed good enough, particularly regarding airtightness. On the other hand, sliding door/window systems were considered applicable in terms of drainage, airtightness, water intrusion and wear and tear due to frequent use. Timber and aluminum sliding systems were evaluated. Eventually, an aluminum system was chosen due to smaller frame sizes.

A purpose-built frame profile for the application as dynamic insulation could not be developed within the framework of the project. Therefore, the experiment was drawn on proven technology—a marked-available, simple and slender existing sliding system developed for glazed balconies and interior sliding doors. The chosen system is however poorly insulating with U-values of the frame around 7 W/m^2K.

2.1.2. Panel

Vacuum insulation panels (VIP) were chosen for the numerical and laboratory studies due to superior insulating properties, while maintaining thin dimensions. To increase robustness, the VIP is mounted in a sandwich element between two layers of plywood in a rather low-tech manner. VIP with dimensioning thermal conductivity of 0.007 W/mK and typical insulation with thermal conductivity of 0.035 W/mK—for example mineral wool—both with thicknesses of 20 mm and 40 mm have been considered and evaluated in numerical studies. Figure 2 shows the overall design of the panel.

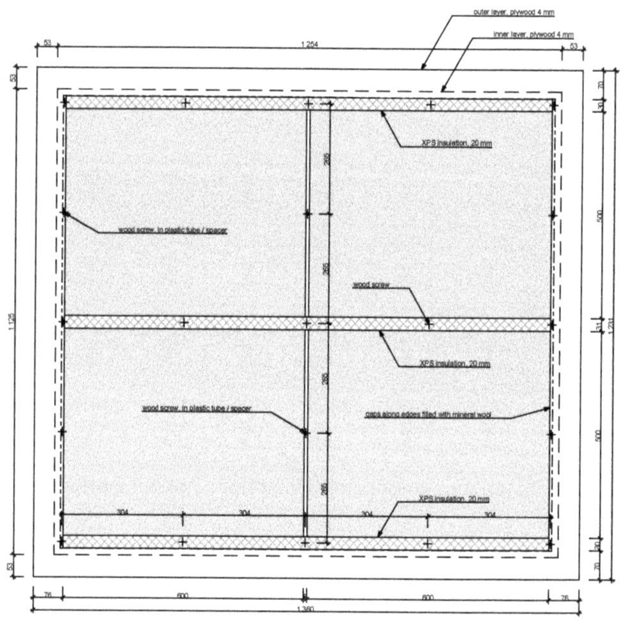

Figure 2. Construction of the shutter panel (VIP marked with grey).

Only standard panel sizes were available that did not match the shutter size. Consequently, gaps were filled with strips of XPS and loose mineral wool, which may slightly impact the insulating capability of the panel.

2.1.3. Joints

The sandwich panel is fixed with glazing beads in the frame like an IGU (see Figure 1). There has been special focus on detailing of the integration of panel and frame to minimize the heat loss linked to the spacers and the poorly insulated frame. Therefore, the outer layer of the sandwich panel is extended over the aluminum frame and the resulting distance between frame and plywood layer is filled with thermal insulation.

Another important detail with respect to thermal bridges and airtightness (and potentially drainage) is the mounting of the system on the facade. The system was only intended as a temporary installation during the experiments. Therefore, the system was only designed to be attached on the outside of the facade in a frame of wooden battens.

It was expected that the mid jamb is crucial to achieve sufficient airtightness. Modified parts of the existing sliding system were utilized. These consist of two elements which interlock to create an airtight joint between closed door leaves. One of these elements was mounted on the outer surface of the wall. In order to achieve good performance, the chosen installation requires a high level of precision.

2.2. Implementation

Originally, the system was designed for field testing, where the system was configured to fit a window of 1170 × 1170 mm (outer frame dimension). This size can be considered common for windows in Nordic conditions. The size of the sliding system was chosen so that no part of the system intrudes a 45° angle from the edge of the glass to disturb access to daylight. The resulting dimension of the outer frame of the sliding system was 2650 × 1300 mm and of the sliding panel 1360 × 1231 mm.

However, the field study was not conducted due to administrative changes in the project and the system was tested instead in the Hot-Box test facility at SINTEF Building and Infrastructure in Trondheim.

3. Numerical Studies

3.1. Method

Numerical studies were performed to assess design aspects and the theoretical thermal performance of the developed system. According to NS-EN ISO 10077-1 [12], the U value of windows with closed shutters is calculated as:

$$U_{WS} = \frac{1}{1/U_W + \Delta R}$$

with:

U_{WS} Thermal transmittance of a window with closed shutter
U_W Thermal transmittance of the window without shutter
ΔR Additional thermal resistance due to the air layer between window and shutter and the shutter itself.

ΔR depends on the thermal properties of the shutter and its air permeability. NS-EN 13125 [13] provides the five classes of air permeability of which classes 3 ("average"), 4 ("low") and 5 ("air tight") can be considered relevant for the system with respect to the expected gap sizes of the system.

The thermal resistance of the shutter R_{sh} can be obtained by measurement, or by calculation according to NS-EN ISO 10077-1 and -2 [14], based on the equation for the thermal transmittance of the shutter U_{sh}:

$$U_{sh} = \frac{\sum A_p \cdot U_p + \sum A_f \cdot U_f + \sum l_p \cdot \Psi_p}{\sum A_p + \sum A_f}$$

with:

A_p Area of the panel
U_p Thermal transmittance of the panel
A_f Area of the frame
U_f Thermal transmittance of the frame
l_p Length of the panel edge
Ψ_p Linear thermal transmittance of the panel edge

The calculations were performed with THERM [15]. Appendix A shows the used materials properties and boundary conditions.

In order to determine U_f and Ψ_p, the calculation procedure includes simulations with the panel with actual material properties and with a calibration panel, that is, an insulation panel with the same dimensions as the actual panel but a thermal conductivity of 0.035 W/mK. These calculations presume an IGU or panel fixed with a glazing bead within the size of the shutter frame. However, in the present case the sandwich panel extends over the frame. Therefore, two alternatives for modelling the calibration panel were investigated. In method A, the sandwich panel is reduced to the width held by the glazing bead similar to an IGU. In method B, the calibration panel replaces the entire sandwich panel. Figure 3 shows the sill of the shutter as example for the two alternative methods in THERM.

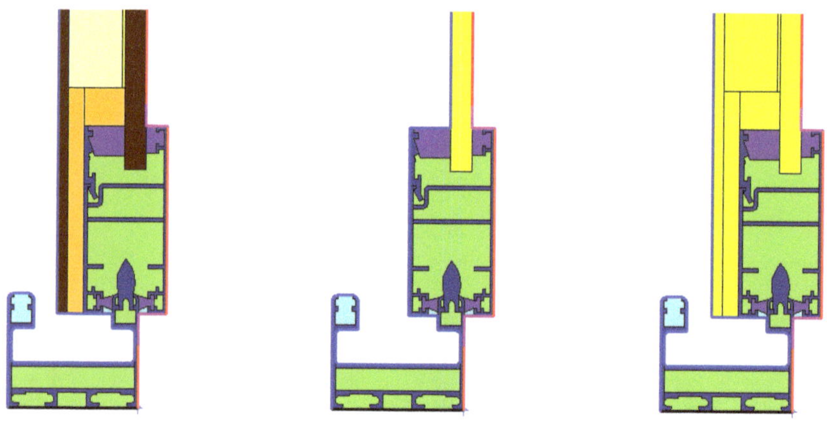

Figure 3. THERM model with actual panel materials (**left**), with calibration panel according to method A (**centre**), and with calibration panel according to method B (**right**).

Moreover, the calculation does not consider a difference between window size and shutter size. Figure 4 illustrates the four possible cases: (a) the shutter does not exceed the window, (b) the shutter, both frame and panel, exceeds the window, (c) the shutter frame exceeds the window, (d) the shutter frame exceeds the window partly.

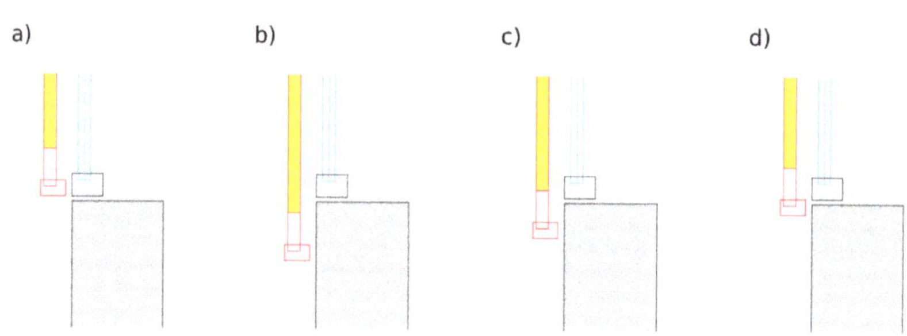

Figure 4. Possible positions of the shutter relative to the window.

For cases (b), (c) and (d) it can be assumed that the influence of the shutter frame represented by A_f will be overestimated in the calculations since the window is more covered by the panel with better insulation properties. Using NS-EN ISO 10077-2 is therefore a conservative approach for the present case, where both jambs correspond to case (b) and head and sill to case (c).

3.2. Results

Table 1 shows a comparison between the results of the two methods for the investigated system. U_{sh} and R_{sh} are calculated for a size of 1360 × 1300 mm and a shutter panel with 20 mm VIP. "Jamb,r" is the mid jamb of the sliding panel with an effective projected width of zero and without U_f value.

Table 1. Calculated thermal transmittances of the shutter components and the resulting thermal resistance R_{sh}.

Element	Value	Method A	Method B
Panel	U_p (W/m²K)	0.318	0.318
Head	b_f (mm)	103.5	103.5
	U_f (W/m²k)	6.847	5.756
	Ψ_p (W/mk)	−0.083	0.029
Sill	b_f (mm)	103.5	103.5
	U_f (W/m²k)	6.891	5.891
	Ψ_p (W/mk)	−0.072	0.032
Jamb,l	b_f (mm)	103.5	103.5
	U_f (W/m²k)	6.847	5.757
	Ψ_p (W/mk)	−0.087	0.025
Jamb,r	b_f (mm)	–	–
	U_f (W/m²k)	–	–
	Ψ_p (W/mk)	0.049	0.026
Total	U_{sh} (W/m²k)	1.66	1.63
	R_{sh} (W/m²k)	0.43	0.45

Method B gives slightly better results than method A, results are however comparable. Adding insulation outside the frame, although only 6 mm thick, is evident in both methods, either with negative Ψp-values in method A, or with lower U_f-values in method B and improves the thermal resistance of the poorly insulated frame profiles notably. Generally, the thermal performance of the total system is significantly lower than the super-insulated panel alone due to the negative effect of the poorly insulated frame.

Figure 5 shows the resulting U_{WS}-values for systems with 20 mm VIP panels with R_{sh} of 0.45 m²K/W. In addition, this base case, results are presented for 40 mm VIP with R_{sh} of 0.52 m²K/W, 20 mm typical insulation with R_{sh} of 0.28 m²K/W and 40 mm typical insulation with R_{sh} of 0.37 m²K/W. As the air permeability class for the system is unknown, results are shown for air permeability classes 3 ("average"), 4 (low") and 5 ("tight") to show the expected range.

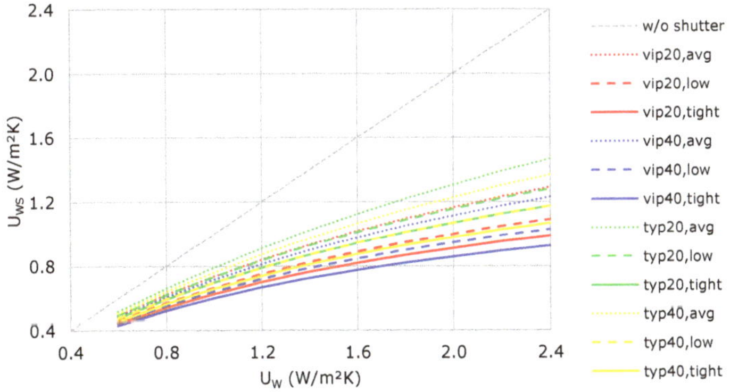

Figure 5. Resulting U_{WS} values of the window with closed shutter for a common range of U-values of windows U_W. "vip" indicates the use of vaccum insulation, "typ" indicates typical insulation.

The expected improvements of the U-value are considerable for the base case with 20 mm VIP, also for highly insulated windows. For common U-values of 1.60, 1.20 and 0.80 W/m^{2}K, the improvement is between 36 to 49%, 30 to 42%, respectively 22 to 32% depending on the air permeability of the system. Generally, the differences between the panel thicknesses and insulation materials are small compared to the influence of the air permeability of the shutter. Differences are only significant for windows with high U_W-values and little airtight shutter systems. A mineral wool insulated shutter with the same panel thickness in class "low" can even have a better thermal performance than a vacuum insulated shutter in air permeability class "average". Therefore, it was decided to continue with the 20 mm VIP in the laboratory tests to achieve best performance.

4. Laboratory Tests

4.1. Set-up

To determine the actual thermal performance, a prototype was built and tested in real scale at the guarded hot box facility of SINTEF Building and Infrastructure according to procedures described in ISO 8990:1994 and ISO 12567-1:2010. The different configurations were measured in two series to reduce systematic uncertainties. Two cases were measured three times due to unexpected results in the first series.

A triple-glazed aluminum frame window was chosen for the test. However, the measurements revealed, that the U-value of the window was not better than 1.98 W/m^{2}K. A second test with a state-of-the-art window with a U-value around 0.8 W/m^{2}K was considered. Unfortunately, such window was not available in the laboratory in the required size. The window was mounted in a template wall constructed as a sandwich element consisting of a 100 mm thick Extruded Polystyrene (XPS) core, clad with 12 mm plywood on the faces exposed to the hot and cold sides.

The shutter system was mounted in a frame of battens 73 × 48 mm with taped joints as close to the template wall as possible to avoid convection in the cavity between window and shutter. The system was mounted so that the shutter is placed centered over the window. Figure 6 shows the mounted system in the template wall of the Hot box.

Figure 6. The shutter system in the hot box. The photo shows test case "3 tapes".

Four test cases were established to investigate the influence of airtightness on the thermal performance of the system (Figure 7):

- "no tape": the joints between the shutter and sliding system respective walls are not sealed
- "3 sides taped": the joints between shutter and sliding system are sealed
- "1 side taped": only the joint to the wall is sealed
- "all sides taped": all 4 sides of the shutter are sealed

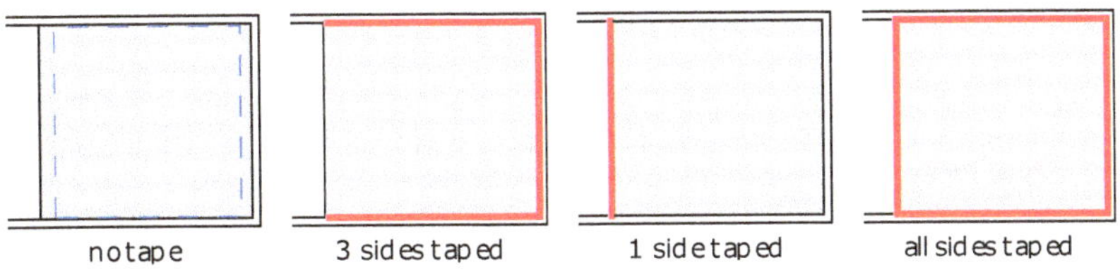

Figure 7. The 4 test cases in the laboratory experiment (tape marked with red). Case "no tape" also shows the centered position of the shutter in relation to the window behind (light blue).

4.2. Results

Table 2 presents the results of the laboratory experiments in the hot box apparatus.

Table 2. Measures U_{WS}-values for the different test series.

Case	Series 1	Series 2a	Series 2b	Average
Window only	1.979 ± 0.005			1.98
No tape	0.840 ± 0.003	0.862 ± 0.004	0.872 ± 0.004	0.86
3 sides taped	0.868 ± 0.004	0.850 ± 0.004	0.850 ± 0.005	0.86
1 side taped	0.791 ± 0.004	0.795 ± 0.003		0.79
All sides taped	0.777 ± 0.004	0.779 ± 0.003		0.78

The thermal performance of the window alone is rather poor with a U-value of 1.98 W/m^{2}K but is significantly improved, when the shutter system is mounted. U-values for cases "no tape" and "3 sides taped" are both 0.86 W/m^{2}K. Case "1 side taped" reaches a U-value close to case "all sides taped" with 0.78 W/m^{2}K. The resulting U-values of the cases with shutter are ca. 60% lower than the U-value of the window.

Figure 8 shows a comparison of the experimental results with the numerical study.

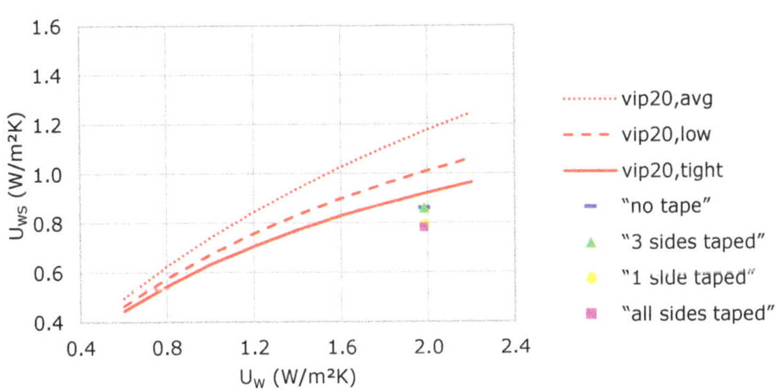

Figure 8. Calculated and measured U_{WS}-values.

Results of the investigated laboratory measurements exceed the results from the numerical studies. Considering the U_W-value of the window of 1.98 W/m^{2}K, a U-values with shutter U_{WS} between 0.91 and 1.17 W/m^{2}K was expected depending on the air permeability class. However, measured U_{WS}-values are 6 to 15% lower than expected for a very air tight system.

Noticeably, "1 side tape" and "all sides taped" give better results than the other two cases, indicating that they are more airtight. Comparing case "1 side taped" to cases "3 sides taped" and "all sides taped," the critical detail appears to be the mid jamb, whereas additional sealing of the other 3 sides shows little improvement. Assuming that cases "no tape" and "3 sides taped" correspond to air permeability class "low" and cases "1 side taped" and "all sides taped" to air permeability class "tight," then R_{sh} of the shutter can be estimated to 0.64 m^{2}K/W. These values cannot be reached in numerical calculations even when using new vacuum insulation with a thermal conductivity of 0.004 W/mK (resulting R_{sh} 0.48 m^{2}K/W), for instance. It is assumed that the difference between calculated and measured U-values is mainly a result of the sliding system extending over the window which is beyond the scope of calculations in NS-EN 10077-1 and -2.

5. Discussion

Numerical and experimental investigations indicate significant improvement of the thermal performance for all types of windows. For state-of-the-art double and triple-glazed windows, U-values are reduced by approximately 40% and 25% respectively. An even larger potential can be foreseen for retrofitting of older windows. For the tested poorly insulated window, an improvement of approximately 60% was reached.

The results verify that the type of frame, detailing of the joints and airtightness of the system are the most critical parameters for thermal performance. The right choice of the frame, that is moving mechanism and frame profiles, is crucial in order to achieve sufficient overall thermal resistance of the shutter. Using a sliding system has proven adequate, as such moving mechanism achieves good airtightness by design. The frame was supposed to be as slim and simple as possible but was therefore poorly insulating and diminished the thermal performance of the entire system. Using a better insulating, thermally broken frame could result in a thicker and more advanced system than desired. However, extending the insulating sandwich panel over the sliding frame profiles improved the thermal resistance of the sliding frame. Despite concerns of an additional thermal bridge and possible air leakage along the edges, no negative effect could be assigned to installing the sliding system in a simple frame of battens.

The laboratory tests with several cases of sealing configurations showed that thermal performance is noticeably improved by better airtightness. However, the experiments also showed that it does require sealing only selected, crucial places, in this case the mid jamb. Results of the laboratory experiments indicate that the reworked interlocking mechanism of the original sliding door system worked sufficiently well to achieve a good level of thermal performance.

On the other hand, little difference between insulation types and thicknesses of the shutter panel was found in the numerical studies. VIP was used in the laboratory experiments but is costly and usually only available in standard panel sizes which can be problematic considering the variety of window sizes. Numerical studies showed that typical insulation materials could achieve good thermal performance as long the system is airtight.

Discrepancies between calculated and measured results show that the calculation procedures in the standards may not cover all aspects of a dynamic insulation system. Further studies should therefore focus on real scale laboratory and field testing.

6. Future Work

From the results of the present studies, the scope for future system design can be outlined:

- The profiles of components should be customized to fit the purpose of a dynamic insulation system. This would also include considerations regarding structural integrity, material strength, drainage and so forth, which is beyond the scope of this project.
- As an alternative to the horizontal sliding system, upward retracting shutters are potentially more airtight as the weight of panel would aid to seal the shutter at the bottom and could work as gradually adjustable shading device.
- Mechanical actuators will need to be integrated in the frame, particularly to allow operation of the system when the room is not occupied.
- A control system including control strategies/algorithms for deployment has to be developed. The challenge is to find the balance between improved thermal performance and reduced solar gains and daylighting. Potentially, sensors should be integrated in the system.
- The environmental impact of system should be considered in a holistic way. A life cycle assessment (LCA) should be conducted to assess if the additional material use for the dynamic insulation system is balanced by potential energy savings.

7. Conclusions

A novel dynamic insulation system has been developed based on an existing poorly insulating sliding door system. The results from laboratory testing are promising and showed improvement of the thermal performance of a poorly insulated window to passive house level. However, more investigation is necessary to test the feasibility for new windows. As the calculation procedures in the relevant standards cannot cover all aspects of such a dynamic insulation system, more laboratory experiments should be conducted with state-of-the-art windows. This would indicate the potential field of application of the dynamic insulation system.

The developed system is still in an experimental state and not market ready. Further development is necessary, both regarding the system design and testing in real scale.

Author Contributions: Conceptualization, B.S.M.; Methodology, B.S.M. and M.G; Software, M.G.; Validation, B.S.M., M.G.; Formal Analysis, M.G.; Investigation, M.G. and B.S.M.; Resources, B.S.M.; Data Curation, M.G.; Writing-Original Draft Preparation, M.G.; Writing-Review & Editing, M.G. and B.S.M.; Visualization, M.G.; Supervision, B.S.M.; Project Administration, B.S.M.; Funding Acquisition, B.S.M.

Acknowledgments: The dynamic insulation system was developed in the frame of the HOME research project, sponsored by the Norwegian Research Council. The authors express their gratitude to colleagues from SINTEF Building and Infrastructure, especially to Egil Rognvik, who performed the testing of the device in the Hot-box, to Schüco Norge for providing the sliding door system and especially Yau Luong for technical assistance and Christian Olavesen at Vacunor for providing the VIP.

Conflicts of Interest: The authors declare no conflict of interest.

Appendix A

Table A1. Used materials and properties in THERM-simulations.

Material	Conductivity (W/mK)	Emissivity (−)
Aluminium alloy, painted	160.00	0.9
EPDM	0.25	0.9
PVC, rigid	0.17	0.9
Mineral wool, loose	0.035	0.9
Mineral wool, stuffed	0.04	0.9
Vacuum insulation	0.007	0.9
Timber	0.13	0.9
Plywood	0.13	0.9
Frame cavity, simplified	...	(0.9)
Frame cavity, slightly ventilated	...	(0.9)

Table A2. Used boundary conditions (according to NS-EN 10077-1).

Boundary Condition	Temperature	Surface Transfer Coefficient
outside	0 °C	0.04 m^2K/W ∣ 23 W/m^2K
inside	20 °C	0.13 m^2K/W ∣ 8 W/m^2K
inside, reduced	20 °C	0.20 m^2K/W ∣ 5 W/m^2K

References

1. Grynning, S.; Gustavsen, A.; Time, B.; Bjørn Petter, J. Windows in the buildings of tomorrow: Energy losers or energy gainers? *Energy Build.* **2013**, *61*, 185–192. [CrossRef]
2. Jelle, B.P.; Hynd, A.; Gustavsen, A.; Arasteh, D.; Goudey, H.; Hart, R. Fenestration of Today and Tomorrow: A State-of-the-Art Review and Future Opportunities. *Sol. Energy Mater. Sol. Cells* **2012**, *96*, 1–28. [CrossRef]
3. Grynning, S. *Transparent Facades in Low Energy Office Buildings—Numerical Simulations and Experimental Studies*; NTNU: Trondheim, Norway, 2015.
4. Schittich, C.; Staib, G.; Balkow, D.; Schuler, M.; Sobek, W. *Glasbau Atlas*; Birkhäuser: Berlin, Germany, 1998.
5. Matusiak, B.; Fridell Anter, K.; Angelo, K. Colour Shifts behind Modern Glazing. Available online: https://www.konstfack.se/SYN-TES/ (accessed on 4 December 2012).
6. Arbab, S.; Matusiak, B.S.; Martinsen, F.A.; Hauback, B.C. The impact of advanced glazing on colour perception. *J. Int. Colour Assoc.* **2017**, *17*, 50–68.
7. Van der Aa, A.; Heiselberg, P.; Perino, M. *Designing with Responsive Building Elements*; Aalborg University: Aalborg, Denmark, 2011.
8. Langdon, W.K. *Movable Insulation: A Guide to Reducing Heating and Cooling Losses through the Windows in Your Home*; Rodale Press: Paris, France, 1980.
9. Shurcliff, W.A. *Thermal Shutters and Shades: Over 100 Schemes for Reducing Heat Loss Through Windows*; Brick House Publishing Co.: Andover, MA, USA, 1980.
10. Ruyssevelt, P.A.; Littler, J.G.F. *Movable Roller Insulation*; OPOCE: Luxembourg, 1985.
11. Bülow-Hübe, H.; Lundgren, M. Fönsterluckor/window shutters. In Proceedings of the 3rd Nordic Passive House Conference on Sustainable Building, Aalborg, Denmark; 7–8 October 2010.
12. Standard Norge. *NS-EN ISO 10077-1:2007 Thermal performance of Windows, Doors and Shutters—Calculation of Thermal Transmittance—Part 1: General*; Standard Norge: Lysaker, Norway, 2007.
13. Standard Norge. *NS-EN 13125:2001 Shutters and Blinds—Additional Thermal Resistance—Allocation of a Class of Air Permeability to a Product*; Standard Norge: Lysaker, Norway, 2001.
14. Standard Norge. *NS-EN ISO 10077-2 Thermal Performance of Windows, Doors and Shutters—Calculation of Thermal Transmittance—Part 2: Numerical Method for Frames*; Standard Norge: Lysaker, Norway, 2012.
15. LBNL. *THERM*; Lawrence Berkeley National Laboratory (LBNL): Berkeley, CA, USA, 2014.

© 2018 by the authors. Licensee MDPI, Basel, Switzerland. This article is an open access article distributed under the terms and conditions of the Creative Commons Attribution (CC BY) license (http://creativecommons.org/licenses/by/4.0/).

sustainability

MDPI

Article

Study on Policy Marking of Passive Level Insulation Standards for Non-Residential Buildings in South Korea

Yeweon Kim * and Ki-Hyung Yu

Building and Urban Research Institute, Korea Institute of Civil Engineering and Building Technology, Ilsan 10223, Korea; raytrace@kict.re.kr

* Correspondence: yeweon.kim@kict.re.kr; Tel.: +82-31-9100-782

Received: 5 July 2018; Accepted: 18 July 2018; Published: 20 July 2018

Abstract: This study presented a methodology and process to establish a passive level for policy making of building energy in South Korea. A passive level in Korea specified in the 2017 Roadmap for non-residential buildings, which was 15 kWh/m^2·year, was defined as the heating energy requirement to strengthen the building energy saving design standards, which were typical building energy regulations in Korea. This study also presented insulation standards of roofs, floors, outer walls, and windows in Pyeongchang, Seoul, Gwangju, and Jeju, which were represented cities of four zones in Korea (Middle 1, Middle 2, Southern, and Jeju). Furthermore, the study results were extended to 66 cities around the nation to calculate the heating energy requirements and a severely cold region was added to existing three regions (Middle, Southern, and Jeju) to extend this to four regions (Middle 1, Middle 2, Southern, and Jeju). Afterwards, insulation standards for four represented regions were presented to derive a measure that minimized an energy loss through outer walls or windows in buildings. Finally, this study derived that a return of investment can be achieved in 10 years, which was determined through the comprehensive economic feasibility analysis due to strengthening insulation performances, proving the rationalization of the legal strengthening.

Keywords: passive building; building energy policy; insulation standards; building energy; building energy requirement; energy benefit

---

1. Introduction

The energy and environmental improvements have become the national goals and tasks as exhibited in the provision of a new greenhouse gas reduction system and the national establishment of a greenhouse gas reduction plan by 2030 (the Paris Agreement in December 2015) after the Kyoto Protocol, which will expire in 2020 [1]. South Korea has planned to reduce greenhouse gas emissions by 37% from business-as-usual (BAU) levels by 2030 and decided to reduce 18.1% in the building sector only [2]. The insulation standards have been strengthened continuously in a stepwise manner for the gradually changing market since the mandatory zero-energy building (by 2025) was announced in the Green Building Promotion Measure (the Sixth Green Committee Report: November 2009) [3]. Although voluntary participation has been induced through constant education about energy cost saving of buildings through advertisement of the need to reduce greenhouse gases and insulation improvements, the initial building cost burden has increased due to rising construction costs of using highly efficient building materials (insulators and high-efficiency insulation windows), which has prevented the voluntary participation by building owners. To overcome these realistic limitations, it is necessary to strengthen the insulation standards to provide superior energy performance while ensuring acceptable policies with minimized burdens on citizens. The passive buildings in Germany are defined as buildings whose annual primary energy consumption is 10 kWh/m^2 or smaller and annual heating energy consumption is 15 kWh/m^2 or smaller. The core of the standards is to strengthen insulation performance thereby minimizing the heating energy requirements and to reduce the role of active facilities such as boilers [4,5]. It will contribute to building energy saving by reducing volumes of active facilities and minimizing basic energy loads through increases in insulation performance of envelopes, which are the largest heat loss areas from the initial design phase. Yoon et al. [6] analyzed a difference in heating energy requirements through the PHPP (Passive House Planning Pakage) program in Germany assuming that demonstration buildings in South Korea are constructed in seven regions, thereby studying the effect of climate characteristics on building energy performance. Lee et al. [7] proved that the adoption of energy saving elements in the passive house standards in detached houses was useful in terms of energy saving. Ko et al. [8] aimed to identify the characteristics of the passive technology according to climates with regard to techniques used in terrain and layout shape and structures according to climates, floor plan, and envelope plan, and material composition.

The passive house concept in Germany is to satisfy thermal comfort requirements without radiation imbalance by preventing drafts or dew condensation in buildings through increases in thermal performance in outer walls and windows. Accordingly, the heating energy requirement standard is specified to 10 kWh/m^2·year [9], and passive houses are defined as buildings whose heating energy requirement in Central Europe is 15 kWh/m^2·year or smaller generally. In addition, for the coefficient of overall heat transmission (U-value) for each part specified in the Passive House, it is specified as 0.15 W/m^{2}K for the covering and as 0.8 W/m^{2}K for the window considering [10] the climate in Central Europe as a criterion. Generally, for the definition, the layout of buildings, envelope thermophysics, building geometry and infiltration and air-tightness are taken into consideration [11].

However, the standards of the passive houses in Korea and other nations are based on detached houses, whereas national mandatory passive standards are not yet presented for non-residential buildings. In addition, no studies have been conducted to establish passive insulation conditions for national policy making that satisfies the passive standards in different types of weather. Thus, this study aims to provide a methodology that presents an insulation level of buildings to satisfy the passive standard (15 kWh/m^2·year of heating energy requirements for non-residential buildings by applying Korean use profiles of non-residential buildings, thereby proposing the insulation standards and measures that take construction cost increase and energy cost saving into consideration comprehensively in case the standards are strengthened.

2. Building Energy Policy in Korea

2.1. Building Energy Roadmap in Korea

Korea has changed the national policy direction from passive response to greenhouse gas reduction in the past to a pro-active response called Low Carbon Green Growth, and the Basic Act of Low Carbon Green Growth took effect in 2010 and green growth has been started and promoted in earnest. Thus, the Ministry of Land, Infrastructure, and Transport in Korea has the goal of lower energy consumption in buildings in 2012 compared to the energy consumed in houses in 2009, and further reduction using passive building levels by 2017, as shown in Figure 1. According to the Green Building Act, and the press release of Measure or Early Promotion of the Response to Climate Change and Zero-Energy Buildings the energy level reduction of buildings will be strengthened step by step due to mandatory zero energy regulations by 2025, which sets the goal as mandatory zero energy performance [3].

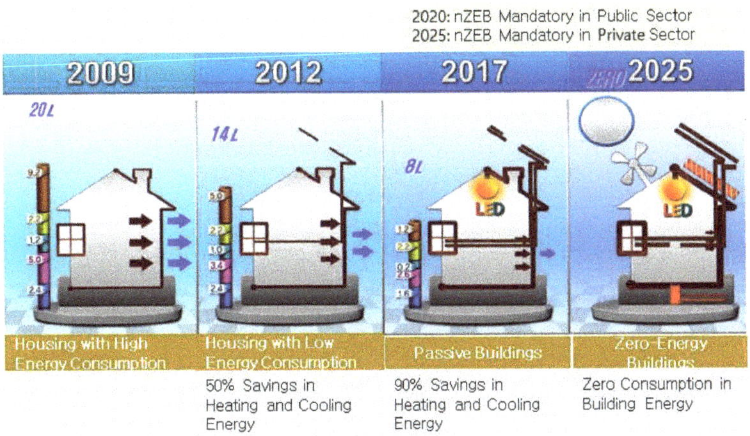

Figure 1. Building energy roadmap in Korea [3].

In addition, a short-term goal of greenhouse gas reduction was set to have a policy goal that reduced the greenhouse gas emitted in the building sector by 26.9% by 2020 compared to that of 2014, which was announced as the First Basic Plan for Green Buildings [3]. As part of the policies, the government has strengthened various energy reduction policies gradually such as building energy efficiency rating certification standards and building energy saving design standards to reduce building energy consumption.

2.2. *Design Standard of Energy Saving in Buildings*

The Building Energy Conservation Code (BECC) is a standard of mandatory energy-related minimum requirements to specify mandatory and energy performance indices of energy saving design including heat loss prevention and energy saving facility use for efficient energy management of buildings. The Korea Energy Agency has started investigation on an energy saving plan document based on the ordinance enacted by local governments in 2003. The energy saving planning document is submitted in accordance with the BECC (notified by the Ministry of Land, Infrastructure, and Transport) and the Green Building Creation Support Act at the time of new building permit application over 500 m^2 for this regulation. The main details of this standard are divided into: (i) building sector—energy saving designs including mean thermal transmittance, air-tight windows, rooftop landscaping, (ii) mechanical and electrical sector: adoption of high-efficiency certified products and energy saving control techniques, and a (iii) new and renewable sectors: cooling and heating, hot water, and electric capacity into new and renewable energy. Based on the above details, a building permit is determined if 65 or higher points out of 120 full mark in the energy performance index (EPI) are acquired (74 points for public institutions) and all mandatory items are adopted. Table 1 summarizes the mandatory items of the standards in buildings, machines and electric parts [12]. In particular, it presents thermal transmittance for building an outer wall and window in the middle, southern, and Jeju regions that are divided based on the Heating Degree Days (HDDs) in relation to passive condition-related insulation standards, and a mean thermal transmittance is presented to limit a ratio of window area, which is excessively large.

The BECC Standards of Thermal Transmittance for Building Parts by Region regulate and strengthen the envelope thermal transmittance according to a building type and region. The thermal transmittance for walls was strengthened from 0.58 W/(m^2·K) in 1987 to 0.26 W/(m^2·K) in 2016 and thermal transmittance for windows from 3.37 W/(m^2·K) to 1.5 W/(m^2·K) on the basis of the middle region.

Table 1. Mandatory energy design criteria of building energy conservation plan.

Sector	Mandatory Criteria
Architecture	Comply with U-value for walls, roofs, floors, windows, doors, etc. Comply with area-weighted average U-value for walls, windows, and doors Comply with insulation method for underfloor heating/installation of vapor barriers Install airtight windows Install solar control devices (only for public buildings)
Mechanical system	Comply with design outdoor temperature and humidity for HVAC(Heating, Ventilation and Air-Conditioning system) load calculation Install Korean Standard (KS)-certified pumps or high-efficiency pumps Comply with insulation requirements for ducts and pipes Install alternative power cooling system (only for public buildings above 1000 m^2)
Electrical system	Install high-efficiency transformers Install power-factor improvement condensers Install high-efficiency lighting/circuits for partial lighting Install certified high-efficiency LED lighting for parking lots and for emergency exit signs Install automatic illumination control systems in entrances of each residential unit Install whole house-off switch Install automatic standby power cut-off devices for more than 30% of electric outlets

2.3. Process to Determine Building Energy Policies in Korea

A number of processes are needed to determine policies to make building energy-related regulations mandatory in the Ministry of Land, Infrastructure, and Transport. In these processes, technical levels of building material suppliers, construction costs rise in case of standard strengthening, energy cost reduction is comprehensively taken into consideration, and the acceptability of the amendment is increased through an advisory conference with building companies, related experts, and material suppliers to facilitate the regulation compliance by the regulated (citizens). Thus, a revision draft is provided after reflecting discussion contents comprehensively based on expert advisory conferences and industry meetings, and regulation examinations such as cost analysis and verification and self-regulation examination (including cost–benefit analysis) are conducted followed by seeking the consultation of related institutions' administrative notices about the revised draft. Finally, opinions of stakeholders such as related institutions are fully gathered through public hearings and the act is amended.

3. Simulation Configuration

This study calculated an insulation level that satisfies a heat quantity required to maintain an indoor temperature constantly according to the regional monthly mean outdoor air temperature based on envelope conduction heat loss, ventilation conduction heat loss, internal heat generation acquisition, and solar heat gain with regard to heating energy requirements as a methodology to calculate building energy requirements of ISO 13790 [13]. Figure 2 shows the process to meet the insulation level required for passive heating requirements. This process is conducted as follows: first, reference building conditions are set according to building use, characteristics, and envelope conditions and then the insulation level that satisfies 15 kWh/m^2·year, which is a heating energy requirement of passive insulation level, is obtained by reflecting the meteorological data by region. In fact, ISO 13790 is an international standard, which is unlikely to produce errors. However, it is also true that it can produce a significant difference in energy results depending on the local use conditions. This study does not aim to evaluate performances of individual buildings in detail conceptually. This study conducts simulations at the standard set conditions in Korea by utilizing statistical data of energy efficiency rating certification and SEUMTER (SEUMTER is the National Standard Information System to apply for permission through the digitized construction administrative system for all construction administrative works in the Republic of Korea), an administration system for building-related public services, for represented regions by region. Thus, it may produce a difference in measurement results of individual buildings. However, these errors compared to actual measured data will be reduced once the local-based set data are constructed constantly in the future.

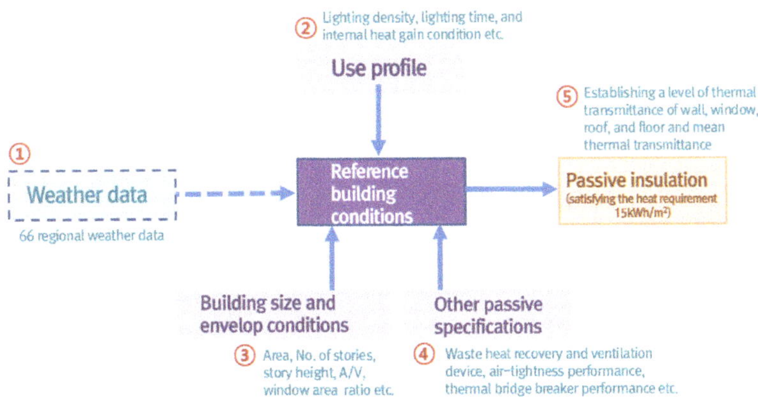

Figure 2. Process to calculate an insulation level to satisfy the passive heating requirement.

3.1. Meteorological Data Conditions

The methodological data used in this study employed 66 data sets [14] in Korea constructed with dry bulb temperature, humidity, solar radiation, and wind volume based on ISO 15927-4 to present thermal transmittance that satisfied the passive level in each of the regions.

3.2. Buiding Sizes, Envelope Conditions, and Other Passive Specifications

An envelope conduction loss and solar heat gain vary in buildings according to building characteristics (size, envelope conditions, and a ratio of envelope area with floor area, etc.). In particular, they are also affected by the parameter zone condition according to a ratio of envelope area with floor area. There is a difference in energy requirements in buildings ultimately due to a sum of heat loss and gain of internal heat generation, envelope loss, and solar radiation according to the characteristics of buildings. This is illustrated in Figure 2. Accordingly, 34,000 buildings that had a permit of the BECC from January 2014 to August 2018 were analyzed to determine the reference conditions of non-residential buildings constructed in Korea and building size and envelope conditions were defined. Note that since a simulation model configuration that satisfied all mean values of the certification data was impossible, some of the variables were referred from existing study 10 based on the certification data. For other passive specifications, passive house standards and DIN (Deutsche Industric Normen) standards in Germany were followed. Table 2 summarizes them.

Table 2. Building sizes, envelope conditions, and other passive specifications.

Building Size and Envelop Conditions		Other Passive Specifications	
No. of stories	4.27 [stories]	Waste heat recovery and ventilation device	75 [%]
Story height (mode)	3.6 [m]	Air tightness performance	0.6 [times, based on n50]
Heating area of ground floor	7060 [m^2]	Thermal bridge	0.1 [W/m·K]
Based on outside insulation Wall area ratio to (A)	0.577 [-]	Area ratio of roof	0.2365 [-]

3.3. Use Profile

Since lighting hours, heat generation by residents, and indoor heat generation can vary according to the use profile due to room configuration of buildings, internal heat generation acquisition is determined for the inside of buildings. The standard setup conditions were produced by calculating the area weight use profile according to a room configuration ratio through building investigations on 881 cases of preliminary certification of building energy efficiency rating in 2016 and 2017 to calculate the conditions of the reference building in Figure 3. Accordingly, this study reflected the use profile according to DIN 18599 [15] in Germany and their setup conditions are presented in Table 3.

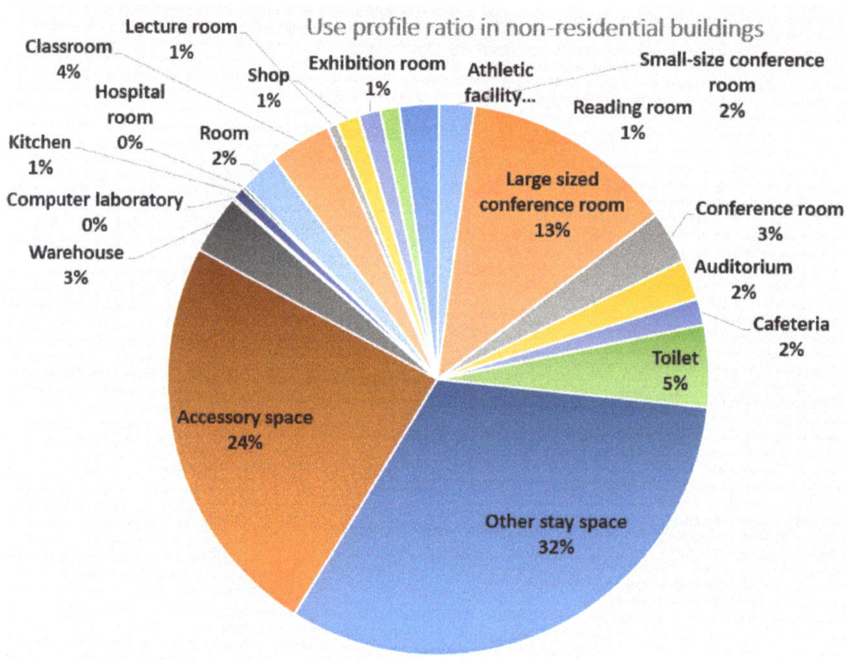

Figure 3. Room configuration ratio through building investigation on 881 cases.

Table 3. Setup conditions of non-residential buildings according to room configuration ratios.

Category	Unit	Buildings Other Than Residential Purpose	Business Facility (Efficiency Grade)	Business Facility (Total Energy Consumption)
Minimum outdoor air volume	[$m^3/(m^2 \cdot h)$]	6.15	5.93	5.66
Lighting time	[h]	10.25	10.4	10.14
Person	[$Wh/(m^2 \cdot day)$]	58.3	59.3	57.02
Work auxiliary equipment	[$Wh/(m^2 \cdot day)$]	38.4	42.4	49.18
Indoor heat emission amount	[$Wh/(m^2 \cdot day)$]	96.7	101.7	106.2

4. Study on the Insulation Standard Level by Region

4.1. Heating Energy Requirment by Case Region According to the Current Energy Saving Design Standards

The heating energy requirement in DIN 18599 is marked as a sum of envelope conduction heat loss, ventilation heat loss, internal heat generating acquisition, and solar radiation acquisition, and it indicates a heat generating value that maintains a temperature according to the set indoor temperature in the condition of monthly mean outdoor air temperature during the heating period.

This study analyzed the annual heating energy requirements using four regional meteorological data of Pyeongchang, Seoul, Gwangju, and Jeju by applying thermal transmittance of the aforementioned reference building conditions and current energy saving design standards [16] (2016) in the building parts by region. In Figure 4, each of Pyeongchang, Seoul, Gwangju and Jeju represent a city in the zone of Middle 1, Middle 2, Southern and Jeju. Pyeongchang, though in the same latitude as Seoul, shows the coldest regional characteristic as the mountainous climate due to the Taebaek mountain range. Seoul, Gwangju and Jeju show the characteristics of a warm climate as they go southward.

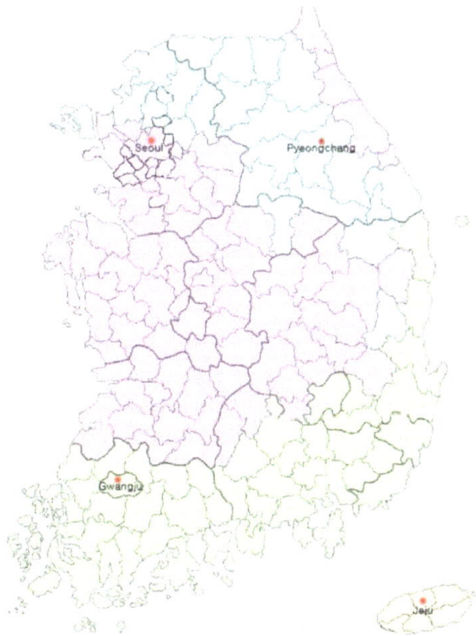

Figure 4. Location of representative cities by area.

The comparison results of regional heating energy requirements in Table 4 showed that, as an outdoor air temperature was decreased, the requirements in most regions were increased. However, despite higher outdoor air temperature in Jeju in April than that of Seoul, a small heating energy requirement was generated. This result indicated that a solar energy gain [16] in Jeju was small due to the meteorological characteristics of Jeju Island despite the high mean temperature, which affected the heating energy requirements. The comparison results of annual heating energy requirements by region showed that Pyeongchang had the highest annual heating energy requirement followed by Seoul, Jeju, and Gwangju. Despite the difference in thermal transmittance in building parts by region due to the insulation performance of the current energy saving standards, buildings in Gwangju and Jeju were close to the passive building level, whereas Pyeongchang and Seoul had relatively high annual heating energy requirements. In particular, Pyeongchang had the same middle region standards applied to Seoul although Pyeongchang was a very cold place whose outdoor air temperature was significantly different from that of Seoul region, which was why Pyeongchang had a higher heating energy requirement than that of Seoul. A very cold place such as Pyeongchang needed improvements in thermal transmittance of more than existing insulation performance for building parts to satisfy the passive level or balance between regions.

Table 4. Heating energy requirements according to monthly outdoor temperature by region.

Category			January	February	March	April	October	November	December	Heating Requirement kWh/(m^2·Year)
Pyeongchang (Middle 1)	Outdoor air temperature	°C	−9.0	−5.0	−0.3	7.7	8.8	1.9	−4.3	29.8
	Heating energy requirement	kWh/(m^2·month)	9.9	5.5	3	0.1	0.1	3.5	8.0	
Seoul (Middle 2)	Outdoor air temperature	°C	−1.5	2.3	5.1	13.1	15.2	8.3	1.9	17.5
	Heating energy requirement	kWh/(m^2·month)	7.0	3.0	1.7	0	0	1.2	4.4	
Gwangju (Southern)	Outdoor air temperature	°C	1.1	3.9	8.3	14.2	15.7	10.2	2.8	14.4
	Heating energy requirement	kWh/(m^2·month)	5.8	2.4	0.9	0	0	0.4	5.0	
Jeju	Outdoor air temperature	°C	5.5	6.6	9.7	13.8	18.3	12.5	8.5	16.7
	Heating energy requirement	kWh/(m^2·month)	6.3	3.7	1.6	0.3	0	0.8	4.1	

Building Insulation Conditions That Satisfy the Passive Condition

Among heat losses that make up the building energy heating requirements, a part of envelope conduction occurs at roofs, floors, windows, and walls of building overall. Thus, it is important to consider an insulation level in each of the parts compositely that satisfies 15 kWh/m^2·year, which is a heating energy requirement of passive standard for reference buildings. Figure 2 shows the correlation of thermal transmittance between window and envelope by region that satisfies the passive level. For Pyeongchang and Seoul, when a mean thermal transmittance 0.2 W/m^{2}K of roof and floor, which is close to the current energy saving standard level, is satisfied, the heat loss through envelopes is large due to low outdoor air temperature in the case of Pyeongchang. Thus, the minimum window thermal transmittance shall satisfy 0.9 W/m^{2}K at least and thermal transmittance of walls shall be lower than 0.1 W/m^{2}K to satisfy the passive level. For Seoul, this belongs to the middle region, although Seoul has a lower outdoor air temperature than that of Pyeongchang, when thermal transmittance of roof and floor is 0.2 W/m^{2}K and window thermal transmittance is 1.5 W/m^{2}K, which is the current middle-region standard, the wall thermal transmittance shall be lower than 0.14 W/m^{2}K to satisfy the passive standard. This result implies that additional regional segmentation is needed for places where outdoor air temperature is significantly different even if those regions belong to the same middle region. In contrast, Gwangju satisfies the passive standard already even in terms of the current regulation. For Jeju, when a mean thermal transmittance of roof and floor is 0.25 W/m^{2}K, which is close to the current standard, and a window thermal transmittance is 2.4 W/m^{2}K, the minimum of the wall thermal transmittance of the passive level should be at least 0.34 W/m^{2}K. Thus, better thermal transmittance than 0.43 W/m^{2}K, which is the current legal standard, is required.

Thus, it is important to strengthen each of the levels that can be acceptable in the market by taking insulation standards of roof, floor, window, and wall into consideration by region compared to the current standards in order to present an insulation standard for building parts in consideration of heating energy requirements with regard to the passive building level.

4.2. Study on the Passive Building Level by Region

4.2.1. Regional Heating Energy Requirement According to the Current BECC

Figure 5 shows the regional heating energy requirements that satisfy the current insulation standards based on the 66 regional standard meteorological data, which indicate the HDDs in the *y*-axis. Table 5 presents the cities and fractions that satisfy the regional passive standards according to Figure 5. The purpose of this study is to provide a measure that satisfies a certain level and minimizes energy loss through outer walls or windows of buildings by applying the insulation standard of the passive building level with regard to all buildings around the nation. As a result, Figure 5 shows that the annual heating energy requirements are not met based on the current insulation standards for regions where outdoor air temperature is low (the HDDs are large).

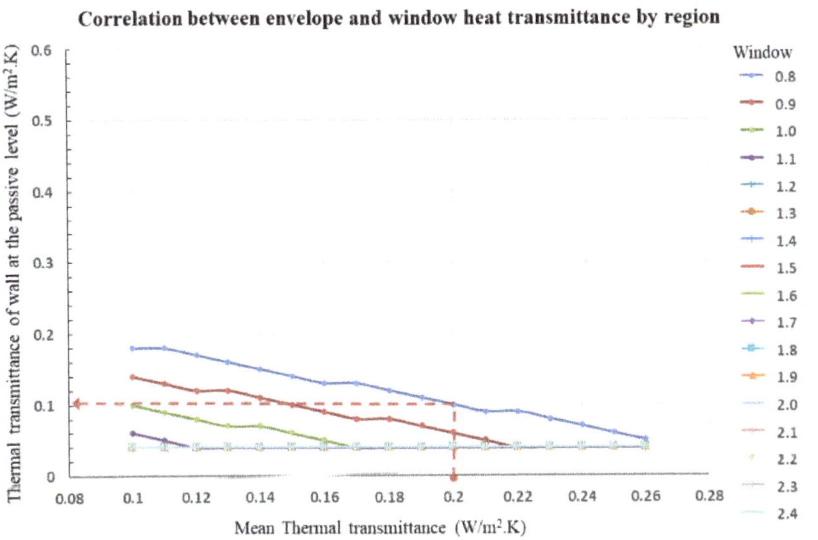

(**a**) Pyoungchang

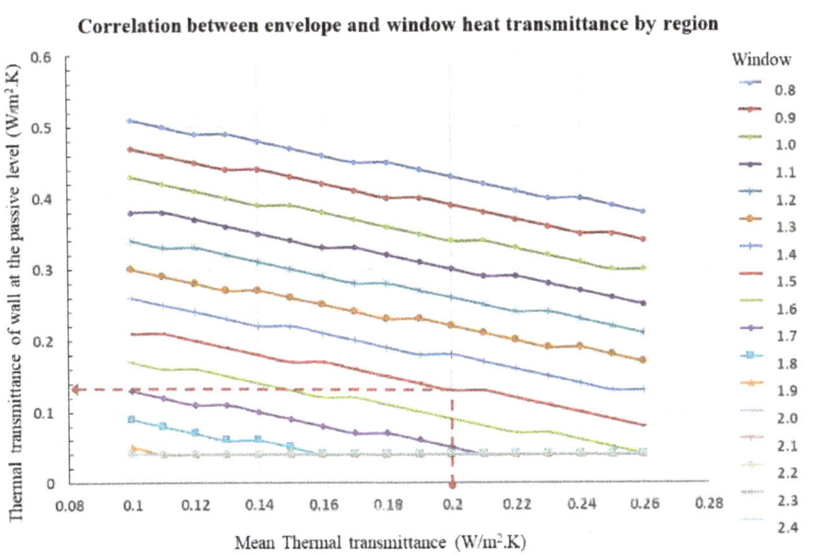

(**b**) Seoul

Figure 5. *Cont.*

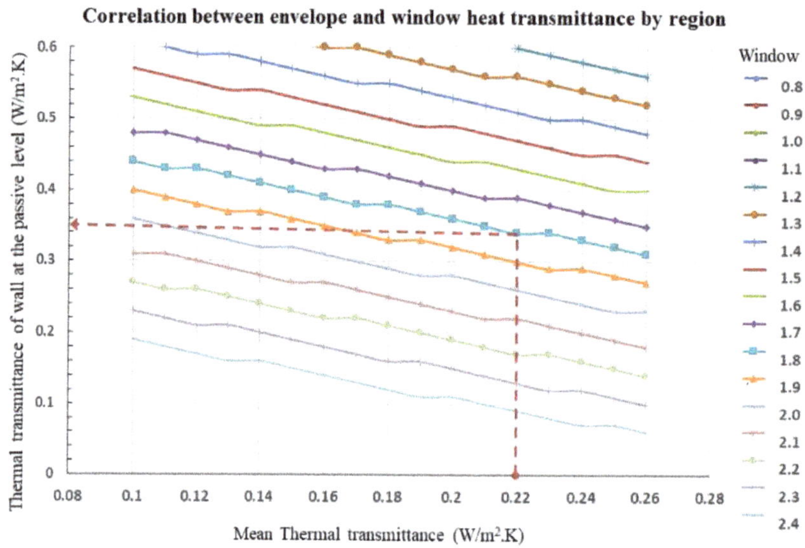

(c) Gwangju

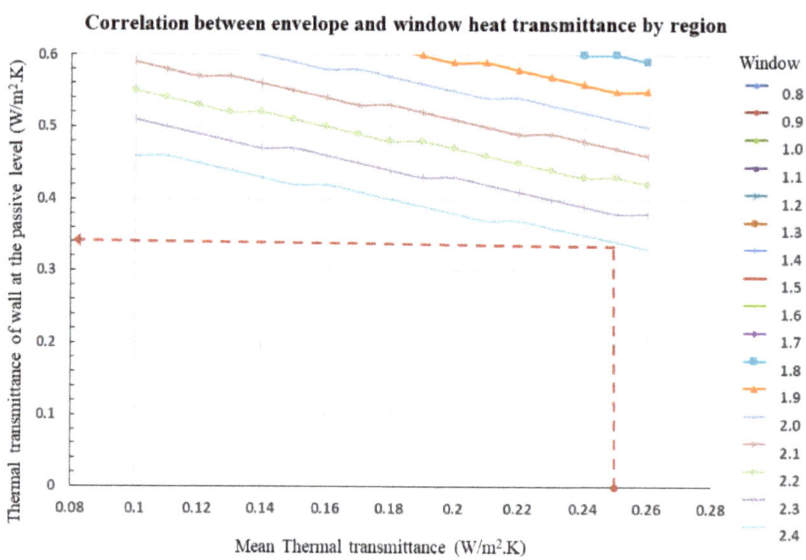

(d) Jeju

Figure 5. Correlation between envelope and window thermal transmittances by region that satisfies the passive level.

Table 5 presents that a proportion of cities that satisfy the passive level in Middle 1 region is 0%, and 10 cities satisfy the passive level out of 31 cities in Middle 2 region (32%), and 18 out of 21 cities (86%) satisfy the passive level in the Southern region. The results in Table 5 indicate that the insulation standards for Middle 1 and 2 regions where an outdoor air temperature is low should be strengthened more, while most of the southern regions except for Jeju satisfy the passive level, which implies that the current standard seems relatively reasonable.

Table 5. Fraction that satisfies the passive conditions for buildings except for apartments by region according to the current insulation standards.

No. of Cities in Region	Middle 1	Middle 2	Southern	Jeju
No. of cities that satisfies the passive conditions	12	31	21	1
Passive fraction	0	10	18	0

4.2.2. Insulation Level by Building Part That Satisfies the Passive Condition by Region in Korea

Figure 6 shows a graph of mean thermal transmittances of window, floor, and roof according to the level of wall thermal transmittance for each region. Using the above method, Table 6 presents the revised version to make Middle 1 and 2 regions satisfy the passive level around 50% based on the current insulation standards [17] and most of the southern and Jeju regions satisfy the passive level, in which a strengthened ratio for each of the thermal transmittances is summarized. Overall, an insulation standard in the Middle 1 region was strengthened by 17% to 35% and that in Middle 2 and Jeju regions by 5% to 8% while the southern region maintained the current level. The thermal transmittance for roofs facing the outdoor air was not strengthened. This was because the thermal transmittance of existing roofs has already been strengthened due to the effect of cold radiation of perforation or winds compared to that of the floor and nearly reached the passive level already. Thus, cities that satisfied the passive level according to the proposed thermal transmittance by building parts exhibited 50% for Middle 1, 47% for Middle 2, 95% for Southern, and 100% for Jeju in Table 7.

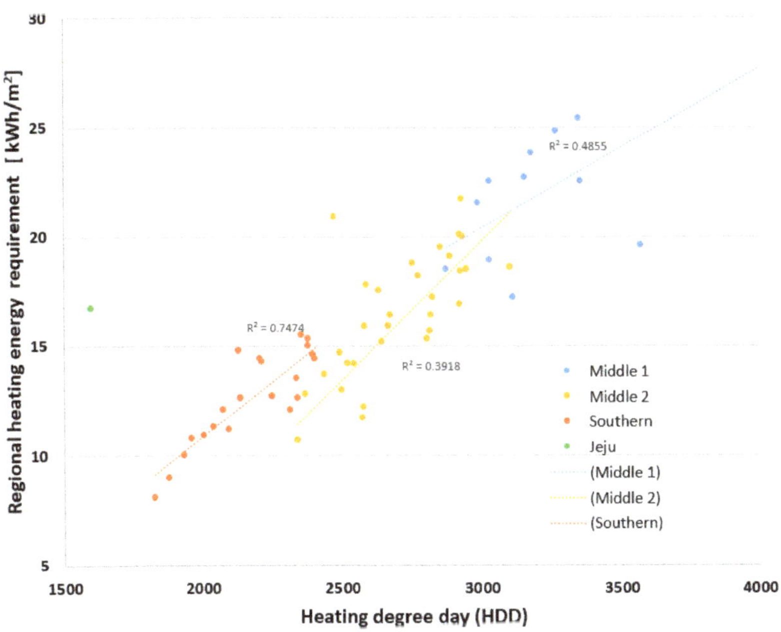

Figure 6. Regional heating energy requirement by region according to the current insulation standard.

Table 6. Strengthened level of insulation standards.

		Middle 1	Middle 2	Southern	Jeju
Existing buildings Thermal transmittance [W/m²·K]	Windows facing the outdoor	1.5		1.8	2.4
	Roof facing the outdoor	0.15		0.18	0.25
	Floor facing the outdoor	0.22		0.22	0.33
	Outer wall	0.26		0.32	0.43
Revised Thermal transmittance [W/m²·K]	Windows facing the outdoor	1.2	1.5	1.8	2.2
	Roof facing the outdoor	0.15	0.15	0.18	0.25
	Floor facing the outdoor	0.15	0.17	0.22	0.29
	Outer wall	0.17	0.24	0.32	0.41
Strengthening ratio [%]	Windows facing the outdoor	20%	0%	0%	8%
	Roof facing the outdoor	0%	0%	0%	0%
	Floor facing the outdoor	17%	6%	0%	0%
	Outer wall	35%	8%	0%	5%

Table 7. Fraction that satisfies the passive conditions for buildings except for apartments by region according to the strengthening of the insulation standards.

	Middle 1	Middle 2	Southern	Jeju
No. of cities in region	12	32	21	1
No. of cities that satisfy the passive conditions	6	15	19	1
Passive fraction	50%	47%	95%	100%

5. Analysis on the Effect of Regulation According to the Passive Level

In South Korea, when a citizen's right is regulated legally, whether the regulation is reasonable or not has to be examined through many procedures by the nation. The energy cost–benefit obtained through the direct cost added due to the standard strengthening and energy reduction rate is compared and regulation examination is recommended to have a net benefit within 10 years.

5.1. Cost–Benefit Analysis for Regulated General Citizens

Based on the aforementioned conventional and proposed insulation conditions, costs due to strengthening insulation standards on walls, roofs, floors, and windows were calculated by region. All costs are expressed in US Dollars.

5.1.1. Envelope Insulation

A ratio of the envelope area to the total area was referred to from 2016 building statistical data of building energy efficiency rating certification, and added material cost per unit area [18] was calculated by referring to the consumer price information and supplier's quotes. For the cost calculation of the insulators used due to insulation strengthening, the use of Graphite Expandable Polystyrene (G-EPS) foam (Rating A) was assumed. Table 8 presents the rising cost added due to insulator construction per 1000 m^2 of total floor area.

Table 8. Calculation of rising construction costs of added insulators per 1000 m^2 of total floor area.

Item	Area	Additional Construction Cost ($)			
		Middle 1	Middle 2 (Existing Middle)	Middle 2 (Existing Southern)	Jeju
Wall	575.5	2.3	0.4	1.2	0.2
Roof	236.5	0	0	0.6	0
Floor	231.6	0.6	0.2	0.5	-

5.1.2. Window Insulation

By referring the above efficiency rating statistical data, additional costs were calculated with regard to a window facing the outdoor air directly (if an indirectly-faced window is applied, additional construction costs are reduced. Thus, this study chose a window facing the outdoor air directly in the conservative calculation). Table 9 presents the additional construction cost calculated based on the strengthened window thermal transmittance.

Table 9. Applied values of additional construction cost according to the window insulation levels per 1000 m^2 of total floor area.

Item	Additional Construction Cost Per 1000 m^2 ($)		
	Current	Revision	
	Window Specifications	Window Specifications	Additional Construction Cost
Middle 1	28T low-e double glazing AL insulation bar window	42T low-e argon triple glazing AL window	6326
Middle 2 (Existing Southern)	24T low-e double glazing AL window	28T low-e double glazing AL insulation bar window	11,678
Jeju	22T double glazing AL window	24T low-e double glazing AL window	2433

5.1.3. Calculation of Total Cost

Table 10 presents the added construction cost with regard to rising construction cost per 1000 m^2 of total floor area based on the previously calculated results in Table 9. This can be converted to the cost that occurred nationally as follows: the government of Korea has developed a database of building permits every year by operating SEUMTER by the Ministry of Land, Infrastructure, and Transport. The additional cost for the whole regions was calculated based on the mean building permit area data for seven years (2010 to 2016), which were statistically processed by building use among the data in SEUMTER. The derived equation is as follows:

$$\text{Middle 1 region } 7{,}576{,}271\ m^2 \times \$9.3 + \text{Middle 2 region (previously Middle region) } 41{,}121{,}076\ m^2 \times \$0.6 + \text{Middle 2 region (previously Southern region) } 18{,}564{,}911\ m^2 \times \$41 + \text{Jeju region } 2{,}498{,}341\ m^2 \times \$2.7 = \$424{,}845{,}000$$

Table 10. Calculation of rising construction cost per 1000 m^2 of total floor area.

Item	Additional Construction Cost ($)				
	Middle 1	Middle 2	Middle 2	Southern	Jeju
Windows	6325.8	-	11,678.4	-	2433.0
Outer wall	2319.3	356.8	1248.8	-	178.4
Roof	-	-	574.4	-	-
Floor	646.1	215.4	502.6	-	71.8
Subtotal	9291.2	572.2	14,004.2	0	2683.2

5.2. *Reduction in Heating Cost Due to Energy Requirement Reduction*

The heating cost reduction benefit was calculated according to the energy cost saving by building owner as a result of improvements in new building energy performances. The benefit analysis was conducted based on buildings constructed in accordance with the conventional design standards. The energy reduction of the building was calculated using the calculation method of heating energy requirements in ISO 13790, and the setup conditions were referred to through the use profile in the building energy efficiency rating certification scheme. Table 11 presents the comparison results after calculating the heating energy requirements according to the strengthened insulation standards based on the buildings located in Middle 1, Middle 2, Southern, and Jeju regions. A unit cost per 1 Mcal was referred to from the Korea Energy Handbook published by the Korea Energy Agency [19].

Table 11. Comparison results after calculating the heating energy requirements according to the strengthened insulation standards based on different region.

Item	Additional Construction Cost ($)				
	Middle 1	**Middle 2**	**Middle 2**	**Southern**	**Jeju**
Windows	6325.8	-	11,678.4	-	2433.0
Outer wall	2319.3	356.8	1248.8	-	178.4
Roof	-	-	574.4	-	-
Floor	646.1	215.4	502.6	-	71.8
Subtotal	9291.2	572.2	14,004.2	0	2683.2

The results showed that the annual energy usage per unit area in non-residential buildings in Korea was 258 kWh/m^2·year and a ratio of heating energy was 35.5% approximately according to the 2014 Energy Consumption Survey [19]. Accordingly, the energy savings per unit area by zone, and the resulting savings are summarized in Table 12.

Table 12. The energy savings per unit area by zone, and the resulting savings.

	Middle 1	**Middle 2**	**Middle 2 (Existing South)**	**Southern**	**Jeju**
Energy savings per unit area-year [Kwh/m^2·year]	21.9	3.6	26	-	8.9
Energy saving amount per unit area-year [$/m^2]	1.4	0.2	1.7	-	0.6
Amount saved by region [$]	$13.207 billion US	$11.656 billion US	$38.458 billion US	-	$1.765 billion US

The average national building permit area from 2010 to 2016 was 101,558,000 m^2. Accordingly, the energy reduction benefit of buildings except for apartments can be calculated for buildings that satisfied the passive level insulation standards based on the areas of the regions presented in Section 5.1.3 considering a discount rate for 30 years of building service life as follows: $13,207,000,000 for Middle 1, $11,656,000,000 for Middle 2 (previously Middle), $38,458,000,000 for Middle 2 (previously Southern) and $1,765,000,000 for Jeju. Thus, a total of $65,086,000,000 can be reduced.

5.3. Analysis on ROI

The return on investment (ROI) for the Middle 1 region was analyzed when the passive level insulation standards are satisfied as follows: for the payback period, assuming that the moving-in starts two years after meeting the passive-level insulation criteria, the initial payback period by applying the discount rate and the generated annual cost savings per zone of residential/non-residential buildings since moving into the buildings (2020) are as shown in Table 13.

The relative ratio of average building permit area in 2010 to 2016 is as follows: Middle 1: 7.79%, Middle 2 (previously Middle): 43.36%, Middle 2 (previously Southern): 19.57%, Southern: 26.68%, and Jeju: 2.60%. The national average ROI is 6.2 years after residents move.

Table 13. The energy savings per unit area by zone, and the resulting savings.

	Middle 1	Middle 2	Middle 2 (Existing South)	Southern	Jeju
Cost Saving Effect [$]	$1.0576 billion US	$0.9334 billion US	$0.3080 billion US	-	$1.4 million US
Increase of Amount in Construction Cost [$]	$7.5778 billion US	$5.8074 billion US	$28.1859 billion US	-	$0.7282 billion US
Payback period	10.2 years	6.6 years		-	6.2 years

6. Conclusions and Policy Implications

This study calculated energy requirements that satisfied the insulation standards of the BECC for building parts by region according to regional climate conditions using the methodology of ISO 13790 and presented the insulation standards that satisfied the passive building standards for each of the regional climate conditions. The conclusions of the study results are summarized as follows:

(1) The regional monthly heating energy requirements had a close correlation mostly with the monthly average outdoor air temperature during the heating period (October to March). The heating energy requirement for each region was as follows: 29.8 kWh/m^2·year in Pyeongchang, 17.5 kWh/m^2·year in Seoul, 14.4 kWh/m^2·year in Gwangju, and 16.7 kWh/m^2·year in Jeju.

(2) The thermal transmittance of building part by region should take the outdoor air condition of the region into consideration compositely to satisfy the passive level of buildings. For Pyeongchang and Seoul, which belonged to the same middle region, when a thermal transmittance of roof/floor of buildings was 0.2 W/m^{2}K, the minimum window thermal transmittance should satisfy 0.9 W/m^{2}K and the minimum wall thermal transmittance should satisfy 0.1 W/m^{2}K in the case of Pyeongchang. On the other hand, the minimum window thermal transmittance should satisfy 1.5 W/m^{2}K and the minimum wall thermal transmittance should satisfy 0.14 W/m^{2}K in the case of Seoul. Buildings in Gwangju, which belonged to the Southern region, satisfied the passive level already. For Jeju regions, the minimum wall thermal transmittance should satisfy 0.34 W/m^{2}K to meet the passive level of buildings if the roof/floor and window thermal transmittances were close to the level specified in the current BECC.

(3) This study investigated the regional meteorological conditions based on 66 meteorological datasets to strengthen energy saving up to the passive building level by region in Korea and proposed a requirement of 50% or higher passive insulation standards for Middle 1 and 2 regions, and nearly 100% passive insulation standards for Southern and Jeju regions considering the market and technical conditions.

(4) The energy benefit obtained through the additional direct cost and energy reduction rate was analyzed due to the strengthening of the insulation standards that satisfied 0.17 W/m^{2}K for Middle 1, 0.24 W/m^{2}K for Middle 2, 0.32 W/m^{2}K for Southern region, and 0.41 W/m^{2}K for Jeju in the case of outer walls, and 1.2 W/m^{2}K for Middle 1, 1.5 W/m^{2}K for Middle 2, 1.8 W/m^{2}K for Southern region, and 2.2 W/m^{2}K for Jeju in the case of windows. The results showed that the national average ROI was approximately 6.2 years after moving to new building.

Author Contributions: Y.K. analyzed the study outcomes. K.-H.Y. conceived the concept.

Funding: Please add: This research was funded by [Infrastructure and Transport of the Korean government] grant number [17CTAP-C116666-03].

Acknowledgments: This research was supported by a grant (17CTAP-C116666-03) from the Architecture and Urban Development Research Program funded by the Ministry of Land, Infrastructure and Transport of the Korean government.

Conflicts of Interest: The authors declare no conflict of interest.

References

1. International Energy Agency. *Towards Sustainable Urban Energy Systems*; ETP (Energy Technology Perspective) Annual Report; IEA: Paris, France, 2016.
2. UNFCCC, United Nations Framework Convention on Climate Change. INDCs as Submitted by Parties. 2015. Available online: http://www4.unfccc.int/Submissions/INDC/Published%20Documents/Republic%20of%20Korea/1/INDC%20Submission%20by%20the%20Republic%20of%20Korea%20on%20June%2030.pdf (accessed on 18 July 2018).
3. International Energy Agency. *Energy Policies of IEA Countries—The Republic of Korea*; 2012 Review; International Energy Agency: Paris, France, 2012.
4. Jang, M.; Kim, J.A. Development and evaluation of laws and regulation for the Low-Carbon and Green Growth in Korea. *Int. J. Urban Sci.* **2010**, *14*, 191–206. [CrossRef]
5. Rhee, S.K.; Jang, D.C.; Chung, Y. A critical review and new policy framework of Low-Carbon, Green-Growth strategy of Korea. In *Green Growth: Managing the Transition to a Sustainable Economy—Learning by Doing in East Asia and Europe*; Vazquez-Brust, D.A., Sarkis, J., Eds.; Springer: Dordrecht, The Netherlands, 2012; Volume 1, pp. 27–42.
6. Yoon, Y.S.; Kim, E.J.; Lee, S.J.; Sim, M.H. A study on the Correlation of Domestic Local Climate and Annual Heating Energy Demand. *J. Archit. Inst. Korea* **2014**, *30*, 209–216.
7. Lee, M.J.; Kim, W.S.; Lee, W.J.; Lee, W.T. A study about Reduction Rates of Building Energy Demand for a Detached House according to Building Energy Efficient Methods. *J. Archit. Inst. Korea Plan. Des.* **2012**, *28*, 275–282.
8. Koh, G.H.; Kim, J.G. Study on Technical Analysis of Regional Climate Sustainable Building-Focused on passive techniques. *J. Archit. Inst. Korea* **2015**, *35*, 327–328.
9. Feist, W.; Schnieders, J.; Dorer, V.; Haas, A. Re-inventing air heating: Convenient and comfortable within the frame of the Passive house concept. *Energy Build.* **2015**, *37*, 1186–1203. [CrossRef]
10. Feist, W. *Lernzielkatalog "Zertifizierter PassivhausPlaner"*; Passive Haus Institut: Darmstadt, Germany, 2015.
11. EMSD. Hong Kong Energy End-Use Data. Electrical & Mechanical Service Department. Available online: http://www.emsd.gov.hk (accessed on 18 July 2018).
12. Park, D.J.; Yu, K.H.; Yoon, Y.S.; Kim, K.H.; Kim, S.S. Analysis of a building energy efficiency certification system in Korea. *Sustainability* **2015**, *7*, 16086–16107. [CrossRef]
13. *EN ISO 13790: 2008 Thermal Performance of Buildings—Calculation of Energy Use for Space Heating and Cooling*; ISO: Geneva, Switzerland, 2008.
14. Kim, Y.; Jang, H.K.; Yu, K.H. Study on Extension of Standard Meteorological Data for Cities in South Korea Using ISO 15927-4. *Atmosphere* **2017**, *8*, 220. [CrossRef]
15. Aldaloro, A.P.E.; Salomone, R.; Ioppolo, G.; Andaloro, L. Energy certification of buildings: A comparative analysis of progress towards implementation in European countries. *Energy Policy* **2010**, *28*, 5840–5866. [CrossRef]
16. Korea Energy Agency. *Regulation on Operation of Building Energy Efficiency Grade Certification System*; Korea Energy Agency: Yongin-si, Korea, 2016.
17. Korea Meteorological Administration. Available online: http://www.kma.go.kr (accessed on 18 July 2018).
18. Korea Price Information. Available online: www.kpi.or.kr (accessed on 18 July 2018).
19. Korea Energy Agency Handbook. Available online: www.kemco.or.kr (accessed on 18 July 2018).

© 2018 by the authors. Licensee MDPI, Basel, Switzerland. This article is an open access article distributed under the terms and conditions of the Creative Commons Attribution (CC BY) license (http://creativecommons.org/licenses/by/4.0/).

sustainability

MDPI

Article

Consumers' Willingness to Pay for Net-Zero Energy Apartment in South Korea

Ju-Hee Kim, Hyo-Jin Kim and Seung-Hoon Yoo *

Department of Energy Policy, Graduate School of Energy & Environment, Seoul National University of Science & Technology, 232 Gongreung-Ro, Nowon-Gu, Seoul 01811, Korea; jhkim0508@seoultech.ac.kr (J.-H.K.); hjinkim@seoultech.ac.kr (H.-J.K.)

* Correspondence: shyoo@seoultech.ac.kr; Tel.: +82-2-970-6802

Received: 8 March 2018; Accepted: 10 May 2018; Published: 14 May 2018

Abstract: In South Korea, buildings account for more than 25% of the nation's total greenhouse gas emissions. Therefore, the government aims to make zero energy buildings mandatory from 2025, onward. For the housing sector, the government has recently built and operated a pilot net-zero energy apartment (NZEA) and plans to expand it to several cities. This article attempts to obtain information about the consumers' willingness to pay (WTP) for the NZEAs. To this end, households' additional WTP for the NZEAs over a conventional apartment was investigated, applying the contingent valuation (CV) approach. The data on the WTP were gathered from a CV survey of 1000 interviewees and analyzed, employing a dichotomous choice question and the spike model, respectively. The mean value of the additional WTP is obtained as KRW 0.46 million (USD 424) per m^2, which is statistically significant. This value corresponds to 17.0% of the conventional apartment price per m^2 (KRW 2.76 million, USD 2436). It can be concluded that the households in South Korea place a significant value on a NZEA over a conventional apartment. In addition, we have investigated the characteristics of consumers which affect the probability that consumers accept additional payment of an amount of money for NZEA over a conventional apartment, finding that higher income earners, higher education, and male consumers have a higher probability.

Keywords: net-zero energy apartment; willingness to pay; contingent valuation

1. Introduction

Abating greenhouse gas (GHG) emissions has already become a crucial issue around the world. The average global temperature has increased by 0.85 in the past 133 years (1880–2012), according to current trends, the global mean temperature is expected to rise by 2.6 to 4.8 °C at the end of the 21st century and the sea level to rise by 45 to 82 cm [1]. As a result, the world has set specific goals and made efforts through the Kyoto protocol in 1997 and the Paris climate change accord in 2015 to reduce GHG emissions, which are considered one of the major factors of climate change.

South Korea has been ranked the second in growth rate of GHG emissions in the OECD countries since 2000 and is currently the fifth largest GHG emitter [2]. Therefore, South Korea pledged to mitigate GHG emissions by 37% from business as usual (BAU) by 2030 in the Paris Climate Change Accord in 2015. Presently, the South Korean government has implemented various policies to reduce GHG emissions. In particular, a quarter of the nation's GHG emissions are from the building sector [3]. The government announced a policy of reducing GHG emissions and energy consumption by mandating zero energy building (ZEB) construction from 2025, onward.

According to United States Department of Energy and International Energy Agency, a ZEB is a residential or commercial building that minimizes energy requirements through building energy efficiency. As buildings consume 36% of the world's energy consumption [4], countries around the world recognize ZEB as a solution to energy saving and various studies are being carried out [5–9].

Zero energy apartments (ZEAs) are residential ZEBs, energy-independent apartments designed to reduce carbon emissions from energy consumption to zero. ZEAs maximize the insulation performance of the building, minimize the energy consumption of the home, and utilize renewable energy to supply the energy needed for building functions. As a result, ZEAs' energy consumption and GHG emissions are expected to be reduced by 90% and 75%, respectively, compared to conventional apartments [3]. However, it is not possible to supply enough energy by time and season, year by year, of renewable energy production facilities alone, or it can cause very large cost increase.

In South Korea, an apartment system has been developed that connects conventional urban energy supply networks, uses non-renewable energy at necessary times, and transmits energy to the supply network at the time when the generated energy is unnecessary. As a result, the apartment's annual energy balance contrast is zero, and the energy requirement is reduced by 60% compared to the conventional apartment. It is called a net-ZEA (NZEA). Currently, the South Korean government has built and piloted NZEA, and plans to expand it to several new cities. However, the NZEA requires greater construction costs than conventional apartments. Moreover, government officials are asking for information about the value that the enforcement of the NZEA construction policy produces to the consumer, which is of great help to obtain some implications concerning whether the NZEA construction should be performed or not.

There are many studies on technological improvements of ZEBs. For example, see Mohammad et al. [5], Bühler et al. [6], Koller et al. [7], Marszal et al. [8], and Wang et al. [9]. However, a few studies analyzed consumers' preferences for ZEBs. For instance, Li et al. [10] examined market acceptability of net-zero energy homes (NZEH) in Canada, proposed a holistic framework to analyze stakeholders for NZEHs, and presented strategies that could be applied to promoting NZEHs. In addition, the study reveals that 77% of the homebuyer survey respondents indicated an interest in buying an NZEH and half of the potential NZEH homebuyers showed a willingness to accept a 5% increase in price compared to conventional homes. The other half of respondents are made up of four categories: (i) those who reported a willingness to pay (WTP) up to 10% more (10%); (ii) those who reported a WTP up to 15% more (10%); (iii) those who reported a WTP up to 20% more (20%); and (iv) those who answered that their decision depends on return (10%).

There are four differences between Li et al.'s [10] study and ours. First, the former dealt with the Canadian case, but the latter will use the specific case of South Korea. Second, the former gathered just 69 responses using an Internet-based survey, while the second collected 1000 responses from a nationwide survey using person-to-person interviewing implemented from a professional survey company. Third, although the former presented just a descriptive distribution of the WTP, our study will estimate the distribution function of the WTP and report the estimates for its parameters. Finally, the former did not derive the mean WTP for NZEH, though the second will compute the mean WTP for NZEAs, which is necessary for analyzing consumers' welfare arising from expanding the supply of the NZEA, employing a random utility maximization model.

Information on the WTP for NZEA is useful for decision-making in building NZEAs. This is because nations continue to make investments in NZEAs, which constitute one of the most important components of their budget for public infrastructure. Economic evaluation is important because it aids in ascertaining whether the public favors proposed projects for the constructing NZEA and in estimating the degree to which it is willing to pay for benefits. In the prevalent context of constrained public budgets, conceptually correct and empirically valid estimates of the consumers' WTP for NZEAs are essential for making economically sound investment decisions.

The main objective of this paper is to estimate the consumer's WTP for NZEAs in South Korea. For this purpose, this study strives to derive household additional WTP for a NZEA over a conventional apartment. The rest of this article consists of four sections. Section 2 describes the methodology that this article adopts. Modeling of the WTP data is explained in Section 3. The fourth section provides the results are the discussion of them. The final section presents conclusions.

2. Methodology

2.1. Object to Be Evaluated

As addressed above, increasing GHG emissions in South Korea have become a critical problem and effective and rigorous policies should be taken to reduce the GHG emissions. Thus, the government seeks to expand the supply of the NZEA. NZEAs use high-performance insulation materials and sunshades to minimize the use of household cooling and heating energy. Additionally, the use of the eco-friendly energy more than conventional fossil fuels makes it possible to dramatically reduce GHG emissions.

In addition, the government is trying to expand the supply of the NZEAs. The betterment goal of the presented policy is to expand the construction of the NZEAs using various policy instruments and mandate it from 2025, onward. The main policy instrument, which was conveyed and explained to the respondents through using newspaper articles, color pictures, and well-made presentation materials during the contingent valuation (CV) survey is to build a NZEA town in a new city, such as Sejong Special Self-governing City, based on the currently-operated NZEA demonstration complex. Thus, the government requires the quantitative information about the consumers' WTP for the NZEA. This article attempts to evaluate the consumers' additional WTP for the NZEA over a conventional apartment in terms of the price of an apartment per m^2.

2.2. Method: CV

From the literature review, it is found that stated preference (SP) methods have been usually applied to carrying out the measurement of the consumers' WTP. The SP methods usually ask people to state their WTP for consuming the goods or services concerned. Two representative approaches belonging to SP methods are the CV approach and choice experiment (CE) approach e.g., [11–17]. The former elicits the WTP response directly. However, the latter derives the WTP responses indirectly. This study will employ the CV approach instead of the CE approach because the first is much simpler to apply than the second and the attributes required in using the CE approach are not well defined in this study.

Our research can be compared with the previous studies in four points. First, the studies that measured people's WTP for NZEA remain scarce. Most of related studies have tackled energy saving and renewable energy [18–20]. In this regard, this study can contribute to the literature on the consumers' evaluation of a NZEA. In particular, there is no former research that measured the consumers' WTP for a NZEA in other countries as well as South Korea.

Second, our application of the CV technique coincides with the practice adopted in the former studies dealing with this kind of research topic. Moreover, the CV technique is based on microeconomics and is, thus, theoretically sound [21]. Since the findings obtained in this article can be used in policy-making and analysis, it is crucial to use reasonable and sound methodology. The CV technique is not only practically useful, but also theoretically robust.

Third, we tried to follow several guidelines recommended for applying the CV approach in the literature. They include the use of a dichotomous choice (DC) question, a minimum sample size of 1000, the announcement of the possible presence of substitutes for the goods to be investigated in the CV survey, and so on. More details will be presented in the next subsections.

Fourth, when eliciting the WTP responses this study paid more attention to not only mitigating the response bias, but also to augmenting the statistical efficiency. The respondents are asked just one question in the single-bounded (SB) DC method. Hence, it may suffer from low statistical efficiency. The double-bounded (DB) DC method, which requires two WTP questions, may suffer from the response bias. As an alternative, Cooper et al. [22] contrived a one-and-one-half-bounded (OOHB) DC method. It can produce more efficient results than the SB DC method and cause less response bias than the DB DC method. Furthermore, the spike model proposed by Kriström [23] is combined with the OOHB DC method in order to model the WTP data with zero observations.

2.3. Sampling and Survey Instrument

As mentioned above, the government will mandate ZEB and NZEA for all buildings built nationwide from 2025, onward. Therefore, 1000 households were sampled nationwide based on the demographic characteristics reported by Statistics Korea [24]. In this regard, our sampling method was stratified random sampling. The entire process of sampling and carrying out the survey was administered by a professional polling company during September 2017. The firm sough to make sure that the sample characteristics represents the population characteristics well. An experienced specialist at the firm ran the whole process.

A pretest using a focus group of thirty persons was implemented with an earlier version of the survey instrument to examine whether it is understandable and clear enough for the interviewees to finish filling in the survey questionnaire or not. The outcomes of the in-depth interviews with the focus group have been utilized to make the questionnaires fully corrected for the use in the main survey. The final version of the survey questionnaire is made up of four parts. The first part presents the background and objective of the survey. The second part includes several questions deriving the interviewees' opinions and judgment regarding the policy of building NZEAs. The third part deals with the questions about the additional WTP for living in a NZEA rather than a conventional apartment. Some questions about respondents' characteristics are given in the final part.

2.4. Elicitation of WTP

As explained above, this article adopted an OOHB DC question for the purpose of eliciting the WTP responses. The DC question was originally recommended for the use in the field CV survey by a number of studies. The main reason for the recommendation is that it can reduce the respondents' burden of answering the WTP question and derive an incentive-compatible response from interviewees [25]. The DC question is quite simple. The only work for a respondent must do is to state "yes" or "no" to a given bid amount. The respondent will report "yes" if her/his additional WTP for purchasing a NZEA over a conventional apartment is more than, or equal to, an offered bid and "no", otherwise. On the other hand, an open-ended question of directly asking the WTP value is not preferred to the DC question in the literature because the former can induce a number of protest WTP responses [26,27]. The OOHB DC format has the merit of taking only the preferential features of the SB and DB DC formats.

One complication involved in applying the CV is that it puts people in a hypothetical situation and, thus, the respondents can have difficulties in stating their true WTP. An appropriate payment can help the respondents confronted with the hypothetical situation to report their WTP making them feel as if they were in the real world. Some examples of the payment vehicle include a tax, such as income tax or property tax, a donation, a fund, a usage fee, and so on. The payment vehicle should be directly related to the object to be evaluated and be familiar to people.

We decided that the payment vehicle meeting the two conditions is the additional price of a NZEA built in Sejong Special Self-governing City, a recently-developed and typically new city in South Korea over a conventional apartment built in the same area. The price is expressed as Korean won per pyeong, which is same as 3.3 m^2. The unit of 'pyeong' is more frequently used in South Korea than that of 'm^2'. Consequently, the WTP question presented to the respondents was "Would your household accept an amount of increase in the price of an apartment per pyeong (=3.3 m^2) for the purpose of living in a NZEA rather than a conventional apartment?"

3. WTP Model

3.1. OOHB DC Model

Cooper et al. [22] proposed an approach to model the OOHB DC CV data. A_j is defined as a bid presented to respondent j. Before implementing the field CV survey, we need to determine several sets of two bids, A_j^L and A_j^U ($A_j^L < A_j^U$). A set is randomly selected of the several sets and presented to each interviewee. Each set is composed of A_j^L and A_j^U. About half of respondents in a group who receive the same set are asked to state "yes" or "no" to the payment of A_j^L. If the response is "yes", an additional question about whether they are willing to pay A_j^U or not is asked. If the response to the payment of A_j^L is "no," the additional question is needless. The other half of the respondents are confronted with a question about the payment of A_j^U. If the response is "no", a follow-up question about whether they are willing to pay A_j^L or not is asked. If the response to the payment of A_j^U is "yes," the follow-up question is not required.

Let Y_j be the interviewee's WTP. Three responses, "yes-yes" ($Y_j > A_j^U$), "yes-no" ($A_j^L < Y_j < A_j^U$), and "no" ($Y_j < A_j^L$) can emerge from the situation where A_j^L is offered at first. One of three responses, "yes" ($Y_j > A_j^U$), "no-yes" ($A_j^L < Y_j < A_j^U$), and "no-no" ($Y_j < A_j^L$), can occur in the case that A_j^U is provided at first. Therefore, there can be six kinds of responses. Let I_j^{YY}, I_j^{YN}, I_j^N, I_j^Y, I_j^{NY}, and I_j^{NN} be binary variables which correspond to the six kinds of responses. For instance, I_j^{YY} is one if jth interviewee reports "yes-yes" and zero otherwise.

$$\begin{aligned} I_j^{YY} &= \mathbf{1}\,(j\text{th respondent's response is "yes-yes"}) \\ I_j^{YN} &= \mathbf{1}\,(j\text{th respondent's response is "yes-no"}) \\ I_j^{N} &= \mathbf{1}\,(j\text{th respondent's response is "no"}) \\ I_j^{Y} &= \mathbf{1}\,(j\text{th respondent's response is "yes"}) \\ I_j^{NY} &= \mathbf{1}\,(j\text{th respondent's response is "no-yes"}) \\ I_j^{NN} &= \mathbf{1}\,(j\text{th respondent's response is "no-no"}) \end{aligned} \tag{1}$$

where $\mathbf{1}(\cdot)$ is an indicator function whose value is one when the argument is true and zero otherwise.

3.2. Combination of the OOHB DC Question and Spike Model

An additional question, "Would your household agree to pay anything?", was given to the respondents who reported "no" response to the lower bid or "no-no" response to the upper bid. Her/his WTP is less than the lower bid and more than zero if the answer is "yes." Her/his WTP is zero if the answer is "no." One more binary variable, I_j^{TY}, is defined as one if the answer is "yes" and zero otherwise. Thus, there are eight outcomes:

- "yes-yes" ($Y_j > A_j^U$),
- "yes-no" ($A_j^L < Y_j < A_j^U$),
- "no-yes" ($0 < Y_j < A_j^L$),
- "no-no" ($Y_j = 0$),
- "yes" ($Y_j > A_j^U$),
- "no-yes" ($A_j^L < Y_j < A_j^U$),
- "no-no-yes" ($0 < Y_j < A_j^L$), and
- "no-no-no" ($Y_j = 0$).

where the first four outcomes are achieved when A_j^L is offered first and the latter four outcomes are obtained when A_j^U is supplied first.

As will be explained below, out of the 1000 respondents, 595 said they had no intention of paying a penny. Thus, the spike model can be usefully employed to deal with the WTP data. Considering that the most-frequently used distribution in analyzing the DC CV data is logistic distribution, we specify the WTP distribution function, $F_Y(\cdot)$, as:

$$F_Y(A;\gamma_0,\gamma_1) = \begin{cases} [1+\exp(\gamma_0-\gamma_1 A)]^{-1} & \text{if } A > 0 \\ [1+\exp(\gamma_0)]^{-1} & \text{if } A = 0 \\ 0 & \text{if } A < 0 \end{cases} \quad (2)$$

where γ_0 and γ_1 are the parameters of $F_Y(\cdot)$, A means the independent variable of $F_Y(\cdot)$, and $\exp(\cdot)$. is exponential function.

The log-likelihood function we deal with is:

$$\begin{aligned} \ln L = \sum_{j=1}^{S} \{ & (I_j^Y + I_j^{YY}) \ln[1 - F_Y(A_j^U;\gamma_0,\gamma_1)] \\ & + (I_j^{YN} + I_j^{NY}) \ln[F_Y(A_j^U;\gamma_0,\gamma_1) - F_Y(A_j^L;\gamma_0,\gamma_1)] \\ & + I_j^{TY}(I_j^N + I_j^{NN}) \ln[F_Y(A_j^L;\gamma_0,\gamma_1) - F_Y(0;\gamma_0,\gamma_1)] \\ & + (1 - I_j^{TY})(I_j^N + I_j^{NN}) \ln F_Y(0;\gamma_0,\gamma_1) \} \end{aligned} \quad (3)$$

where S is the sample size, γ_0 and γ_1 are the parameters of $F_Y(\cdot)$, and I_j^{YY}, I_j^{YN}, I_j^N, I_j^Y, I_j^{NY}, and I_j^{NN} are defined in Equation (1).

We can obtain the estimates for γ_0 and γ_1 by finding the values for γ_0 and γ_1 maximizing Equation (2), that is, using the maximum likelihood estimation method. When using Equation (1) and the estimates for γ_0 and γ_1, the average WTP can be obtained as:

$$E(Y) = \int_0^\infty [1 - F_Y(A;\gamma_0,\gamma_1)]dA - \int_{-\infty}^0 F_Y(A;\gamma_0,\gamma_1)dA = (1/\gamma_1)\ln[1+\exp(\gamma_0)] \quad (4)$$

where γ_0 and γ_1 are the parameters of $F_Y(\cdot)$, A means the independent variable of $F_Y(\cdot)$, and $\exp(\cdot)$ is exponential function.

4. Results and Discussion

4.1. Data

USD 1.0 was equal to KRW 1132 when the survey was performed. The list of sets of A_j^L and A_j^U used in the CV survey is 250/350, 300/400, 350/450, 450/550, 550/650, 650/750, and 800/900 ten thousand KRW per pyeong (=3.3 m^2). They were determined through the focus group interview of thirty individuals as follows: first, we asked the WTP for the enforcement and obtained a set of WTP values; second, we deleted zero WTP values and then sorted the remaining positive WTP values to look into empirical distribution; third, some bids were selected from the distribution. One of the seven sets of A_j^L and A_j^U was randomly offered to the interviewees.

Finally, 1000 usable observations were obtained from the CV survey. Table 1 reports a summary of the interviewees' responses to each set of bids. Overall, 29, 45, 126, and 300, 52, 24, 129, and 295 interviewees gave "yes-yes", "yes-no", "no-yes", "no-no", "yes", "no-yes", "no-no-yes", and "no-no-no" responses, respectively. Out of the 1000 respondents, 595 said they had no intention of paying a penny ("no-no" and "no-no-no" responses).

Table 1. Summary of the interviewees' responses to each set of bids.

Bid Amount [a]		Lower Bid Is Offered at First (%) [b]				Upper Bid Is Offered at First (%) [b]				Sample Size
		"Yes-Yes"	"Yes-No"	"No-Yes"	"No-No"	"Yes"	"No-Yes"	"No-No-Yes"	"No-No-No"	
250	350	4 (2.8)	10 (7.0)	16 (11.2)	42 (29.4)	10 (7.0)	3 (2.1)	16 (11.2)	42 (29.4)	143 (100.0)
300	400	6 (4.2)	12 (8.4)	17 (11.9)	36 (25.2)	4 (2.8)	5 (3.5)	14 (9.8)	49 (34.3)	143 (100.0)
350	450	6 (4.2)	10 (7.0)	12 (8.4)	43 (30.1)	10 (7.0)	2 (1.4)	18 (12.6)	42 (29.4)	143 (100.0)
450	550	3 (2.1)	4 (2.8)	21 (14.7)	44 (30.8)	7 (4.9)	5 (3.5)	20 (14.0)	39 (27.3)	143 (100.0)
550	650	4 (2.8)	6 (4.2)	17 (12.0)	44 (31.0)	13 (9.2)	6 (4.2)	19 (13.4)	33 (23.2)	142 (100.0)
650	750	4 (2.8)	1 (0.7)	18 (12.7)	48 (33.8)	5 (3.5)	1 (0.7)	18 (12.7)	47 (33.1)	142 (100.0)
800	900	2 (1.4)	2 (1.4)	25 (17.4)	43 (29.9)	3 (2.1)	2 (1.4)	24 (16.7)	43 (29.9)	144 (100.0)
Totals		29 (2.9)	45 (4.5)	126 (12.6)	300 (30.0)	52 (5.2)	24 (2.4)	129 (12.9)	295 (29.5)	1000 (100.0)

Notes: [a] The unit is tens of thousands of Korean won (USD 1.0 = KRW 1132 at the time of the survey) per pyeong (=3.3 m^2). [b] A percentage of sample size is given in parentheses beside the number of responses.

4.2. Estimation Results of the Model

Table 2 presents the estimation results of the model. The estimates for γ_0 and γ_1 are all statistically significant. In particular, a negative sign of the estimate for γ_1 means that higher bid amount induces a lower probability of saying "yes" to an offered bid. From Equation (1), the spike is derived as $[1+\exp(\gamma_0)]^{-1}$. The estimate for the spike is calculated as 0.594 and statistically significant. Since the spike implies the possibility of the interviewees having zero WTP, the estimated spike should not be significantly different from the sample ratio of zero WTP (59.5%). This is the case with our study.

Table 2. Estimation results of the model.

Variables	Estimates [d]
Constant	−0.379 (−5.90) #
Bid amount [a]	−3.411 (−19.07) #
Spike	0.594 (38.27) #
Mean additional WTP per m^2	KRW 0.46 million (USD 424)
t-value	15.73 #
95% confidence interval [b]	KRW 0.41 to 0.53 million (USD 364 to 455)
99% confidence interval [b]	KRW 0.40 to 0.55 million (USD 364 to 485)
Sample size	1000
Log-likelihood	−1057.30
Wald statistic (*p*-value) [c]	247.41 (0.000)

Notes: [a] The unit is tens of thousands of Korean won and the exchange rate was USD 1.0 = KRW 1132 at the time of the survey. [b] It is calculated by the use of the parametric bootstrapping method given in Krinsky and Robb [28]. [c] It is computed under the null hypothesis of all the parameters' being jointly zero. [d] The values reported in parentheses beside the coefficient estimates are *t*-values. # implies statistical meaningfulness at the 1% level.

Table 2 also provides an estimate of average WTP calculated using Equation (3). The average WTP has the value of KRW 0.46 million (USD 424) per m^2 and statistical meaningfulness. It is desirable to calculate its confidence interval to explicitly take into account the uncertainty concerning the calculation of the point estimate. The 95% confidence interval computed adopting the parametric bootstrapping method presented in Krinsky and Robb [28] is KRW 0.41 to 0.53 million (USD $364 to $455) per m^2.

4.3. Reflection of Covariates

Covariates mean the factors that can have an effect on the probability of saying "yes" to a given bid. Usually, the interviewees' characteristics are used as covariates. Usually, the covariates are reflected in the model by inserting them into γ_0 in Equation (1). We consider four variables: education, income, head, and gender. The variables are explained in Table 3. Therefore, a positive sign of the coefficient for a variable implies that the greater the value of the variable, the higher the probability of reporting "yes" to a presented bid.

Table 3. Description of the variables used for covariates.

Variables	Definitions	Mean	Standard Deviation
Education	The respondent's educational level in years	14.04	2.27
Income	The respondent's household's monthly income before tax deduction (unit: million Korean won)	4.80	5.99
Head	Whether the respondent is the head of household (1 = head; 0 = otherwise)	0.55	0.50
Gender	The respondent's gender (0 = male; 1 = female)	0.50	0.50

The estimation results of the model with covariates are shown in Table 4. The estimate for γ_1 is expected to be negative and statistically meaningful. The coefficient estimates for other variables are all statistically significant. The respondent's education level is positively related to the probability of reporting "yes" to a given bid. Similarly, wealthier interviewees are more inclined to accept the payment of a proposed bid than less wealthy interviewees.

Table 4. Estimation results of the spike model with covariates.

Variables [a]	Estimates	*t*-Values
Constant	−2.6080	−6.21 *
Bid amount [b]	−3.5066	−19.12 *
Education	0.1424	4.75 *
Income	0.0581	2.05 *
Head	0.5515	1.80 *
Gender	−0.7127	−2.32 *
Spike	0.4679	29.47 *
Mean additional WTP per m^2	KRW 0.45 million (USD 394)	
t-value	15.63 **	
95% confidence interval [c]	KRW 0.40 to 0.51 million (USD 364 to 455)	
99% confidence interval [c]	KRW 0.38 to 0.53 million (USD 333 to 464)	
Wald statistic (*p*-value) [d]	491.94 (0.000)	
Log-likelihood	−1036.54	
Number of observations	1000	

Notes: [a] Table 3 explains the variables. [b] The unit is tens of thousands of Korean won and the exchange rate was USD 1.0 = KRW 1132 at the time of the survey. [c] It is calculated by the use of the parametric bootstrapping method given in Krinsky and Robb [28]. [d] It is computed under the null hypothesis of all the parameters' being jointly zero. *and ** indicate statistical meaningfulness at the 10% and 1% levels, respectively.

4.4. Discussion of the Results

To look into consumers' preferences for NZEAs, we estimated a mean additional WTP for a NZEA over a conventional apartment. The most important issue in the estimation process is whether the sample represents the population. Thus, whether some variables for the sample are similar to those for the population or not should be examined. In this regard, the ratio of female respondents, and the household's monthly income are looked into here. The sample averages for the variables were 50.0%, and KRW 4.80 million. The population averages were 50.0%, KRW 4.50 million when the survey was conducted [24]. Interestingly, it seems that there are no significant gaps between the two values for each variable. This finding makes the representativeness of our sample even stronger.

How to select covariates may influence the mean WTP estimate. Therefore, the mean WTP estimate found in the model with no covariates is used. The mean additional WTP for a NZEA is estimated to be KRW 0.46 million (USD 424) per m^2, which is statistically meaningful at the 1% level. This corresponds to 17.0% of the price of a conventional apartment per m^2 (KRW 2.76 million or USD 2436). Consumers' WTP for a NZEA over a conventional apartment in South Korea is illustrated in Figure 1. The 95% and the 99% confidence intervals for the mean additional WTP for NZEA are KRW 0.41 to 0.53 million (USD $364 to $455) and KRW 0.40 to 0.55 million (USD 364 to 485) per m^2, respectively. In this regard, it can be concluded that the households in South Korea place a significant value on a NZEA over a conventional apartment. A summary of the procedures of conducting this study is shown in Figure 2.

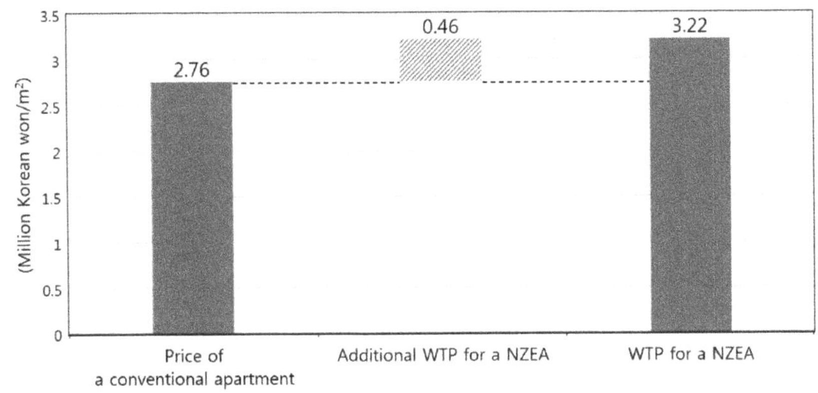

Figure 1. Consumers' willingness to pay (WTP) for a net zero energy apartment (NZEA) over a conventional apartment in South Korea.

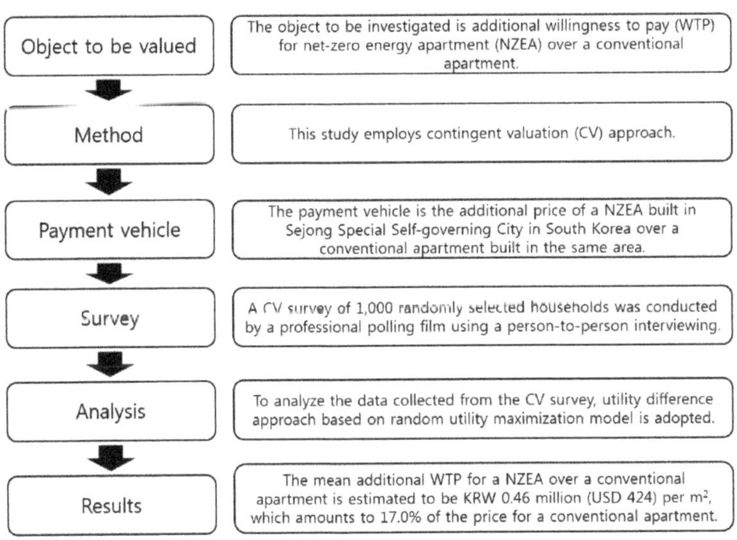

Figure 2. Summary of the procedure of conducting this study.

The additional WTP reflects the present values of the consumers' future cost savings related to energy expenditures. If the additional cost involved in building a NZEA that can significantly mitigate GHG emissions by adding energy saving technology and renewable energy production facility is less than the additional WTP plus the environmental benefits concerning the reduction of GHG and air pollutants emissions, expanding the supply of the NZEAs can be successfully implemented. If not, further action is needed for successful expansion of the NZEAs. For example, it is related to granting subsidies for renewable energy production facilities, extending the lending limits of housing city funds. In addition, the governments can offer a variety of tax credits and other incentives to expand the NZEAs. For instance, it is recommended that subsidies be paid as an incentive to companies that develop renewable energy production facilities and use construction materials that reduce energy consumption. Furthermore, the information on the WTP found in the article can be used to determine the level of tax relief, tax deductions, or subsidies needed to expand the supply of the NZEAs.

5. Conclusions

In South Korea, as GHG emission has increased recently, there is concern about climate change. Particularly, one quarter of the nation's GHG emissions are from the building sector. For the purpose of abating GHG emissions, we should reduce the energy consumption of buildings. The South Korean government, therefore, has planned to expand the supply of a NZEAs. This study applied a CV technique to assessing the consumers' additional WTP for a NZEA over a conventional apartment. The estimate for the mean WTP for the NZEA was KRW 0.46 million (USD 424) per m^2. It has statistical meaningfulness at the 1% level and the sample also represented the population well. Considering that the apartment sale price was KRW 2.76 million per m^2 at the time of the survey [24], consumers were willing to pay 17.0% more to live in the NZEA.

We also presented the mean WTP estimate from the model with covariates and its 95% confidence interval in Table 4. How much does the WTP change when considering the covariates provide more details about policies and, in particular, for informing the population about policy measures as they can be more targeted towards different societal groups? The mean WTP obtained was KRW 0.45 million (USD 394) and its 95% confidence interval was estimated as KRW 0.40 to 0.51 million. To check whether the mean WTP estimate from the model without the covariates is different from that from the model with the covariates, we applied a simple test. The null hypothesis is that they do not differ. Given that the 95% confidence intervals for the mean WTP estimates from the models with the covariates and without the covariates do overlap, thus, the null hypothesis cannot be rejected at the 5% level, which implies that they do not significantly differ.

These results are a useful baseline for understanding the potential implications of NZEAs. The government and industry should be aware of how much potential consumers' additional WTP for a NZEA is. For example, it is encouraged to expand the supply of the NZEAs because the consumers gave higher value to a NZEA than a conventional apartment. This article added a contribution to the current literature by deriving the household WTP for living in a NZEA rather than a conventional apartment. The study provided empirical evidence that the CV approach, theoretically grounded in microeconomics, could be successfully utilized in measuring the additional WTP.

Furthermore, the factors affecting the public trust in the policy need to be identified and investigated. This is because if people do not have trust in the policy enforcement, it cannot succeed. Thus, the government should consider which factors influence and how much the factors affect people's trust when designing a policy of expanding a supply of NZEAs. It may be useful to examine how the public value varies as time passes by conducting the CV survey every year for some years and analyzing the CV data. Investigating how much the value changes across regions and identifying other geographic factors which affect the value are also a good research topic. Comparing the findings from this study with those from other studies for foreign countries and analyzing the gap between the two enable us to obtain new insight into the WTP estimate. These works can provide us with a new point of view concerning the additional WTP.

Author Contributions: All of the authors played an important role in the preparation of this paper. J.-H.K. wrote most of the paper; H.-J.K. conducted the empirical analysis; and S.-H.Y. took charge of making the survey questionnaire and gathering the data.

Acknowledgments: This work was supported by the Korea Institute of Energy Technology Evaluation and Planning (KETEP) and the Ministry of Trade, Industry and Energy (MOTIE) of the Republic of Korea (no. 20164030201060).

Conflicts of Interest: The authors declare no conflict of interest.

References

1. Intergovernmental Panel on Climate Change. *Climate Change 2014: Synthesis Report*; IPCC: Geneva, Switzerland, 2014; p. 151.
2. OECD. *OECD Environmental Performance Review: Korea 2017*; OECD Publishing: Paris, France, 2017.
3. MOLIT. Policy Report: To Cope with Climate Change Zero Energy Building Promotion Plan. 2016. Available online: http://www.molit.go.kr (accessed on 5 March 2018). (In Korean)
4. International Energy Agency. *World Energy Outlook*; OECD/IEA: Paris, France, 2016. Available online: www.iea.org (accessed on 5 March 2018).
5. AbuGrain, M.Y.; Alibaba, H.Z. Optimizing existing multistory building designs towards net-zero energy. *Sustainability* **2017**, *9*, 399. [CrossRef]
6. Bühler, D.; Schuetze, T.; Junge, R. Towards development of a label for zero emission buildings: A tool to evaluate potential zero emission buildings. *Sustainability* **2015**, *7*, 5071–5093. [CrossRef]
7. Koller, C.; Talmon-Gros, M.J.; Junge, R.; Schuetze, T. Energy toolbox-framework for the development of a tool for the primary design of zero emission buildings in European and Asian cities. *Sustainability* **2017**, *9*, 2244. [CrossRef]
8. Marszal, A.J.; Heiselberg, P.; Bourrelle, J.S.; Musall, E.; Voss, K.; Sartori, I.; Napolitano, A. Zero Energy Building–A review of definitions and calculation methodologies. *Energy Build.* **2011**, *43*, 971–979. [CrossRef]
9. Wang, L.; Gwilliam, J.; Jones, P. Case study of zero energy house design in UK. *Energy Build.* **2009**, *41*, 1215–1222. [CrossRef]
10. Li, H.X.; Patel, D.; Al-Hussein, M.; Yu, H.; Gül, M. Stakeholder studies and the social networks of NetZero energy homes (NZEHs). *Sustain. Cities Soc.* **2018**, *38*, 9–17. [CrossRef]
11. Min, S.H.; Lim, S.Y.; Yoo, S.H. Consumer's willingness to pay a premium for eco-labeled LED TVs in Korea: A contingent valuation study. *Sustainability* **2017**, *9*, 814. [CrossRef]
12. Park, S.Y.; Lim, S.Y.; Yoo, S.H. The economic value of the national meteorological service in the Korean household sector: A contingent valuation study. *Sustainability* **2016**, *8*, 834. [CrossRef]
13. Lim, S.Y.; Kim, H.Y.; Yoo, S.H. Public willingness to pay for transforming Jogyesa Buddhist temple in Seoul, Korea into a cultural tourism resource. *Sustainability* **2016**, *8*, 900. [CrossRef]
14. Kim, H.Y.; Park, S.Y.; Yoo, S.H. Public acceptability of introducing a biogas mandate in Korea: A contingent valuation study. *Sustainability* **2016**, *8*, 1087. [CrossRef]
15. Wang, J.; Ge, J.; Ma, Y. Urban Chinese consumers' willingness to pay for pork with certified labels: A discrete choice experiment. *Sustainability* **2018**, *10*, 603. [CrossRef]
16. Yang, H.J.; Lim, S.Y.; Yoo, S.H. The environmental costs of photovoltaic power plants in South Korea: A choice experiment study. *Sustainability* **2017**, *9*, 1773. [CrossRef]
17. Vanstockem, J.; Vranken, L.; Bleys, B.; Somers, B.; Hermy, M. Do looks matter? A case study on extensive green roofs using discrete choice experiments. *Sustainability* **2018**, *10*, 309. [CrossRef]
18. Abdullah, S.; Jeanty, P.W. Willingness to pay for renewable energy: Evidence from a contingent valuation survey in Kenya. *Renew. Sustain. Energy Rev.* **2011**, *15*, 2974–2983. [CrossRef]
19. Guo, X.; Liu, H.; Mao, X.; Jin, J.; Chen, D.; Cheng, S. Willingness to pay for renewable electricity: A contingent valuation study in Beijing, China. *Energy Policy* **2014**, *68*, 340–347. [CrossRef]
20. Lim, K.M.; Lim, S.Y.; Yoo, S.H. Estimating the economic value of residential electricity use in the Republic of Korea using contingent valuation. *Energy* **2014**, *64*, 601–606. [CrossRef]
21. Champ, P.A.; Boyle, K.J.; Brown, T.C.A. *Primer on Nonmarket Valuation*; Kluwer Academic Publisher: Dordrecht, The Netherlands, 2004.
22. Cooper, J.C.; Hanemann, W.M.; Signorello, G. One and one-half bound dichotomous choice contingent valuation. *Rev. Econ. Stat.* **2002**, *84*, 742–750. [CrossRef]
23. Kriström, B. Spike model in contingent valuation. *Am. J. Agric. Econ.* **1997**, *79*, 1013–1023. [CrossRef]
24. Statistics Korea. Available online: http://kosis.kr (accessed on 2 March 2018).
25. Hanemann, W.M. Welfare evaluations in contingent valuation experiments with discrete responses. *Am. J. Agric. Econ.* **1984**, *66*, 332–341. [CrossRef]
26. Hanemann, W.M.; Loomis, J.; Kanninen, B.J. Statistical efficiency of double-bounded dichotomous choice contingent valuation. *Am. J. Agric. Econ.* **1991**, *73*, 1255–1263. [CrossRef]

27. Johnston, R.J.; Boyle, K.J.; Adamowicz, W.; Bennett, J.; Brouwer, R.; Cameron, T.A.; Hanemann, W.M.; Hanley, N.; Ryan, M.; Scarpa, R.; et al. Contemporary guidance for stated preference studies. *J. Assoc. Environ. Resour. Econ.* **2017**, *4*, 319–405. [CrossRef]
28. Krinsky, I.; Robb, A.L. On approximating the statistical properties of elasticities. *Rev. Econ. Stat.* **1986**, *68*, 715–719. [CrossRef]

© 2018 by the authors. Licensee MDPI, Basel, Switzerland. This article is an open access article distributed under the terms and conditions of the Creative Commons Attribution (CC BY) license (http://creativecommons.org/licenses/by/4.0/).

sustainability

MDPI

Article

Analysis of Heating and Cooling Loads of Electrochromic Glazing in High-Rise Residential Buildings in South Korea

Myunghwan Oh [1], Sungho Tae [2,*] and Sangkun Hwang [3]

1 Sustainable Building Research Center, Hanyang University, 55 Hanyangdaehak-ro, Sangrok-gu, Ansan-si 15588, Gyeonggi-do, Korea; bigomh@hanyang.ac.kr
2 School of Architecture & Architectural Engineering, Hanyang University, 55 Hanyangdaehak-ro, Sangrok-gu, Ansan-si 15588, Gyeonggi-do, Korea
3 Research Center of Technology, ALU ENC Co., 10, Baekjegobun-ro 9-gil, Songpa-gu 05562, Seoul, Korea; hsk3111@naver.com
* Correspondence: jnb55@hanyang.ac.kr; Tel.: +80-10-6730-2101

Received: 22 February 2018; Accepted: 31 March 2018; Published: 9 April 2018

Abstract: This study compares the impact of the recently developed electrochromic glazing technology on load reduction by comparing it with the double-glazing and shading devices that are sold commercially for high-rise residential buildings in Korea. These buildings are similar to large office buildings in terms of their high window-to-wall ratio. The energy consumption of such buildings was simulated using an analytical model of a high-rise residential building. The patterns between the heating and cooling loads were found to be similar to that of office buildings, in that the cooling load was considerably higher than the heating load. This study hypothesizes that the load reduction performance of electrochromic glazing with variable solar control and high solar radiation rejection is superior to that of existing double-glazing products and shading devices. This hypothesis was tested by analyzing the cooling and heating loads of buildings with different types of double glazings. Bleached electrochromic glazing exhibited lower transmittance than colored glass double glazing, low-e double glazing, and double glazing with a shading device, and is thus not effective in reducing heating load. Colored electrochromic glazing provided higher solar radiation rejection than colored glass double glazing and low-e double glazing, and thus is effective in reducing cooling load.

Keywords: electrochromic glazing; optical properties; heating and cooling loads; high-rise residential building; curtain wall; EnergyPlus

1. Introduction

1.1. Research Background and Objective

Windows and doors are the only glazing systems that can capture solar energy. They can also serve as a medium that reduces building load and improves the indoor environment through appropriate inflow and the rejection of solar radiation [1]. As glazing systems, windows and doors require separate shading devices for solar control, such as roll shades or blinds. However, smart glass, which does not need any separate shading device to control the optical properties of glass and, thus, solar radiation, has recently been actively researched in the literature. Types of smart glass include electrochromic, PDLC (polymer-dispersed liquid crystal) [2], and SPD (suspended particle device) [3]. Of these, this study analyzes the optical properties of electrochromic glazing and its effect on HVAC (Heating, ventilation, air conditioning) load reduction in buildings [4,5].

Regarding the main characteristics of electrochromic glazing, which are dealt with in this study, transmittance can be controlled in stages depending on the amount of electricity that is supplied to either the cathode (−) or the anode (+), and the scope of transmittance control covers the area of visible and near-infrared rays, which account for the largest portion of energy in the solar spectrum [6]. Therefore, with appropriate control according to changes in the outdoor environment using the characteristics of selective transmittance control, it is possible to contribute significantly to a reduction in heating and cooling loads as well as the energy used for lighting [7,8]. Based on these characteristics, studies have been actively conducted on electrochromic glazing control to reduce the amount of energy consumed by buildings using such data as indoor–outdoor temperatures and the amount of solar radiation and illumination. Moreover, the daylight glare index and lighting energy have been investigated in the context of indoor light to represent the characteristics that can control the range of visible rays [6,9]. According to the results of simulations of office buildings in the United States (US), 10–20% of the energy saving effect is obtained at the perimeters of buildings [10], and at least 54% of the energy can be saved in a Mediterranean climate [11]. A study showed that 37–48% more lighting energy can be saved compared with manual control-type blinds [12]. Furthermore, several simulations and experimental studies are ongoing on the performance of electrochromic glazing technology [13]. A few recent studies have examined electrochromic glazing that can intensively control near-infrared (NIR) rays. It maintains a visually clear state and secures visible rays related to lighting energy while controlling only the NIR rays that influence cooling and heating energy [14]. Moreover, photovoltaic-related electrochromic glazing is being developed, which is colored using power charged on batteries and produced using photovoltaic energy. Thus, it does not require a separate power supply, and is an eco-friendly product that is easy to build and does not consume operation energy [15]. Furthermore, electrochromic glazing can be controlled on smartphones using Wi-Fi in association with the Internet of Things. This provides convenient functions compared with existing shading devices [16]. Further, apart from electrochromic glazing with a glass substrate, electrochromic films with a web-coated flexible polyester (PET) substrate have also been developed, and have various applications [17].

Another advantage of electrochromic glazing is that unlike the blinds or roll shades installed indoors, it blocks solar radiation through glass, which is the outermost part of an insulated glazing unit. As a result, it is more effective at reducing the cooling load, owing to its high solar radiation rejection. Moreover, it has a memory effect whereby it maintains its state once supplied with electric power once, without requiring any further power once it is colored or bleached. This minimizes the operation energy that is needed. However, the drawback of most electrochromic glazing products is their low speed of response. It requires five minutes for a 10 cm × 30 cm glass [18] to be colored, and eight minutes and 12 min for a 1.2 m × 0.8 m glass to be colored and bleached, respectively [19]. On the contrary, a few recently released electrochromic glazing products can be colored within three minutes. These products show promise in enhancing the performance of commercialized products in the future [20].

The objective of this study is to analyze the optical properties of test products for the commercialization of large-area electrochromic glazing for the first time in Korea, to the best of the authors' knowledge, and determine whether the optical properties are effective at reducing the energy consumed by buildings. High-rise residential buildings in Korea with high window-to-wall ratios were assessed to this end. For performance comparison, this study analyzed the building load performance of existing double-glazing and shading devices that are available for commercial use in Korea.

1.2. Research Method and Scope

We created electrochromic glazing specimens to obtain the transmittance, reflectance, and absorptance for each range of spectra through measurements using a spectrum analyzer. Using raw data, we obtained the input data for WindowMaterial:Glazing, which is a component of structural glass on EnergyPlus, by employing the LBNL Optic 5.1 tool. We thus formed the double glazing, and obtained heat and optical data. The data for commercially available glass were obtained from the data for glass that is distributed in Korea on IGDB provided by LBNL Window 7.4.

To analyze building loads using the heat and optical data, we used EnergyPlus8.5, a dynamic building energy analysis program, to model a high-rise residential building with a high window-to-wall ratio because of curtain walls, located in Seoul, Korea. We then applied 10 cases, including colored double glazing, low-e double glazing, double glazing with a shading device, and electrochromic double glazing on EnergyPlus, and comparatively analyzed building load performance through a parametric study.

We focused on window heat gain, window heat loss, and hourly load change on typical days of winter and summer based on a design day. We hence analyzed the effects of the characteristics of heat flow of each glazing on the load. We ultimately analyzed monthly and annual loads, which we used to comparatively analyze the performance of electrochromic glazing and commercially available double-glazing and shading devices in terms of building load. Figure 1 show the research guideline and flowchart.

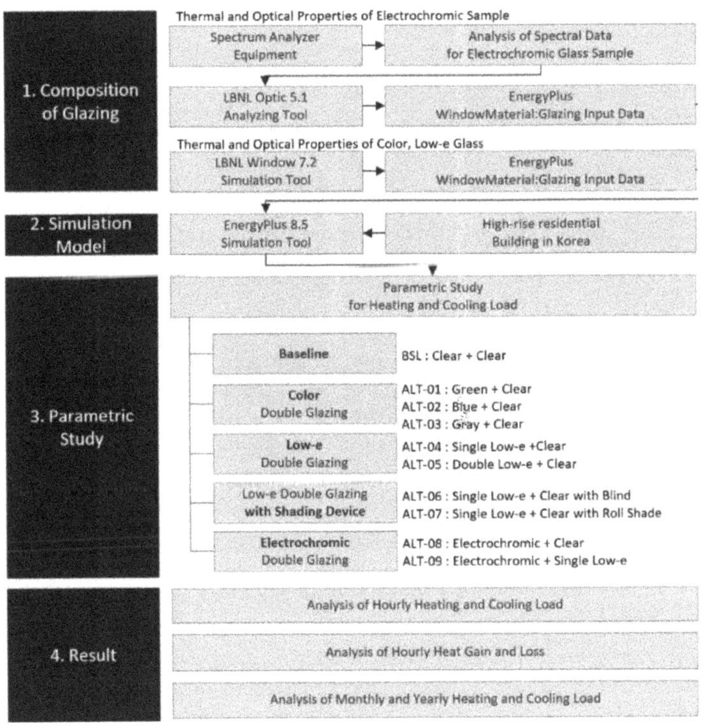

Figure 1. Research guideline and flowchart.

2. Analysis of Optical Properties of Electrochromic Glazing

Electrochromic glazing consists of a transparent conductive object, a colored layer, an ion storage layer, and an electrolyte. In this study, we used TEC 10 (manufactured by Pilkington) as a transparent conductive object with glass on which electric current can be applied. We coated the colored layer with tungsten oxide (WO_3) and the ion storage layer with nickel tungsten (NiW) using a sputtering device that could coat areas of up to 1500 mm × 1800 mm. We used $LICLO_4$ in gel form for electrolytes for ion movement, and formed a silver paste and bus bar electrodes to provide electrical power.

Specimens of size of 50 mm × 50 mm were fabricated as shown in Figure 2 to analyze the optical properties of electrochromic glazing. We then measured transmittance and reflectance at a solar spectrum wavelength of 0.3–2.5 μm (0.005-μm intervals) using a spectrum analyzer. These measurements were performed with electrochromic glazing in the bleached and colored states at a voltage of 2 V in each state.

To enter the raw data of transmittance and reflectance extracted through the spectrum analyzer for the WindowMaterial:Glazing component of EnergyPlus, we imported the data on the LBNL Optic 5.1 program, and extracted the spectral average data of solar transmittance, solar reflectance, visible transmittance, and visible reflectance, as shown in Table 1.

Optical properties were analyzed using spectroscopic analysis and the Optic 5.1 tool, and we found a considerable difference in the solar transmittance (Tsol) and visible transmittance (Tvis) between the bleached and colored states. As shown in Table 1, solar transmittance (Tsol) could be adjusted by 41.6%, from 48.1% when bleached to 6.5% when colored, and visible transmittance (Tvis) could be adjusted by 52.8%, from 64.8% when bleached to 12% when colored.

Figures 3–5 show the results of a spectroscopic analysis of electrochromic glazing and the spectral data of clear glass and single low-e glass provided by Optic5.1.

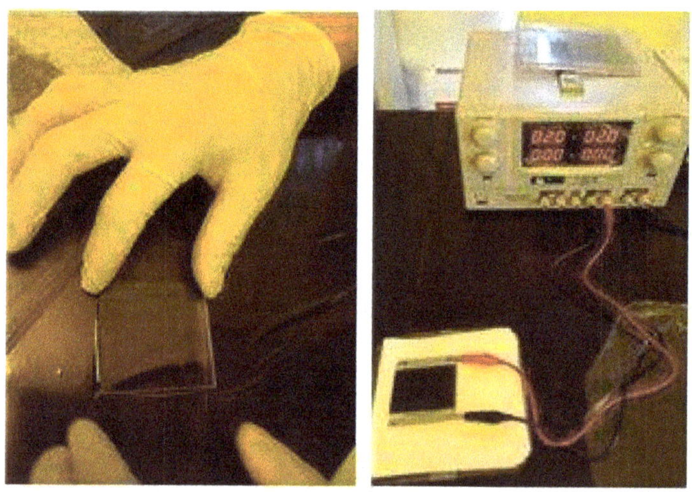

Figure 2. Electrochromic glass cell and colored mode.

Building load can be predicted using spectral data. According to the transmittance graph shown in Figure 3, electrochromic glazing exhibited a higher visible ray rejection in the colored state than in the clear glass or single low-e glass state, and similar rejection to that of single low-e glass in the NIR range. Its total solar transmittance (Tsol) was low, and thus, it is effective in terms of cooling load reduction in the summer. In bleached state, its visible ray and NIR transmittance was higher than that of single low-e glass, and thus it is more likely to reduce heating load. However, its transmittance was lower than that of clear glass, and thus it is less likely to reduce heating load. Moreover, the front and back-side reflectance graphs provided in Figures 4 and 5 show that even though the reflectance of electrochromic glazing was lower than that of single low-e glass in the colored state, transmittance was low primarily due to absorption.

Optical properties and building load can be predicted through spectroscopic analysis. However, for more accurate predictions, it is necessary to analyze the solar heat gain coefficient of the glazing system by considering the penetration, absorption, and reflection of glass, and to conduct analyses based on dynamic building energy simulations, which are described in the following section.

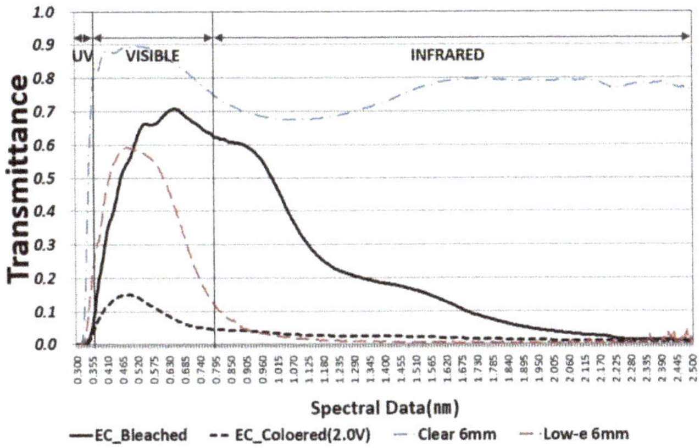

Figure 3. Transmittance of clear glass, single low-e glass, and electrochromic (EC) glass.

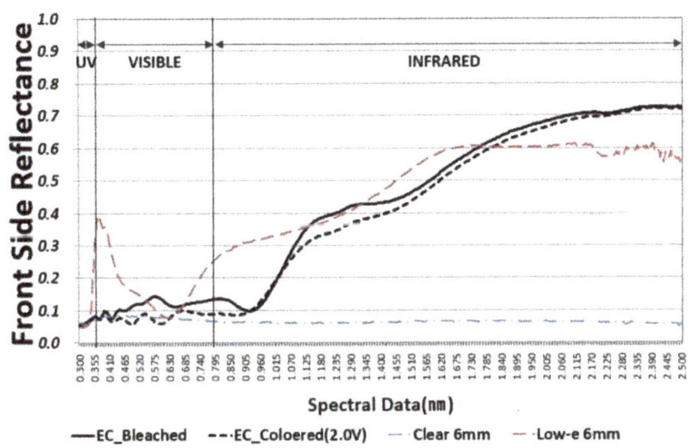

Figure 4. Front-side reflectance of clear glass, single low-e glass, and electrochromic (EC) glass.

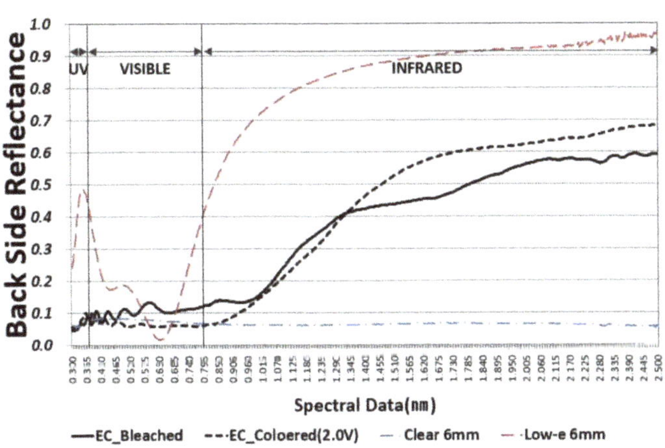

Figure 5. Back-side reflectance of clear glass, single low-e glass, and electrochromic (EC) glass.

Table 1. Input data concerning electrochromic (EC) glass to EnergyPlus.

Division		EC Glass (Bleached)	EC Glass (Colored)
Thickness (mm)		9	9
Solar Transmittance	front	0.481	0.065
	back	0.481	0.065
Solar Reflectance	front	0.214	0.181
	back	0.190	0.168
Visible Transmittance	front	0.648	0.120
	back	0.648	0.120
Visible Reflectance	front	0.128	0.074
	back	0.111	0.062
Front and back side emissivity		0.840	0.840
Conductivity (W/mK)		1.000	1.000

3. Simulation Conditions

3.1. Overview of Analytical Simulation Model

The facades of recently constructed high-rise residential buildings in Korea are similar to those of office buildings, as they use curtain wall windows and doors with high window-to-wall ratios in order to provide a view. We set as our hypothesis the claim that high-rise residential buildings have high cooling loads, and by applying electrochromic glazing with variable solar control and high solar radiation rejection, this load can be significantly reduced in comparison with the reduction obtained by using double-glazing products or shading devices. Accordingly, we selected a high-rise residential building recently constructed in Seoul as an analytical model, as shown in Figure 6.

EnergyPlus 8.5 (developed by the DOE (Department of Energy)) was used as simulation tool [21]. EnergyPlus is based on the heat balance equation, and thus can calculate heat transfer by using the conduction, convection, and radiation of windows and doors as well as the penetration, reflection, and absorption of solar radiation [22–28]. It also allows for a detailed analysis of heat transfer between the shading device, and doors and windows [29].

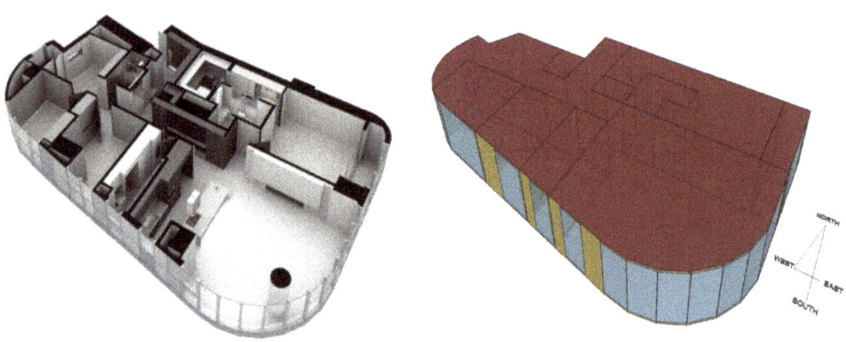

Figure 6. Perspectival view (**left**) and EnergyPlus simulation model by OpenStudio (**right**).

The floor area of the analytical model was 215.12 m^2, the air conditioning area was 181.63 m^2 (84.4%), the height of the living space was 2.9 m, and the area occupied by the windows was 67.7 m^2 on an exterior wall with an area of 106.5 m^2. Thus, the window-to-wall ratio was 63.5%.

The analytical mode was constructed based on a typical floor. The ceilings, floors, and northern side wall were set as adiabatic boundaries. The southern, eastern, and unclear western exterior walls followed the insulation standards set by the Standard for Energy Saving Design in Buildings (2016), as shown in Table 2. The southern, eastern, and western windows had clear double glazing (6 mm clear + 12 mm air + 6 mm clear), with a heat transmission coefficient of 2.685 W/m^2 K, a solar heat gain coefficient of 0.714, and a visible ray transmittance of 0.790. Clear double glazing was applied to analyze the load of the baseline, and the following section describes the application of different types of glass.

The temperatures set for the model were 22 °C and 28 °C, based on the indoor temperature standard for the cooling and heating systems in the Standard for Energy Saving Design in Buildings [30]. The heating period was assumed to span from October to March, and the cooling period was assumed to span from May to September. Weather data for Seoul were used. The city has a continental climate where the annual variation in temperature is as high as 30 °C, as high pressure from the Eurasian continent dominates the winter season, and oceanic air masses at high temperature and humidity significantly influence the weather in the summer. The IdealLoadsAirSystem (provided by EnergyPlus) was applied for load analysis of the HVAC system to eliminate interference by system variables as much as possible. Using the IdealLoadsAirSystem, the cooling and heating loads required in the zone were processed by a virtual air conditioning system with infinite capacity, and load calculation was performed using only the difference in enthalpy between the specified air supply and the mixed (ventilation + outside) air.

The internal heat gain is shown in Table 3 with reference to the standard of ASHRAE (American Society of Heating, Refrigerating and Air-Conditioning Engineers, Inc.) fundamentals (2009) [31]. Multifamily two-zone (bedroom and other rooms) provided by the schedules of the datasets in EnergyPlus was applied for the internal heat gain schedules of the body, the lighting, and the equipment [32]. Outdoor air volume was set to 1.1 m^3/m^2 h, which is in accordance with building energy rating certificate guidelines [30].

Table 2. Composition of simulation base model. SHGC: solar heat gain coefficient.

Division	Materials	Thermal Properties
Exterior wall	200 mm concrete 155 t insulation 19 mm gypsum board	U-value 0.429 W/m^2 K
Interior floor	150 mm concrete 30 mm insulation 40 mm ALC(Autoclaved lightweight concrete) 30 mm mortar	U-value 1.186 W/m^2 K
Glazing	24 mm double glazing (6 clear + 12 air + 6 clear)	SHGC 0.695 Tvis 0.777 U-value 2.543

Table 3. Internal gain.

Internal Load Type	Maximum Value
Overhead lighting (W/m^2)	10.8
Peak occupancy (m^2/person)	17.0
Equipment (W/m^2)	8.6

3.2. *Analysis of Building Load Characteristics of Analytical Model*

As stated earlier, we hypothesize that the facades of high-rise residential buildings in Korea are similar to those of office buildings, and thus have high cooling loads. We found through the building load simulation that the heating load was 4285.1 kWh/year (16.0%), and the cooling load was 22,490.0 kWh/year (84%), as shown in Figure 7, Table 4.

Table 4. Proportions of heating and cooling loads.

Division	Total Load (kWh)	
	Heating	Cooling
Total	4285.1	22,490.0
Rate	16.0%	84.0%

The energy consumption for heating and cooling may be different depending on the operating conditions or efficiency of the systems. However, as the systems are operated using IdealLoadsAirSystem, and their performance and relevant conditions are excluded, the load that is needed to continuously maintain the optimum temperature in light of indoor comfort was larger for cooling, as presumed in the hypothesis.

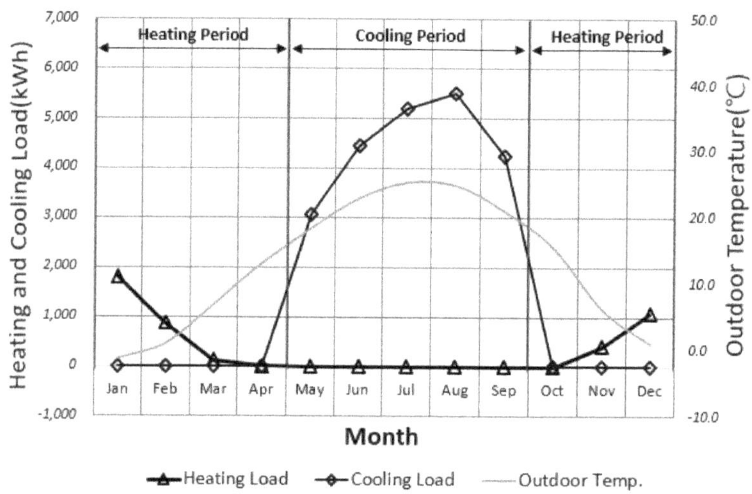

Figure 7. Monthly heating and cooling loads of baseline model.

The cooling load was high, because glass constitutes a large portion of the exterior wall, and thus the internal heat gain owing to solar penetration was dominant. On the contrary, in the analytical model, the heating load was reduced by increasing the internal heat gain through solar penetration. This is explained further in Section 4.

Thus, for Korean high-rise residential buildings with high window-to-wall ratios, solar control may be key to reducing the building's energy consumption. We aim to test the second hypothesis, that electrochromic glazing with variable solar control and high solar radiation rejection is more effective at reducing the load on buildings compared with existing double-glazing and shading devices.

3.3. Composition of Double-Glazing and Shading Devices for Comparative Analysis

We collected data concerning the heat and optical properties of commercialized double-glazing and shading devices to analyze the building performance of electrochromic glazing in terms of reducing the energy consumption of buildings. IGDB (LBNL) was used for data collection, and Window 7.2 (LBNL) was used to obtain the data through double-glazing composition.

For the first category, green, blue, and gray-colored glass was layered with clear glass to form ALT-01, 02, and 03 to compare with the colored glass that was already installed in the given building. To compare its performance with electrochromic glazing on equal terms, the glazing was layered with clear glass to form ALT-08.

To compare with low-e glass, which was the second category, single low-e and double low-e glass was layered with clear glass to form ALT-04 and 05, respectively. "Single" and "double" here represent the number of soft low-e coatings. Emissivity decreased as the number of coatings increased; this lowered the solar heat gain coefficient (SHGC) and the heat transmission coefficient. We used ALT-08 data to compare the performance on equal terms with low-e glass.

We selected Venetian blinds and roll shades as the shading devices for the third category. These devices are used in high-rise residential buildings in Korea. We formed ALT-06 and 07 by integrating them with the double glazing, which was a combination of single low-e glass and clear glass. Both shading devices were assumed to be installed indoor, open in winter (October–March), and closed in summer (May–September), with a focus on energy reduction. Moreover, the slats of the Venetian blinds were completely shut in the closed state. The reflectance of the blinds and roll shades was set to 80%, and their transmittance was set to 0%. For transmittance, we selected the condition where load reduction for cooling was the highest based on the results of past studies [33]. Electrochromic glazing consisted of the double glazing that was a combination of electrochromic glazing and single low-e glass for the comparative analysis of performance on equal terms with the shading devices. It was assumed that the electrochromic glazing was bleached in the winter and colored in the summer for energy reduction. Table 5 and Figure 8 show data concerning the heat and the optical properties of each type of double glazing for the comparative evaluation of load reduction performance.

Table 5. Thermal properties of each type of glazing.

Division				U-Value	SHGC	Tvis
BSL *	Baseline	clear + clear		2.685	0.714	0.790
ALT-01	Colored glazing	green + clear		2.685	0.632	0.671
ALT-02		blue + clear		2.685	0.616	0.511
ALT-03		gray + clear		2.685	0.618	0.393
ALT-04	Low-e glazing	single low-e+ clear		1.630	0.571	0.711
ALT-05		double low-e + clear		1.600	0.325	0.503
ALT-06	Low-e glazing with shading device	interior blind + clear + single low-e	open	1.630	0.571	0.711
			close	1.624	0.333	0.000
ALT-07		interior roll shade + clear + single low-e	open	1.630	0.571	0.711
			close	1.579	0.319	0.000
ALT-08	Electrochromic glazing (controlled)	EC + clear	bleached	2.853	0.487	0.579
			colored	2.853	0.191	0.107
ALT-09		EC + single low-e	bleached	1.619	0.409	0.521
			colored	1.619	0.128	0.096

* BSL: Baseline.

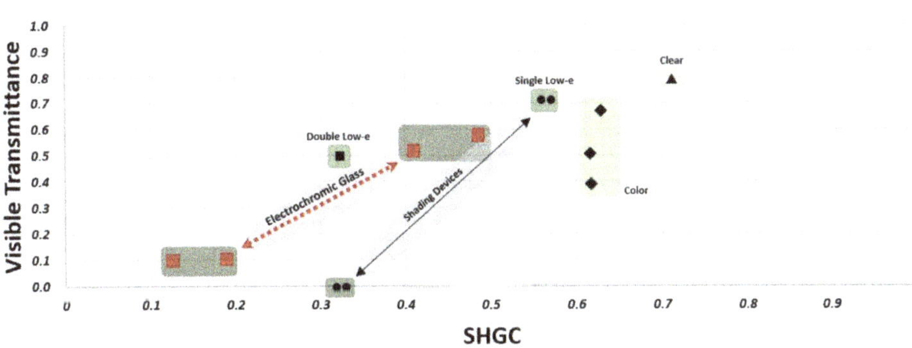

Figure 8. Optical properties of each type of glazing.

4. Results and Discussion

4.1. Performance Evaluation of Hourly Heating and Cooling Loads

For the evaluation of hourly heating and cooling loads, we analyzed the hourly changes in heating load according to the type of double glazing under the conditions obtained on the design day using meteorological data for Seoul. As described above, the indoor cooling and heating temperatures were set to 22 °C and 28 °C, respectively, using the IdealLoadsAirSystem. This was done to calculate the heating and cooling loads that were necessary to reach the indoor temperature, based on which we can check for the reduction in heating and cooling loads according to type of double glazing.

4.1.1. Comparative evaluation of hourly heating load performance

Figure 9 and Table 6 show a comparison of the performance of different types of double glazing that are listed in Table 5 in terms of reducing heating load. Figures 10–12 show the changes in heating load for each type of double glazing. This load is always high when electrochromic glazing is applied, regardless of time.

We analyzed heat flow through glass to determine the factors directly affecting heating load, and the relevant graph is shown in Figure 9, Table 7. We selected Display AdvancedReportVarialbles of Output:Dignostics as the data output of the EnergyPlus simulation tool, which supports a detailed analysis of the heat gain and heat loss of glass. The formula for calculating heat flow is given in Equation (1):

$$Q = \sum_{i=1}^{n} (TS_i + Cv_i + IR_i - SR_i) \quad (1)$$

where Q indicates the accumulated heat flow through the glass over time (i), which is calculated by subtracting the hourly accumulation of short-wave radiation back out (SR_i) from the hourly accumulation of transmitted solar (TS_i), convective heat flow (Cv_i), and net IR heat flow (IR_i). A positive value (+) implied heat gain, whereas a negative value (−) indicated heat loss. SR_i is the transfer of short-wave radiation from the zone back out the window in watts. This is a measure of the diffuse short-wave light (from reflected solar and electric lighting) that leaves the zone through the window.

Table 6. Heating loads of clear, colored, and electrochromic (EC) glazing.

Glass Division		Clear	Green	Blue	Gray	EC
						Bleached
Glazing properties	U-Value (W/m^2 K)	2.685	2.685	2.685	2.685	2.853
	SHGC	0.714	0.632	0.616	0.618	0.487
	Tvis	0.790	0.671	0.511	0.393	0.579
Heating load (kWh)		83.0	89.7	91.0	90.9	105.0

※ Double Glazing—[Clear, Green, Blue, Gray, EC] + Clear Glass

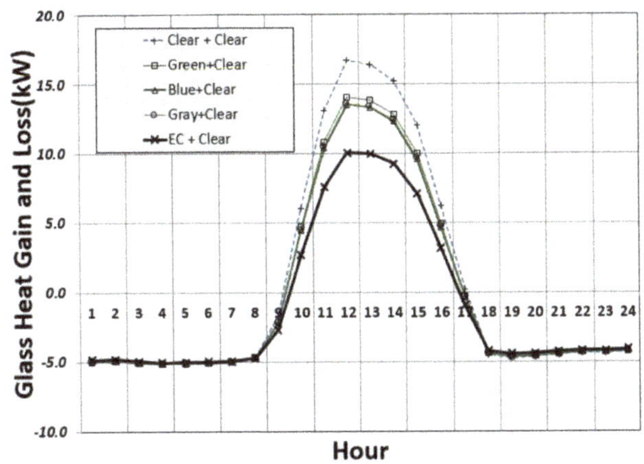

Figure 9. Glass heat gain and loss of clear, blind, roll shade, and electrochromic glazing.

Table 7. Heat flow through clear, colored, and electrochromic (EC) glazing.

Glass Division	Clear	Green	Blue	Gray	EC
					Bleached
Transmitted solar (kWh)	106.1	62.0	52.7	54.2	64.4
Convection heat flow (kWh)	−26.5	−19.7	−17.9	−18.2	−27.9
Infrared heat flow (kWh)	−70.1	−45.8	41.0	−41.8	−61.0
Shortwave heat flow (kWh)	3.3	1.2	0.9	0.9	1.3
Total heat flow (kWh)	6.2	−4.8	−7.0	−6.8	−25.8

Figure 9 shows no significant difference in heat loss for double glazing during the night when there was no solar radiation. This is because the heat transfer coefficient of each type of double glazing was similar. However, the low solar heat gain coefficient of electrochromic glazing resulted in low solar penetration during the day; this had the strongest impact on the increase in heating load.

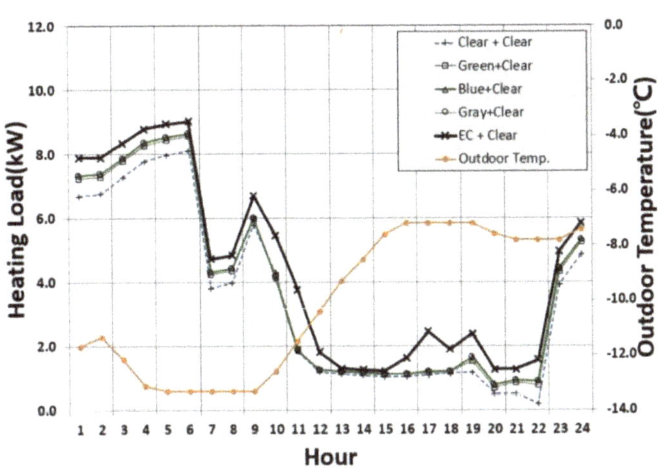

Figure 10. Heating loads of clear, colored, and electrochromic (EC) glazing.

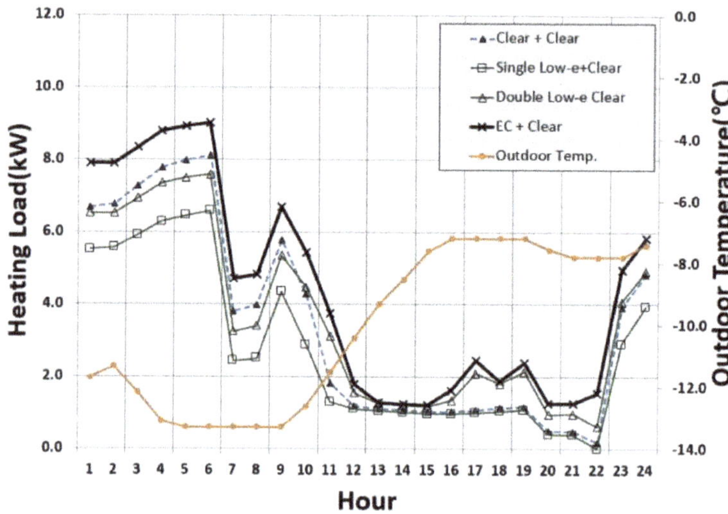

Figure 11. Heating loads of clear, single low-e, double low-e, and electrochromic (EC) glazing.

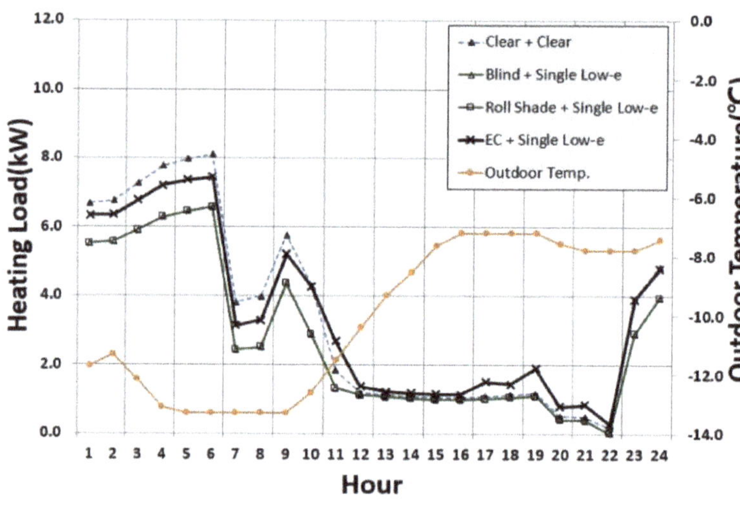

Figure 12. Heating loads of blind, roll shade, and electrochromic (EC) glazing.

In other words, compared with clear double glazing and colored glass, electrochromic glazing has unfavorable heat and optical properties in terms of reducing heating load. Therefore, based on the cumulative heat flow in Table 6, it appears necessary to achieve as much solar penetration as that for clear double glazing, or to increase the heat transmission coefficient to minimize the heat loss and compensate for the heat lost overnight to improve the performance of electrochromic glazing.

Figure 11, Table 8 compares the performance of electrochromic glazing with that of low-e double glazing in terms of reducing heating load, and shows that the heating load of electrochromic glazing was the highest. Table 8 shows the daily cumulative heating load for each type of double glazing.

Table 8. Heating loads of clear, low-e, and electrochromic (EC) glazing.

Glass Division		Clear	Single Low-e	Double Low-e	EC
					Bleached
Glazing properties	U-Value (W/m^2 K)	2.685	1.630	1.600	2.853
	SHGC	0.714	0.571	0.325	0.487
	Tvis	0.790	0.711	0.503	0.579
Heating load (kWh)		83.0	65.7	86.2	105.0
※ Double Glazing—[Clear, Single Low-e, Double Low-e, EC] + Clear Glass					

The heat flow shown in Figure 13 was analyzed to determine the cause of the increase in the heating load of electrochromic glazing. The heat transmission coefficient of single and double low-e double glazings is lower than that of clear double glazing and electrochromic glazing. Thus, heating load was low at night when there was no solar radiation. It should be noted here that after 10:00 a.m. (Figure 11), the heating load of double low-e glazing increased to a greater extent than that of clear glass. This might have been because the heat transmission coefficient of double low-e glazing was low, as shown in Figure 13, Table 9 and, thus, heat loss was low at night. However, the heat gain became the lowest as solar penetration decreased compared with clear glass when the solar radiation began. Thus, it appears that a combination of double glazing with single low-e glass, which can suitably maintain transmittance while increasing heat insulation, is more effective at reducing building energy in the winter in light of the variable solar radiation of electrochromic glazing.

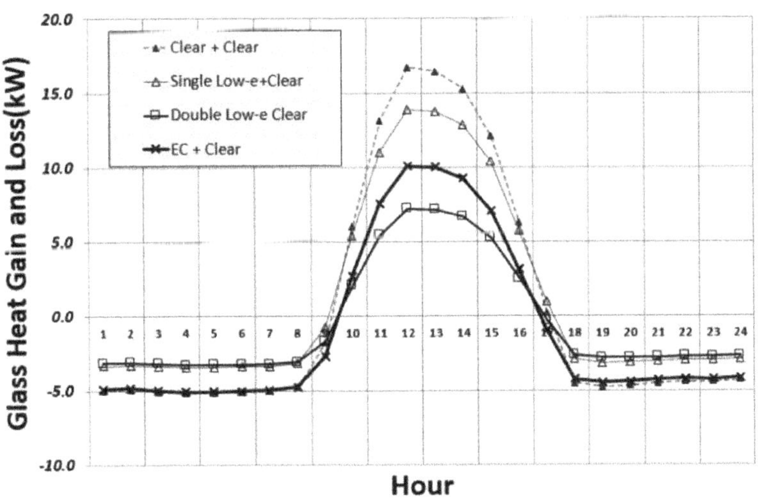

Figure 13. Glass heat gain and loss of clear, single low-e, double low-e, and electrochromic (EC) glazing.

Table 9. Heat flow through clear, low-e, and electrochromic (EC) glazing.

Glass Division	Clear	Single Low-e	Double Low-e	EC
				Bleached
Transmitted solar (kWh)	106.1	68.3	33.3	64.4
Convection heat flow (kWh)	−26.5	−9.8	−11.6	−27.9
Infrared heat flow (kWh)	−70.1	−34.3	−32.1	−61.0
Shortwave heat flow (kWh)	3.3	1.5	0.4	1.3
Total heat flow (kWh)	6.2	22.7	−10.7	−25.8

Figure 12, Table 10 compares the performance of electrochromic glazing with that of shading devices in terms of reducing heating load. The double glazing was formed by combining single low-e glass with all shading devices and electrochromic glazing, based on the results of a comparative analysis with low-e glass. It can be seen that the heating load of electrochromic glazing decreased to a greater extent than that of clear glass, because applying single low-e glass reduced the heat transmission coefficient, which in turn reduced the heat lost overnight, as shown in Figure 14, Table 11.

The blinds and roll shades were kept open in winter and, thus, heating load and heat flow were the same as those for single low-e glazing. As the solar heat gain coefficient of electrochromic glazing is lower than that of single low-e glazing, its heating load was still higher than that of double glazing when the blinds and roll shades were open.

In summary, transmittance was low when the electrochromic glazing was bleached, and hence, was not effective in terms of reducing heating load. Therefore, work is needed to increase transmittance in the bleached state when developing electrochromic glazing in the future.

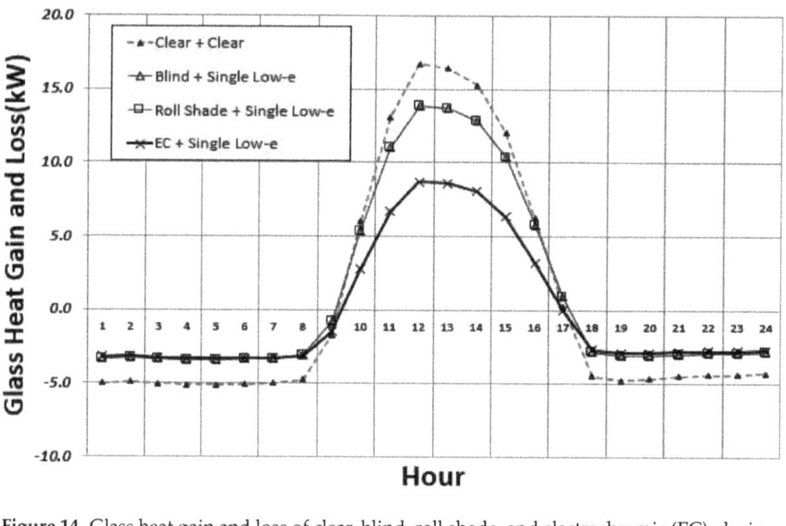

Figure 14. Glass heat gain and loss of clear, blind, roll shade, and electrochromic (EC) glazing.

Table 10. Heating load of clear, shading devices, and electrochromic (EC) glazing.

Glass Division		Clear	Blind	Roll Shade	EC
			Open		Bleached
Glazing properties	U-Value (W/m^2 K)	2.685	1.630	1.630	1.619
	SHGC	0.714	0.571	0.571	0.409
	Tvis	0.790	0.711	0.711	0.521
Heating load (kWh)		83.0	65.7	65.7	81.5
※ Double Glazing—[Clear, Single Low-e, Double Low-e, EC] + Single Low-e Glass					

Table 11. Heat flow through clear, blind, roll shade, and electrochromic (EC) glazing.

Glass Division	Clear	Blind	Roll Shade	EC
				Bleached
Transmitted solar (kWh)	106.1	68.3	68.3	42.4
Convection heat flow (kWh)	−26.5	−9.8	−9.8	−11.9
Infrared heat flow (kWh)	−70.1	−34.3	−34.3	−34.0
Shortwave heat flow (kWh)	3.3	1.5	1.5	0.6
Total heat flow (kWh)	6.2	22.7	22.7	−4.1

4.1.2. Comparative Evaluation of Hourly Cooling Load Performance

Figure 15, Table 12 compares the performance of electrochromic glazing in terms of cooling load with that of colored glass double glazing. Cooling load is always low when electrochromic glazing is applied, regardless of time. The daily cumulative cooling load for each type of double glazing is provided in Table 12. The magnitude of reduction in cooling load was larger than the increase in the heating load.

Such a reduction in the cooling load of electrochromic glazing occurred because, as shown in Figure 16, Table 13, there was no considerable difference between the overnight heat loss for double glazing. This is because the heat transmission coefficients were similar and solar penetration was low owing to the nature of electrochromic glazing, which has a low solar heat gain coefficient. In other words, compared with clear double glazing and colored glass, electrochromic glazing has favorable heat and optical properties in terms of reducing cooling load.

Table 12. Cooling loads of clear, colored, and electrochromic (EC) glazing.

Glass Division		Clear	Green	Blue	Gray	EC
						Colored
Glazing properties	U-Value (W/m^2 K)	2.685	2.685	2.685	2.685	2.853
	SHGC	0.714	0.632	0.616	0.618	0.191
	Tvis	0.790	0.671	0.511	0.393	0.107
Cooling Load (kWh)		262.2	252.6	250.7	251.0	185.8
※ Double Glazing—[Clear, Green, Blue, Gray, EC] + Clear Glass						

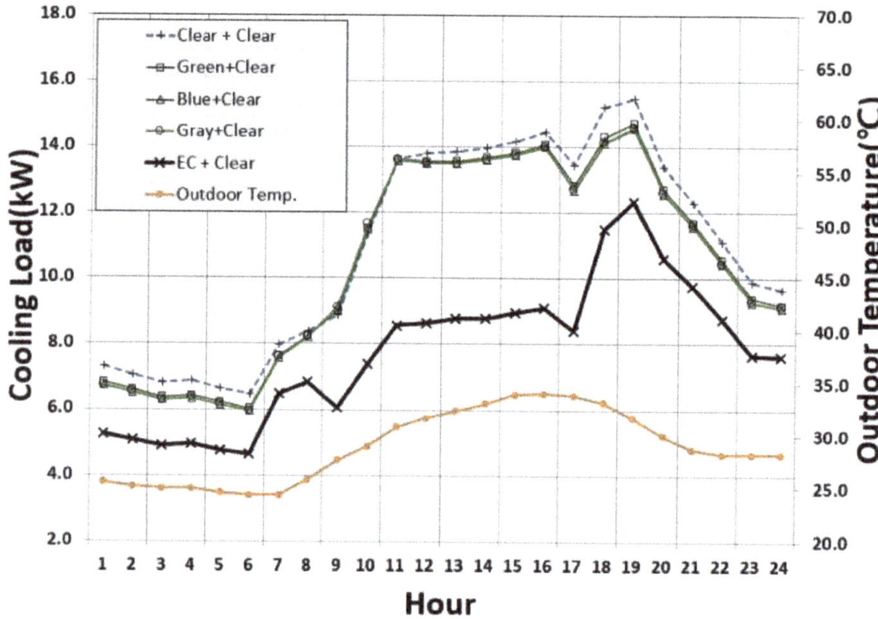

Figure 15. Cooling loads of clear, colored, and electrochromic (EC) glazing.

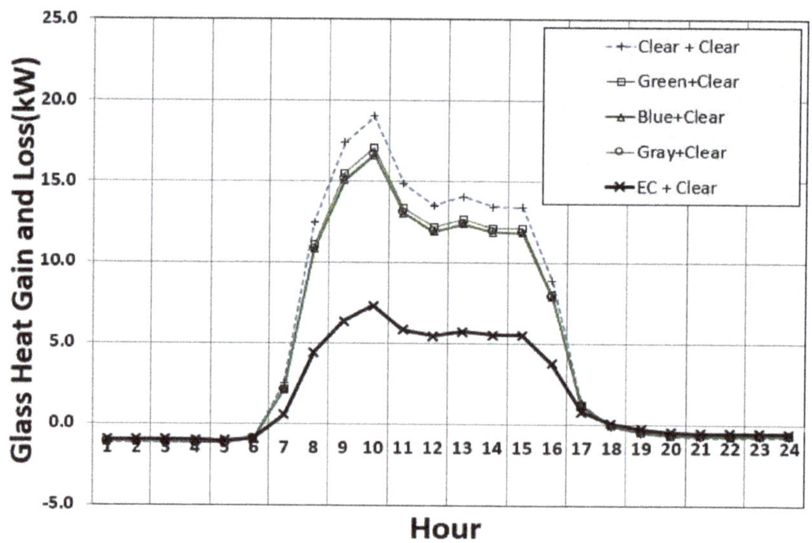

Figure 16. Glass heat gain and loss of clear, blind, roll shade, and electrochromic (EC) glazing.

Figure 17, Table 14 shows a comparison of the performance of electrochromic glazing in terms of cooling load with that of low-e double glazing. It is clear that the cooling load of electrochromic glazing was lower.

The analysis of heat flow in Figure 18, Table 15 shows no significant difference in overnight heat loss due to the difference in the heat transmission coefficient in summer. The heat gain due to the difference in the solar heat gain coefficient during the day had the strongest impact on the reduction in the cooling load. In other words, unlike in winter, where there was a complex impact of the heat transmission coefficient and solar heat gain coefficient, only the effect of the solar heat gain coefficient was dominant in summer. Therefore, as shown in Figure 17, electrochromic glazing with the lowest solar heat gain coefficient was the most effective in terms of reducing the cooling load.

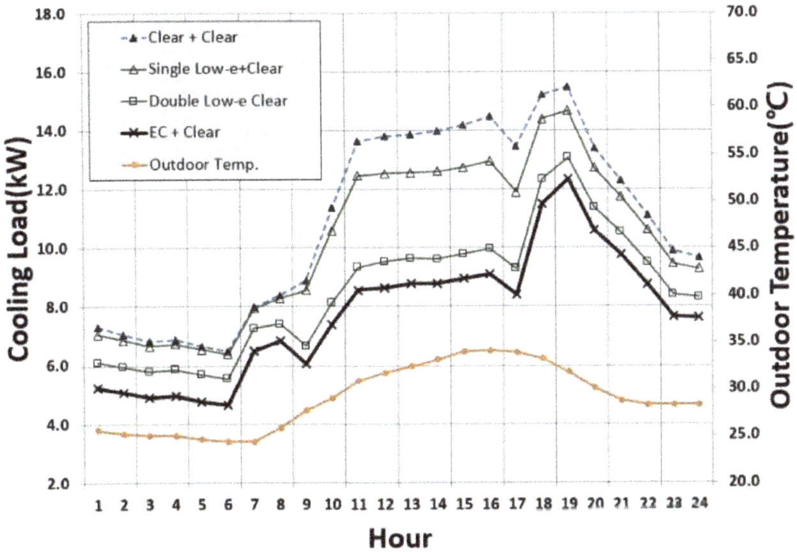

Figure 17. Cooling loads of clear, single low-e, double low-e, and electrochromic (EC) glazing.

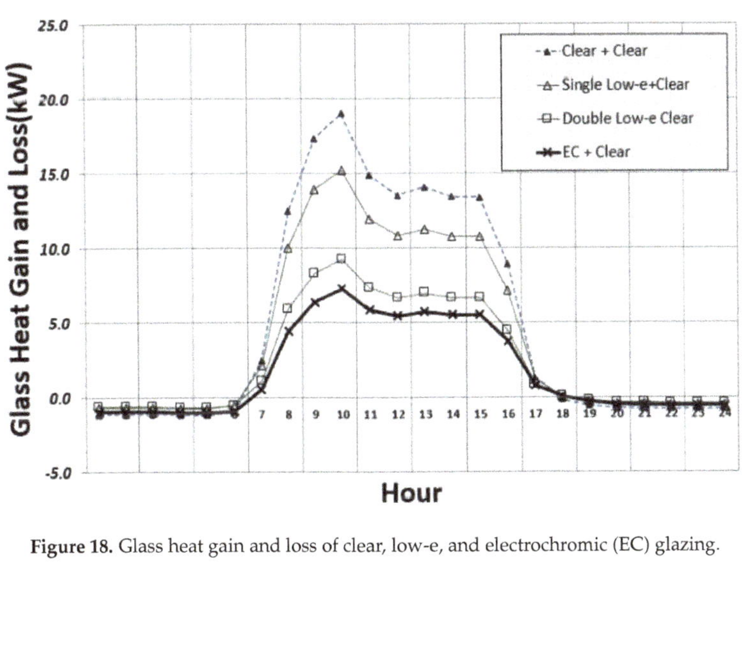

Figure 18. Glass heat gain and loss of clear, low-e, and electrochromic (EC) glazing.

Table 13. Heat flow through clear, colored, and electrochromic (EC) glazing.

Glass Division	Clear	Green	Blue	Gray	EC
					Colored
Transmitted solar (kWh)	106.6	61.4	52.0	53.5	8.7
Convection heat flow (kWh)	9.8	20.3	22.4	22.0	13.4
Infrared heat flow (kWh)	0.7	24.4	29.3	28.4	20.6
Shortwave heat flow (kWh)	2.8	1.0	0.7	0.7	0.0
Total heat flow (kWh)	114.3	105.1	103.0	103.2	42.7

Table 14. Cooling loads of clear, low-e, and electrochromic (EC) glazing.

Glass Division		Clear	Single Low-e	Double Low-e	EC
					Colored
Glazing properties	U-Value (W/m^2 K)	2.685	1.630	1.600	2.853
	SHGC	0.714	0.571	0.325	0.191
	Tvis	0.790	0.711	0.503	0.107
Cooling load (kWh)		262.2	246.2	205.1	185.8
※ Double Glazing—[Clear, Single Low-e, Double Low-e, EC] + Clear Glass					

Figure 19, Table 16 compares the performance of electrochromic glazing with that of the shading devices in terms of the cooling load. Similar to the case in winter, single low-e glass was combined with all of the shading devices and electrochromic glazings to form double glazing; blinds and roll shades were closed, and the electrochromic glazing was bleached. A reflectance of 90% and a transmittance of 0% were applied to the blinds and rolls shades to achieve the highest reduction in cooling load in the summer, and the slats of the blinds were assumed to be completely closed. We observed that electrochromic glazing yielded an excellent performance in terms of reducing cooling load. While shading devices block solar radiation, they are installed indoors, and thus, the solar radiation that penetrates through the glass is partly absorbed by them, and flows indoors in the form of long-wave radiation, thereby affecting the heat gain. Table 17 shows that the heat gain of the blinds and roll shades was high because of the convection and radiation between the double glazing and the blinds. Figure 20 shows that the indoor temperature of the glass surface was higher than that of the electrochromic glazing. Therefore, electrochromic glazing is more effective than the currently used shading devices as a solar control system that blocks solar radiation. As shown in Figures 15, 20 and 21, a comprehensive analysis of hourly heating and cooling loads revealed that the low transmittance values of the colored electrochromic glazing resulted in the largest reduction in solar gain from 07:00 a.m. h to 17:00 p.m., and was thus more effective at reducing cooling loads than the colored, low-e, and shading devices.

Table 15. Heat flow through clear, low-e, and electrochromic (EC) glazing.

Glass Division	Clear	Single Low-e	Double Low-e	EC
				Bleached
Transmitted solar (kWh)	106.6	68.4	33.5	8.7
Convection heat flow (kWh)	9.8	14.7	11.3	13.4
Infrared heat flow (kWh)	0.7	14.7	13.7	20.6
Shortwave heat flow (kWh)	2.8	1.2	0.3	0.0
Total heat flow (kWh)	114.3	96.5	58.2	42.7

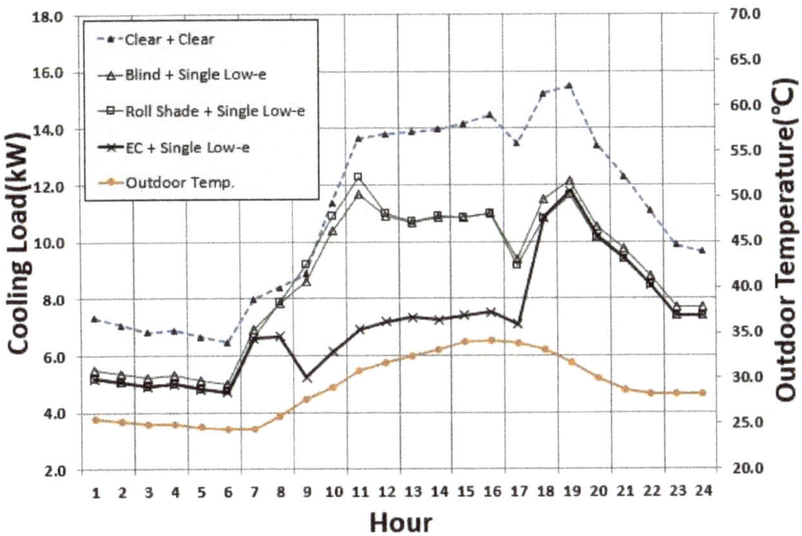

Figure 19. Cooling loads of blind, roll shade, and electrochromic (EC) glazing.

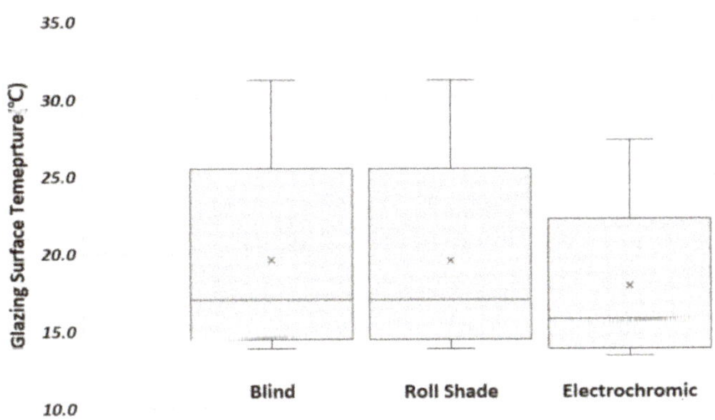

Figure 20. Internal surface temperature of blind, roll shade, and electrochromic (EC) glazing.

Table 16. Cooling loads of clear, blind, roll shade, and electrochromic (EC) glazing.

Glass Division		Clear	Blind	Roll Shade	EC
			Closed		Colored
Glazing properties	U-Value (W/m^2 K)	2.685	1.624	1.579	1.619
	SHGC	0.714	0.333	0.319	0.128
	Tvis	0.790	0.000	0.000	0.096
Cooling load (kWh)		262.2	208.8	205.4	170.6

※ Double Glazing—[Clear, Single Low-e, Double Low-e, EC] + Single Low-e Glass

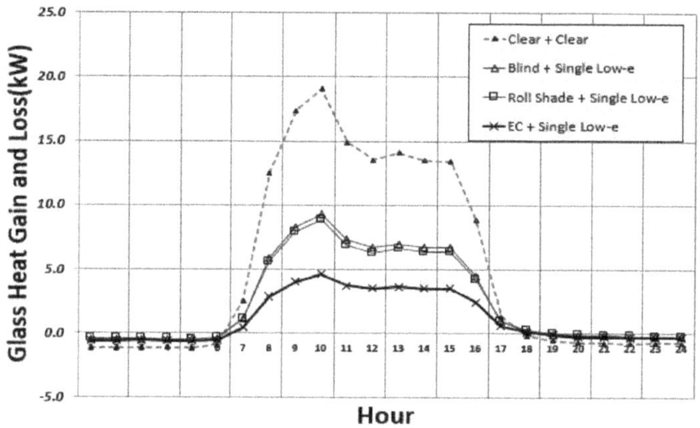

Figure 21. Glass heat gain and loss of clear, blind, roll shade, and electrochromic (EC) glazing.

Table 17. Heat flow through clear, blind, roll shade, and electrochromic (EC) glazing.

Glass Division	Clear	Blind	Roll Shade	EC
				Bleached
Transmitted solar (kWh)	106.6	1.5	0.0	5.7
Convection heat flow (kWh)	9.8	39.5	46.4	8.6
Infrared heat flow (kWh)	0.7	19.5	11.4	13.4
Shortwave heat flow (kWh)	2.8	0.0	0.0	0.0
Total heat flow (kWh)	114.3	60.5	57.7	27.7

4.2. Monthly and Annual Changes in Heating and Cooling Loads

Figure 22 and Table 18 show the total annual cooling and heating loads. Even though electrochromic glazing is sub-optimal at reducing heating loads, it was the most effective at reducing the total annual load in buildings with the cooling load being dominant, which is similar to the analytical model in this study. In particular, compared with shading devices that are mainly used in high-rise residential buildings using the baseline, electrochromic glazing significantly reduced the total load by up to 12.7–14.9%. This implies that electrochromic glazing has potential for use in next-generation structural glass.

creffig:sustainability-280043-f023,fig:sustainability-280043-f024,fig:sustainability-280043-f025 show the monthly cumulative changes in heating and cooling loads, according to the type of double glazing. The same trend as that in the analysis of hourly cooling and heating loads was observed. Compared with all of the other types of double glazing, the heating load of electrochromic glazing was high in winter even when bleached, and this contributed to a reduction in cooling load in the colored state in the summer. In other words, applying double glazing is more effective for buildings with high cooling loads.

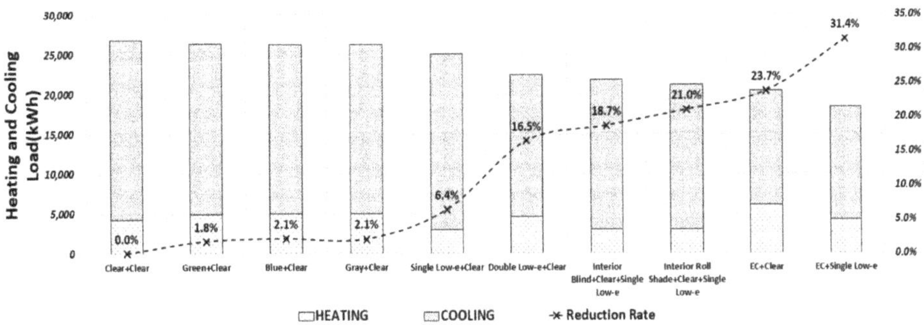

Figure 22. Comparison of annual heating and cooling loads.

Table 18. Heating and cooling loads.

Glass Division	Heating Load	Cooling Load	Total Load	Reduction Rate
Baseline (clear + clear)	4285.1	22,490.0	26,775.1	0.0%
Green + clear	4895.5	21,409.2	26,304.7	−1.8%
Blue + clear	5014.4	21,190.0	26,204.5	−2.1%
Gray + clear	4999.5	21,215.5	26,214.9	−2.1%
Single low-e + clear	2999.4	22,060.6	25,060.0	−6.4%
Double low-e + clear	4654.4	17,715.3	22,369.6	−16.5%
Single low-e + clear with blind	2999.4	18,775.8	21,775.3	−18.7%
Single low-e + clear with roll shade	2999.4	18,148.6	21,148.0	−21.0%
EC + clear	6073.3	14,343.0	20,416.3	−23.7%
EC + single low-e	4294.7	14,082.1	18,376.8	−31.4%

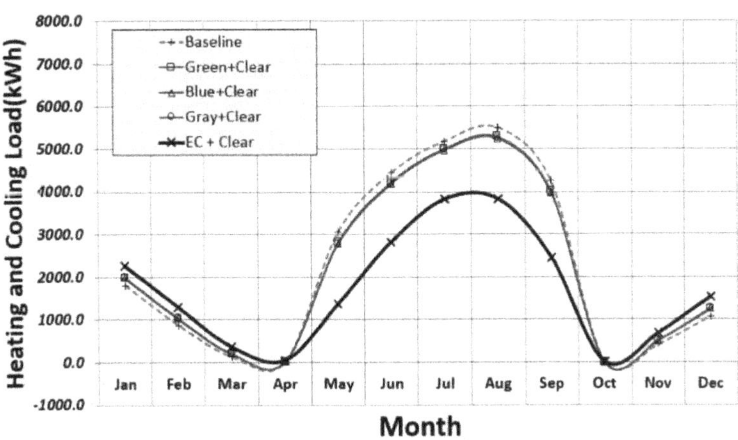

Figure 23. Annual heating and cooling loads of clear, colored, and electrochromic (EC) glazing.

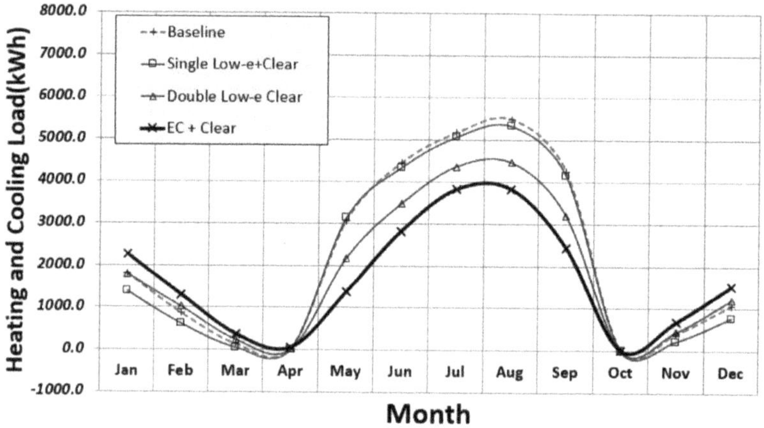

Figure 24. Annual heating and cooling loads of clear, low-e, and electrochromic (EC) glazing.

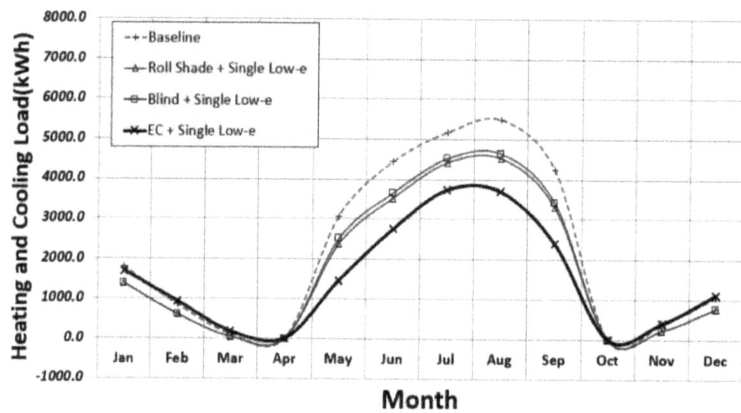

Figure 25. Annual heating and cooling loads of clear, blind, roll shade, and electrochromic (EC) glazings.

5. Conclusions

To assess the load reduction performance of the recently developed domestic electrochromic glazing, this study conducted a comparative analysis of the double-glazing and shading devices that are used for high-rise residential buildings in Korea. The results of the analysis can be summarized as follows:

(1) To test the hypothesis that the cooling load of the high-rise residential buildings was high, we analyzed the annual load using an analytical model on the EnergyPlus simulation tool. Heating and cooling loads comprised 84.0% and 16.0% of the total load, respectively, with the load pattern focused more on cooling.

(2) We hypothesized that the load reduction performance of electrochromic glazing with variable solar control and high solar radiation rejection is better than that of double-glazing products and shading devices, and analyzed the cooling and heating loads for each type of double glazing.

(3) In South Korea, which has different weather conditions in summer and winter, controlling solar radiation is important. The analysis of optical properties of electrochromic glazing showed that the solar transmittance values were 48.1% and 6.5% in the bleached and colored states,

respectively. This indicates that allowing and blocking solar radiation during winter and summer, respectively, are effective for reducing the heating and cooling loads in buildings.

(4) Colored electrochromic glazing had a higher solar radiation rejection than colored glass double glazing and low-e double glazing. Thus, it is effective for reducing cooling load. Moreover, it was observed to be excellent in terms of cooling load reduction compared with shading devices (blinds and roll shades), because such devices are installed indoors. Thus, the solar radiation that penetrates through the glass is partly absorbed by the shading devices, and flows indoors in the form of long-wave radiation, thereby reducing the heat gain due to convection and radiation. However, electrochromic glazing is applied to the outermost part to reduce the inflow of solar radiation.

(5) Electrochromic glazing was found to be more effective for reducing the cooling loads compared with the heating loads. However, in buildings with dominant cooling loads, similar to the analytical model used in this study, it exhibited the best performance in terms of reducing the total annual load.

The limitations of this study are that it failed to analyze the effect of energy consumption based on the use schedule of the HVAC unit, the effect of load reduction on selecting the capacity of the HVAC system, and LCC (Life Cycle Cost) analysis according to initial investment cost and actual energy consumption. Therefore, it is necessary to examine the actual reduction in cooling and heating energy when cooling and heating systems are applied to buildings based on this study. Furthermore, we controlled transmittance by limiting the dependent variables of the control of electrochromic glazing to cooling and heating loads, and categorizing the glazing as in the colored state in summer and bleached state in winter. However, by expanding the scope of dependent variables according to changes in the environment, more significant findings can be obtained in terms of building energy and the construction environment. Furthermore, we limited the shading devices that were used to internal venetian blinds and roll shades, which are widely used in buildings in South Korea. Thus, further research is needed for a comparative analysis between external shading devices and electrochromic glazing. This study also made a simplistic distinction between the summer colored state and the winter bleached state in order to control transmission. This control strategy was chosen with the aim of analyzing the effects of the optical properties of the proposed electrochromic glazing, which exhibited bleached and discolored states, on heating and cooling loads. However, a key advantage of electrochromic glazing is that transmission can be controlled in stages. Further research is needed to examine an optical control method to integrate energy due to cooling, heating, and lighting.

Acknowledgments: This research was supported by a grant (16CTAP-C114806-01) from the Technology Advancement Research Program (TARP), funded by the Ministry of Land, Infrastructure, and Transport of the Korean government.

Author Contributions: All authors contributed substantially to all aspects of this article.

Conflicts of Interest: The authors have no conflict of interest to declare.

References

1. Reilly, S.; Arasteh, D.; Selkowitz, S.E. Thermal and optical analysis of switchable window glazings. *Sol. Energy Mater.* **1991**, *22*, 1–14. [CrossRef]
2. Ghosh, A.; Norton, B.; Duffy, A. Daylighting performance and glare calculation of a suspended particle device switchable glazing. *Sol. Energy* **2016**, *132*, 114–128. [CrossRef]
3. Ghosh, A.; Mallick, T.K. Evaluation of colour properties due to switching behaviour of a PDLC glazing for adaptive building integration. *Renew. Energy* **2018**, *120*, 126–133. [CrossRef]
4. Rezaei, S.D.; Shannigrahi, S.; Ramakrishna, S. A review of conventional, advanced, and smart glazing technologies and materials for improving indoor environment. *Sol. Energy Mater. Sol. Cells* **2017**, *159*, 26–51. [CrossRef]

5. Ihara, T.; Gustavsen, A.; Jelle, B.P. Effect of facade components on energy efficiency in office buildings. *Appl. Energy* **2015**, *158*, 422–432. [CrossRef]
6. Piccolo, A.; Pennisi, F. Daylighting performance of an electrochomic window in a small-scale test-cell. *Sol. Energy* **2009**, *83*, 832–844. [CrossRef]
7. Tavares, P.F.; Gaspar, A.R.; Martins, A.G.; Frontini, F. Evaluation of electrochromic windows impact in the energy performance of buildings in Mediterranean climates. *Energy Policy* **2014**, *67*, 68–81. [CrossRef]
8. Elizabeth, P.; Leidy, J. Passive Design Strategies for Building Envelopes in Different Orientation. *Int. J. Sustain. Build. Technol. Urban Dev.* **2017**, *8*, 201–227.
9. Lee, E.S.; Tavil, A. Energy and visual comfort performance of electrochromic windows with overhangs. *Build Environ.* **2009**, *42*, 2439–2449. [CrossRef]
10. Lee, E.S.; Yazdanian, M.; Selkowitz, S. *The Energy-Savings Potential of Electrochromic Windows in the US Commercial Buildings Sector*; University of California: Oakland, CA, USA, 2004.
11. Papaefthimiou, S.; Syrrakou, E.; Yianoulis, P. Energy performance assessment of an electrochromic window. *Thin Solid Films* **2006**, *502*, 257–264. [CrossRef]
12. Fernandes, L.L.; Lee, E.S.; Ward, G. Lighting energy savings potential of split-pane electrochromic windows controlled for daylighting with visual comfort. *Energy Build.* **2013**, *61*, 8–20. [CrossRef]
13. Ajaji, Y.; André, P. Thermal comfort and visual comfort in an office building equipped with smart electrochromic glazing: An experimental study. *Energy Proc.* **2015**, *78*, 2464–2469. [CrossRef]
14. DeForest, N.; Shehabi, A.; Garcia, G.; Greenbalt, J.; Masanet, E.; Lee, E.S.; Selkowitz, S.; Milliron, D.J. Regional performance targets for transparent near-infrared switching electrochromic window glazings. *Build. Environ.* **2013**, *61*, 160–168. [CrossRef]
15. Cannavale, G.E.; Eperon, P.; Cossari, A.; Abate, H.J.; Snaith, G.; Gigli, P. Photovoltachromic cells for building integration. *Energy Environ. Sci.* **2015**, *8*, 1578–1584. [CrossRef]
16. Sage Glass. Available online: https://www.sageglass.com (accessed on 22 February 2018).
17. Granqvist, C.G.; Bayrak Pehlivan, I.; Niklasson, G.A. Electrochromics on a roll: Web-coating and lamination for smart windows. *Surf. Coat. Technol.* **2017**, 6–11. [CrossRef]
18. Rottmann, M.; Kraft, A.; Heckner, K.H.; Hausler, T.; Fischer, U.; Stenzel, H. Large-area electrochromic safety glass; switching behaviour and transmission control of solar radiation. In Proceedings of the 9th International Glass Conference on Glass Processing Days, Tampere, Finland, 17–20 June 2005.
19. Tavares, P.; Bernardo, H.; Gaspar, A.; Martin, A. Control criteria of electrochromic glasses for energy savings in Mediterranean buildings refurbishment. *Sol. Energy* **2016**, *134*, 236–250. [CrossRef]
20. HALIO Glass, Asahi Glass Co. (AGC). Available online: https://halioglass.com (accessed on 22 February 2018).
21. EnergyPlus. EnergyPlus Engineering Reference. The Reference to EnergyPlus Calculation. Available online: http://www.energyplus.gov (accessed on 22 February 2018).
22. Winkelmann, F. Modeling windows in EnergyPlus. In Proceedings of the 7th International IBPSA Conference, Rio di Janeiro, Brazil, 13–15 August 2001.
23. Arasteh, D.K.; Reilly, M.S.; Rubin, M.D. A versatile procedure for calculating heat transfer through windows, American Society of Heating, Refrigeration and Air-Conditioning Engineers. *ASHRAE Trans.* **1989**, *95*, 755–765.
24. Finlayson, E.U.; Arasteh, D.K.; Huizenga, C.; Rubin, M.D.; Reilly, M.S. *WINDOW 4.0: Documentation of Calculation Procedures, Lawrence Berkeley National Laboratory Report LBL-33943*; Lawrence Berkeley National Laboratory: Berkeley, CA, USA, 1993.
25. Arasteh, D.K.; Kohler, J.C.; Griffith, B.T. *Draft, Modeling Windows in EnergyPlus with only U, SHGC, and Optionally VT. Lawrence Berkeley National Laboratory Report*; Lawrence Berkeley National Laboratory: Berkeley, CA, USA, 2009.
26. Winkelmann, F.; Selkowitz, S. *Day-Lighting Simulation in DOE-2: Theory, Validation and Applications. Lawrence Berkeley National Laboratory Report*; Lawrence Berkeley National Laboratory: Berkeley, CA, USA, 1985.
27. Lomanowski, B.A.; Wright, J.L. Modeling fenestration with shading devices in building energy simulation: a practical approach. In Proceedings of the 11th International IBPSA Conference, Glasgow, Scotland, 27–30 July 2009.
28. Chantrasrisalai, C.; Fisher, D.E. Comparative analysis of one-dimensional slat type blind models. In Proceedings of the SimBuild 2004, Boulder, CO, USA, 4–6 August 2004.

29. International Organization for Standardization. *ISO 15099, Thermal Performance of Windows, Doors, and Shading Devices—Detailed Calculations*; International Organization for Standardization: Geneva, Switzerland, 2003.
30. The Korean Ministry of Land, Transport and Maritime Affairs. *Standard for Energy Saving Design in Buildings*; The Korean Ministry of Land, Transport and Maritime Affairs: Gwacheon, Korea, 2017.
31. American Society of Heating, Refrigerating and Air-Conditioning Engineers, Inc. *ASHRAE Fundamentals Handbook*; American Society of Heating, Refrigerating and Air-Conditioning Engineers, Inc.: Atlanta, GA, USA, 2009.
32. EnergyPlus. EnergyPlus Input Output Reference. The Encyclopedic Reference to EnergyPlus Input and Output. Available online: http://www.energyplus.gov (accessed on 22 February 2018).
33. Oh, M.H.; Lee, K.H.; Yoon, J.H. Automated control strategies of inside slat-type blind considering visual comfort and building energy performance. *Energy Build.* **2012**, *55*, 728–737. [CrossRef]

© 2018 by the authors. Licensee MDPI, Basel, Switzerland. This article is an open access article distributed under the terms and conditions of the Creative Commons Attribution (CC BY) license (http://creativecommons.org/licenses/by/4.0/).

sustainability

MDPI

Article

Phase Change Material (PCM) Application in a Modernized Korean Traditional House *(Hanok)*

Jaewook Lee [1] and Jiyoung Park [2,*]

1 Illinois School of Architecture, University of Illinois at Urbana-Champaign, Champaign, IL 61820, USA; jlee764@illinois.edu

2 Department of Architecture, Inha University, Inharo 100, Namgu, Incheon 22212, Korea

* Correspondence: jypark@inha.ac.kr; Tel.: +82-32-860-7583

Received: 1 February 2018; Accepted: 22 March 2018; Published: 24 March 2018

Abstract: Social and policy interest in the modernization and revitalization of the Korean traditional house (*Hanok*) has increased recently in Korea but its low thermal performance is one of its weaknesses. A feasibility study was conducted to evaluate the suitability of a Phase Change Material (PCM) in a modernized *Hanok*. The research method involved a test of the heating and cooling load reduction and Predicted Mean Vote (PMV) analysis for human comfort using an Esp-r simulation adopting multi variable PCM types as the building wall composite. The influence of PCMs on reducing the building energy load was assessed as a criterion for upgrading materials and infiltration to the passive house regulation. Compared to the base case, the heating and cooling load reduction ratio were as follows: Case 1 (old-*Hanok*), 10%; Case 2 (Korean Building Act), 21%; and Case 3 (passive house regulation), 53%. The optimal phase change temperatures of the PCMs were Case 1 (24–26 °C), Case 2 (23–25 °C) and Case 3 (24–26 °C). PMV analysis showed that the use of a PCM can narrow the comfort range and centralize the optimal point. Therefore, the following contents can be presented as the design and material guidelines. First, the optimal PCM temperature can vary according to the combination of materials and local climate. In addition, the infiltration and insulation should be verified and a certain portion of them should be secured. Finally, the addition of insulation to a passive house level should be considered actively using a PCM as a supplement for net zero energy building (nZEB).

Keywords: Korean traditional house; *Hanok*; PCM (Phase Change Material), ESP-r; energy simulation; building materials

1. Introduction

In Korea, there has been increasing interest in *Hanok*, which is the original type of Korean architecture. As the social demand for *Hanok* is expanding, it has great potential for creating various cityscapes and becoming an alternative to new residential environments. Social and policy interest in the modernization and revitalization of *Hanok* has increased and recently a technical development research project was initiated by the Korean Government. Consequently, a modernized *Hanok*, which includes various reinterpretations and recognition of changes to its layout and performance required for modern life, was constructed. The question is whether to accept and apply the characteristics of the traditional *Hanok*. This is an important part of the modernization of *Hanok* and its application to the future [1]. Figure 1 presents typical images of traditional and modern *Hanok*. The main difference is in their construction method. The old-*Hanok* is built using wet construction methods but the modern *Hanok* uses mainly dry construction methods.

(a) (b)

Figure 1. Typical *Hanok* images: (**a**) old-*Hanok* (*Myeongjaegotak*, Nonsan, Korea, 1709) [2]; (**b**) modernized *Hanok* (*Hwagyeongdang*, Seoul, Korea, 2013) [3].

From this background, some critical issues have emerged, such as narrow space dimensions and the layout for a modern lifestyle, poor insulation, weak fire resistance and expensive construction costs compared to contemporary buildings. In particular, based on a survey dealing with the inconvenience of *Hanok*, "hot and cold" sensations were the biggest issues with a reported rate of approximately 18% [4,5]. Moreover, inefficient energy consumption—particularly energy loss through the building envelope—is one of main concerns with *Hanok* [6–8]. Therefore, if hot and cold problems can be solved, it will help reduce energy consumption and carbon dioxide emissions as well as provide indoor human comfort.

One of the easiest ways to improve the thermal performance is to add insulation. The traditional earthen wall made from loess—called *Hwangtoh,* used to construct the traditional *Hanok* [9]—is impractical because of the high thermal conductivity of approximately 1–2 W/m·K [10], which is approximately 10 times higher than that of insulation [11]. Therefore, an appropriate amount of insulation is needed but various difficulties are encountered as the walls of *Hanok* are thickened because of the limitations regarding the thickness and the aesthetic proportions of the wooden pillars.

An additional option is to install a Phase Change Material (PCM), which is a heat storage material. Its basic principle is to absorb a large amount of heat and energy and release it to the indoors or outdoors by melting and solidifying at certain temperatures [12–16]. In addition, its main advantage is to maintain a certain temperature for a long time and improve the energy efficiency in the building energy load.

In previous studies, the PCM for building applications can be classified into two types: the traditional and encapsulated PCM [17]. Based on the classification, a computer simulation and field study were performed to assess the applications of PCMs in walls [18–25], roofs [26,27], windows [28,29] and floors [30,31]. In particular, there are advantages in installing a PCM in wall systems depending on which part is focused on, such as the materials and combinations, which geographic conditions they are testing and whether the tests are conducted under heating or cooling conditions. For example, finite different numeric calculations of a PCM wall system were performed at different melting temperatures and climates for net zero energy buildings (nZEB) based on characteristic days [19,20]. In addition, a thermal comfort test for a PCM wall system was conducted based on lightweight concrete with an encapsulated PCM for residential housing in Hong Kong. The results showed that the application of PCM to lightweight aggregate improved economic feasibility and was expected to have positive results in an environmental evaluation through CO_2 reduction [21]. In terms of the experiment with a numerical evaluation, two identical rooms were used to assess the performance and effectiveness of a shaped-stabilized PCM [22]. In addition, they selected different locations to determine the climate differences and its impact on the performance of PCM. Numerical and experimental analyses were performed based on a nano-PCM to evaluate the application of a PCM and gypsum wallboard; this system showed better energy performance in maintaining the indoor temperature [23]. Microencapsulated PCM with copper foam as a medium was simulated for thermal comfort in Malaysia and the system reduced the cooling energy load [24]. A performance test of an insulated concrete block with PCM was conducted in the United Arab Emirates to determine the energy consumption in the cooling season with natural or mechanical ventilation [25].

As one of the main concerns around maintaining and improving the thermal performance, the phase change temperature should be considered. A PCM has inherent properties, such as the melting temperature, heat of fusion, thermal conductivity and density [14]. Selecting a PCM with a specific melting temperature is very important because it can absorb a large amount of energy and release it to the buildings at a specific temperature as the melting temperature changes. In particular, the energy performance can vary greatly depending on which PCM is selected because the melting temperature ranges from −30 to +66 °C [32,33].

Therefore, the application of a PCM not only reduces the total energy consumption but also affects the peak load reduction positively [34]. The thermal storage function of the PCM also positively affects human thermal comfort [35,36]. In addition, fire retardation, long term stability and enhanced heat transfer are important.

In this study, the effects of PCM application for the selection of the material property in a modernized *Hanok* (so called new-*Hanok*) were simulated. In previous studies, research on the energy analysis of historical buildings was conducted [37,38]. In addition, the performance of PCM has been evaluated after adding traditional materials, lime mortar with PCM [39], but, few studies have assessed the applicability of PCM to the modernization of traditional architecture and one was initially introduced to conference proceedings in 2001 [40]. Compared to previous studies, there have been major amendments of new contents; the building regulations related to the insulation thickness have changed considerably and the latest codes are reviewed in this paper. In particular, a PCM can play a very important role in the planning of new-*Hanok* because the main material of traditional housing is soil and wood, which have poor thermal performance. Finally, the following are addressed: (1) identify how the energy (total consumption) aspects and human comfort change when applying several different types of PCM to a building envelope using an energy simulation; (2) predict the appropriate phase change setting temperature for various cases through sensitivity analysis; and (3) propose a material and engineering guide for how PCM can effectively reduce energy consumption and increase human comfort.

2. Methods

The analysis is presented in three steps, as shown in Figure 2:

(1) select a new-*Hanok* model, materials and simulation assumptions including infiltration;

(2-1) analyze the basic heating and cooling energy load to determine the adoptable PCMs;

(2-2) diversify temperature variables of PCM, which is aimed at minimizing the energy load and the peak-load shift by sensitive analysis; and

(3) analyze and compare the application of several types of PCM on the building model for minimizing the heating and cooling load and optimize the PMV range for human comfort.

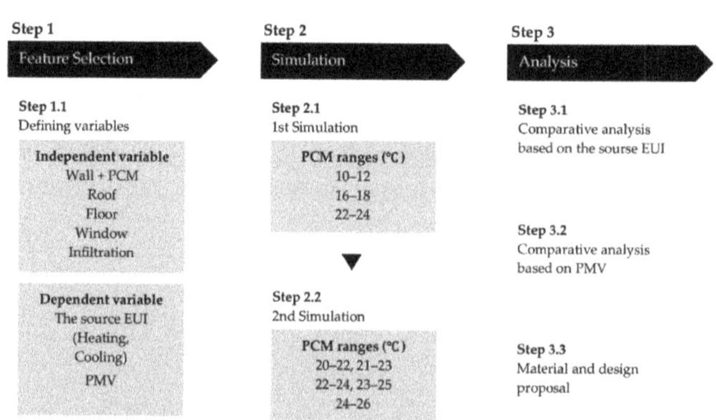

Figure 2. Research diagram.

The procedure for basic design of *Hanok*, material components, simulation settings and criteria for the selection of measures are described below.

2.1. New-Hanok Drawing

As a base model of this study, one of the new-*Hanok* plans by the '*Hanok* Technical Development Research Institute' was selected [41]. Figure 3 shows the plan and perspective drawing for the simulation. The main feature of new-*Hanok* design, '*Daechung*,' which is the living room space, was considered the indoor space while the semi-outdoor space was the unconditioned area in the original *Hanok*. In addition, the room and window size were adjusted to be suitable for modern life and furniture arrangement. This model consisted of three bedrooms, a living room, a kitchen and so forth. The length of the building was 15.6 m; the depth was 13.2 m; the floor to ceiling height was 2.4 m; the area was 102.24 m^2; the volume was 245.38 m^3; and the WWR (Window to Wall Ratio) was 19%.

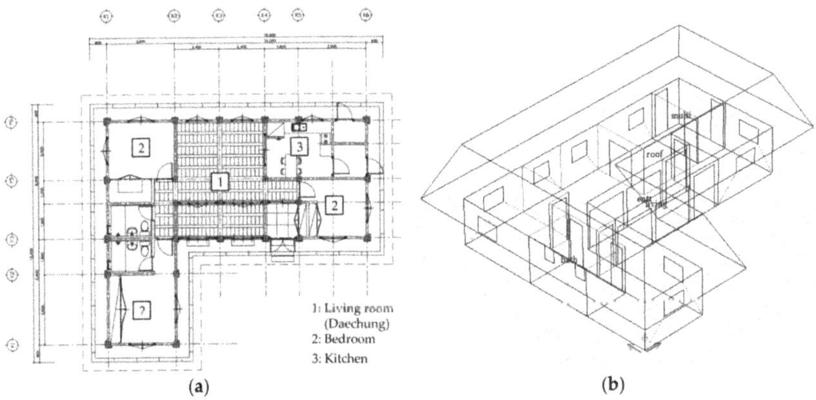

Figure 3. Drawing of the proposed *Hanok* (**a**) Floor Plan [41]; (**b**) 3-d perspective view of the energy simulation modeling.

2.2. Esp-r Computer Simulation and Setting

Esp-r 11.1, which is a dynamic energy simulation tool introduced in 2001, was selected to identify the applications of PCM–*Hwangtoh* (*Hwangtoh* is the traditional earthen wall made by loess) composites for thermal comfort in new-*Hanok* [42,43]. The 'Active Materials' option of Esp-r has been used to conduct research related to the application of PCMs [44]. This research was targeted for integrating the energy simulation with PCM and applying a gypsum board-PCM composite in a building. Therefore, there is a function that enables the application of a PCM to the multilayers of a composite in Esp-r. The basic equation for transient heat conduction with thermophysical properties is shown below [44,45]:

$$\frac{\partial}{\partial t}\rho(T)h(T) = \nabla\cdot[k(T)\nabla T(r^{\rightarrow},t)] + q(r^{\rightarrow},t)$$

where T is the temperature; ρ is the density; h is the enthalpy; k is the conductivity; and q the heat-generation rate [44].

The weather data was supported by the Esp-r default climate file, which was set to Seoul, Korea ('KOR_SeoulAB'_from Korea Institute of Energy Research (KIER)). According to the ASHRAE climate classification, Seoul belongs to Zone 4A, a moderate climate zone. In terms of the annual weather averages, July is the hottest month in Seoul with an average temperature of 24 °C and the coldest month is January at −4 °C. The annual-average daily global solar radiation is 3.58 kWh/m^2·day [46]. As a HVAC control, the set points of cooling and heating were 24 and 21 °C, respectively [47,48].

The internal loads for the equipment were calculated to be 8.07 W/m^2 (0.75 W/ft^2) peak load and the sensible load per person was 297 W, which is 8.07 W sensible/m^2 (0.75 W sensible/ft^2) [49].

In terms of infiltration, the performance in old-*Hanok* appears to be poor due to deformation and joint problems with wood and *Hwangtoh*. Table 1 lists the infiltration input data described by the actual data in old-*Hanok* [50], ASHRAE standard Leakage Class [51,52] and Passive House Planning Package (PHPP) [53]. The formulation converting from Air Changes per Hour (ACH) (50 Pa) to ACH (0 Pa) is followed by EN 13790 [54].

Table 1. Infiltration properties for the simulation.

Types		ACH (50 Pa)	ACH
Passive House Planning Package (PHPP)		0.6	0.038
ASHRAE	A	1	0.064
	B	2	0.128
	C	3	0.192
	D	5	0.321
old-*Hanok*		-	-

Based on the simulation setting, the annual heating and cooling loads were evaluated based on the sum of the load in two seasons; winter and summer. In addition, the reduction ratio, which is expressed as the relationship between the value of the base and target cases was used to compare the energy savings with the base case.

2.3. Material Properties

To select the applicable material properties in a new-*Hanok*, the following three different types were chosen: Case 1, an old traditional *Hanok*, called *Genjaegotaek*, built in 1869 [50] with real data; Case 2, a house that uses materials fulfilling the latest Korean Building Act (revised in 2017); and Case 3, a house that respects the passive house building guideline materials (PHPP 2015 [53]) for nZEB. Table 2 lists the conditional U-value properties of the three cases.

Table 2. Material properties calculated as the U-value (W/m^{2}K).

Types	Wall	Roof	Floor	Window
Old-*Hanok* (Case 1)	3.270	0.788	0.590	5.879
Korean Building Act (Case 2)	0.210	0.150	0.180	1.200
Passive house guideline (Case 3)	0.150	0.150	0.150	0.800

2.4. PCM (Phase Change Material)

Among the many techniques for increasing the thermal comfort, PCM has been used for heating-cooling load reduction and peak load shifting. The thermodynamic features of PCM are known as the absorption and release of the phase change latent heat under isothermal condition [55]. Theoretical and numerical analysis were conducted and the integration of PCM modeling with a building dynamic simulation was performed using ESP-r [44,56].

The most critical properties of a PCM for improving the building thermal performance are the melting point, transition temperature range and latent heat of the phase change. Assuming that the normal temperature distribution in Korea is approximately 0 to 30 °C, three PCMs, which range in ambient temperature were chosen [32,57–59] for the simulation and are listed in Table 3. These three PCMs were tetrabutylammonium bromide, propyl palmitate and paraffin C13–C24, with melting and solidification temperatures of 10–12 °C, 16–18 °C and 22–24 °C, respectively, with a thickness of 25 mm. To reduce the energy load and maintain a uniform indoor temperature, the PCMs were installed in the inner side of the external wall, as shown in Figure 4. In addition, the basic descriptions and abbreviated terms are as follows: T_m, melting temperature; T_s, solidification temperature; latent heat of fusion, which is the heat absorbed as a substance changes phase from liquid to solid; setting temperature; melting and solidification temperatures; and reduction ratio, the energy savings compared to the base case. In terms of thermal conductivity, density and specific heat capacity, the mean or median values based on commercial product details were used [20].

Table 3. Thermophysical properties of the PCMs used in the simulation.

Names	Type	T_m (°C)	T_s (°C)	Latent Heat of Fusion (kJ/kg)	Thermal Conductivity (W/m·K)	Density (kg/m^3)	Specific Heat Capacity (J/kg·K)
Tetrabutylammonium bromide [60]	Organic	10	12	193	0.6	1500	2000
Propyl palmitate [61]	Organic	16	18	186	0.6	1500	2000
Paraffin C13–C24 [62]	Organic	22	24	189	0.6	1500	2000

2.5. Wall Composite with PCM

Three different types of wall configurations were set as the main variables in this experiment, as shown in Figure 4. One common feature of the three cases is that 25 mm PCM-*Hwangtoh* is used as a component of the wall, the main difference being the value of heat transfer (U-value). Case 1 is composed of a wall of old-*Hanok*. This assumes a wall composed of pure *Hwangtoh*. Case 2 assumes a wall that meets the legal standards required by Korean housing. At this time, cellulose which is a pack of cellulose pulp in felt form was used as the thermal insulation material based on the guidelines of the National *Hanok* Center [1] and its thickness was calculated to be 155 mm to meet the current code (Korean Building Act). Finally, in Case 3, the insulation standard was reinforced to a passive house level and an insulation of 260 mm was used with the other materials remaining the same as in Case 2. In Cases 2 and 3, terracotta was applied as an alternative to *Hwangtoh* and waterproofing and steel frames were also used [1].

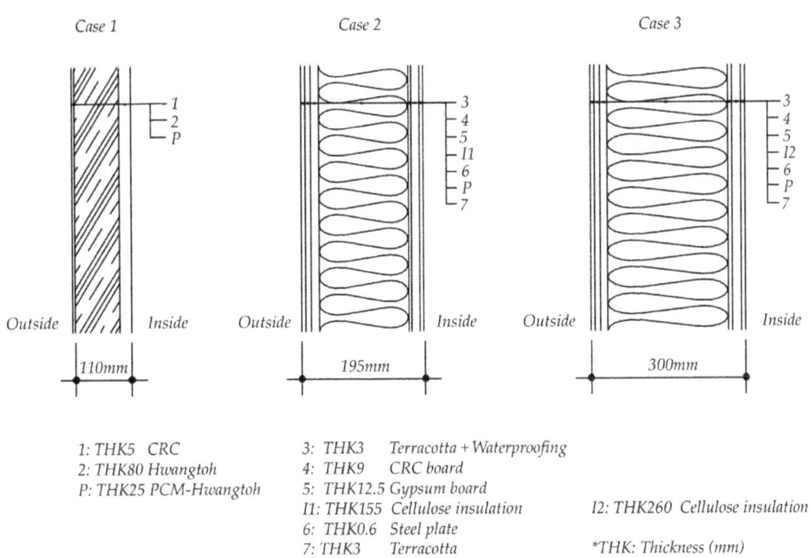

Figure 4. Section drawing detail of PCM-*Hwangtoh* composites installation.

2.6. PMV (Predicted Mean Vote)

Thermal comfort, which is a human's satisfaction with the thermal environment, can be expressed as the PMV, based on Fanger's model [63]. The PMV defines how people would be satisfied in a thermal environment and it is standardized as ASHRAE 55-2004 and used to evaluate the thermal comfort [51,64–66]. The values range from −3 (too cold) to +3 (too warm) and a range from −0.5 to 0.5 is normally recognized as a comfort zone by ASHRAE [51]. In this paper, the number of hours with $|PMV| < 1$, that is, PMV values ranging from −1 to 1, which is the cumulative time comfort over a year, was selected to evaluate the thermal comfort.

3. Results and Discussion

The results can be divided into three parts: energy performance evaluation in terms of the infiltration changes, two simulations with a PCM temperature setting and PMV analysis. In the first part, the simulation was performed to determine how the energy performance changes based on the infiltration uncertainty. Second, two simulations were conducted, which are presented as a process to find the optimal PCM temperature according to various material characteristics through sensitivity analysis. Finally, PMV was applied to determine how thermal comfort is achieved at the junction between the various materials and how much energy reduction and human comfort are similar or different.

3.1. Energy Performance According to the Insulation Material and Infiltration Level Change

The annual heating and cooling loads were drawn by varying the material properties and infiltration without applying a PCM. The three infiltration values were taken from Table 1 and are shown in Figure 5. Considering that the ACH of a typical residential building in Korea is approximately 0.4, 0.321 was selected as the infiltration intermediate value in the simulation [67].

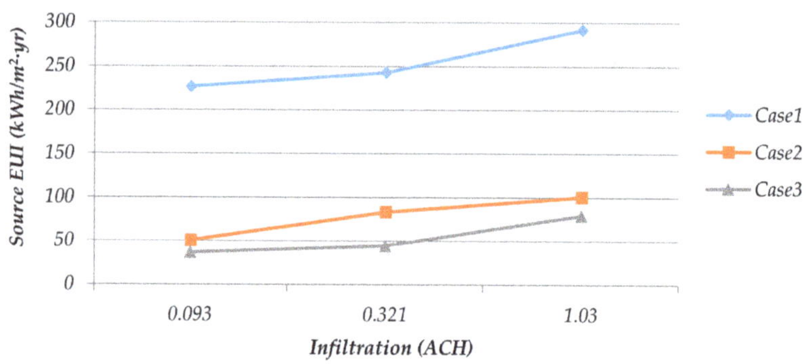

Figure 5. Annual heating and cooling load based on three different cases of insulation.

Figure 5 shows that strengthening of the insulation performance and infiltration in the *Hanok* can reduce the energy load by approximately 66%, 21% and 73% from Case 1 to Case 2, Case 2 to Case 3 and Case 1 to Case 3, respectively. For the material properties for a new-*Hanok*, the addition of thermal insulation materials and strengthening infiltration can improve the energy performance. The next part clarifies how the energy performance (total energy consumption and peak load energy consumption) and thermal human comfort can be improved when a PCM is applied.

3.2. Energy Performance Based on the Variation of the PCM-Hwangtoh Composite

The first simulation was conducted to select the appropriate temperature range in phase change mode. In setting the phase change temperature, three types in the range among the mean ambient temperature 0 to 30 °C were chosen, which are PCMs with a melting and solidification temperature range of 10–12 °C, 16–18 °C and 22–24 °C. A total of 12 simulations were performed based on three types of material properties, Cases 1, 2 and 3, according to the three different types of PCMs and non-PCM. Figure 6 compares the annual heating and cooling loads by applying the three types of PCMs. The reduction ratio is expressed as the relationship between the value of the base case and the target case.

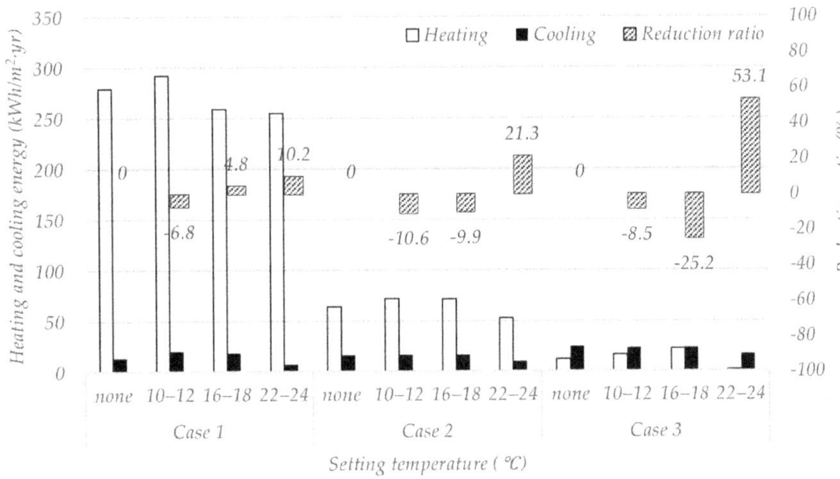

Figure 6. Annual heating and cooling load by the PCMs; 1st simulation.

According to the simulation results, setting the phase change temperature range to 10–12 °C and 16–18 °C increases the total heating and cooling loads slightly compared to the base case. In the case when 22–24 °C was used, the PCM decreased the heating and cooling load reduction ratio by 10.2, 21.3 and 53.1 in Cases 1, 2 and 3, respectively, compared to the non-PCM application. This means that selecting the 22–24 °C PCM temperature range, which is the cooling and heating set point, is efficient.

Based on the first simulation, the next simulation as sensitivity analysis, which ranged from 20–26 °C was conducted and the results are shown in Figure 7. Therefore, the second simulation was performed by further dividing the PCM temperature into six different types: none, 20–22 °C, 21–23 °C, 22–24 °C, 23–25 °C and 24–26 °C. The reduction ratio in the figure was calculated based on the PCM not being included in each case, which is the base case. The simulation results showed that the optimal set points of the PCMs for Cases 1, 2 and 3 are 24–26 °C, 23–25 °C and 24–26 °C respectively. This means that the optimal PCM set point should be applied differently according to the change in material properties. In addition, the heating and cooling load reduction ratio increased when the material properties were upgraded from Case 1 to Case 3. Following these results, for the near future in a passive house, PCM can reduce the heating and cooling loads significantly in a new-*Hanok* by approximately 60% compared to the non PCM applied building.

In terms of the monthly energy reduction, PCM-*Hwangtoh* composites positively affect the condition not only in spring and autumn but also in summer and winter, as listed in Table 4. The monthly heating load reduction ratio in January, which demands the maximum heating loads, was Case 1 at 2.1%, Case 2 at 11.4% and Case 3 at 94% compared to the non-PCM application. This means that the heating load reduction of the PCM can be maximized in a fully insulated situation, so the effect of the PCM and insulation on the heating energy saving is positive.

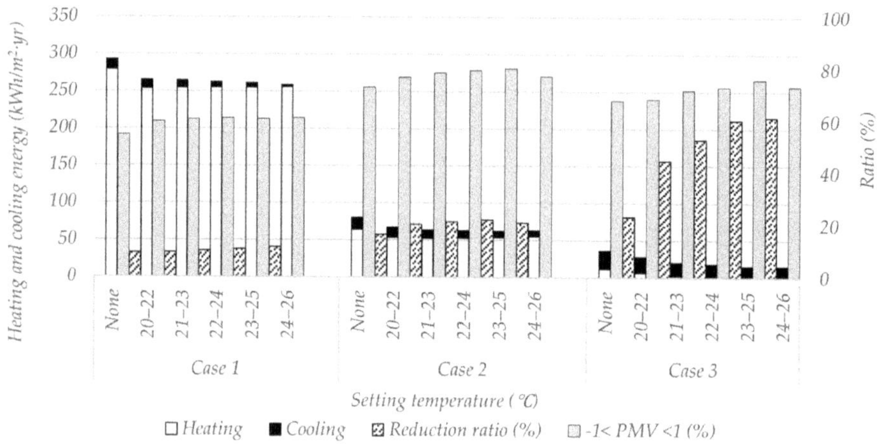

Figure 7. Annual heating and cooling load by the PCMs; 2nd simulation.

Table 4. Monthly heating and cooling load by the PCMs; 2nd simulation.

Types	Heating and Cooling Energy	Jan	Feb	Mar	Apr	May	Jun	Jul	Aug	Sep	Oct	Nov	Dec
Case 1	Heating (kWh)	6511	5328	3604	2018	729	106	2	4	68	1119	3239	5794
	Cooling (kWh)	0	0	0	0	32	157	333	586	198	43	0	0
PCM (24–26)	Heating (kWh)	6370	5163	3247	1468	380	7	0	0	0	768	2948	5783
	Cooling (kWh)	0	0	0	0	1	6	68	200	4	4	0	0
Case 2	Heating (kWh)	1647	1255	785	402	97	8	0	0	1	189	724	1459
	Cooling (kWh)	0	1	26	42	150	234	299	457	257	142	14	0
PCM (21–23)	Heating (kWh)	1459	1106	569	177	47	0	0	0	0	51	542	1328
	Cooling (kWh)	1	2	10	7	31	100	230	442	89	62	14	0
Case 3	Heating (kWh)	382	234	115	37	10	0	0	0	0	9	89	336
	Cooling (kWh)	0	3	65	106	326	384	372	515	423	300	35	0
PCM (24–26)	Heating (kWh)	9	7	7	1	0	0	0	0	0	1	7	9
	Cooling (kWh)	2	5	24	18	77	179	263	454	183	156	35	0

On the other hand, the difference in energy reduction between summer and winter is obvious. In Figures 8 and 9, there are two types of days, the hottest and coldest days of the year and the daily ambient temperature and heating and cooling loads are drawn. The x-axis of the figures represents the time and the y-axis represents the temperature and heating and cooling energy load. In addition, the dotted line in the graph indicates the temperature, the thick line indicates the PCM (24–26) applied case and the thin line indicates the base case, which is the non-PCM case. Compared to the non-PCM energy loads, in summer, the gap, which is shaded in gray (darkened), between non-PCM and PCM (24–26) is minor. On the other hand, in winter, the gap between the two material composites is extremely large. The reason for this is during the hottest day, two major factors, both temperature and solar radiation increase the indoor temperature. On the other hand, in winter, temperature and solar radiation have an opposite effect on the indoor temperature.

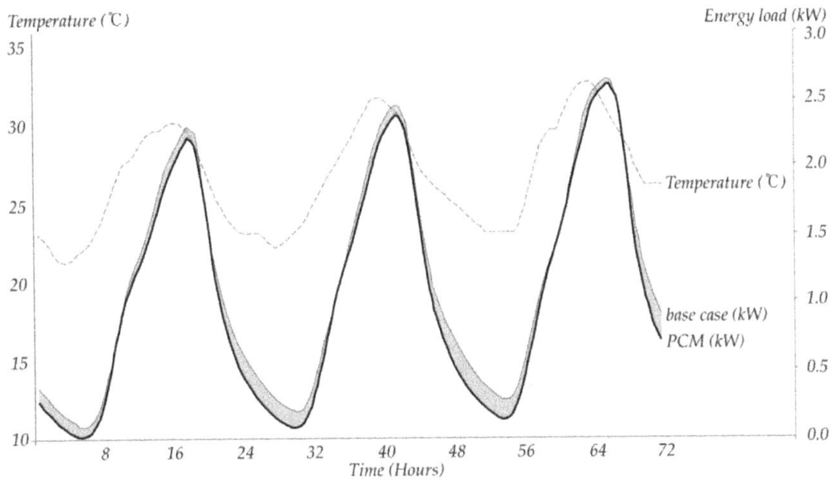

Figure 8. Daily peak load reduction in Case 3 (Hottest days of a year).

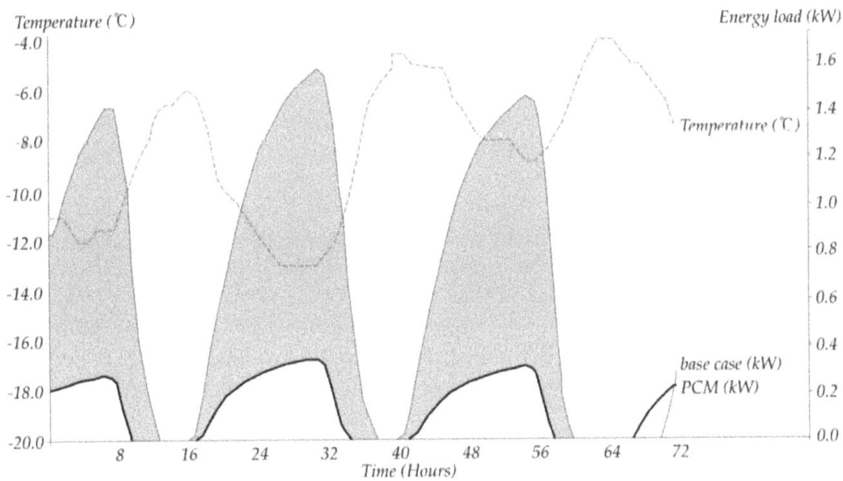

Figure 9. Daily peak load reduction in Case 3 (Coldest days of a year).

3.3. *Thermal Performance with PMV Analysis Based on the Variation of the PCM-Hwangtoh Composite*

Based on the results of the two energy simulations, PMV analysis was performed and the basic input data is as follows: (1) Clothing level, 0.7 (clo); (2) Activity level, 1.46 (MET) and (3) Air velocity, 0.1 (m/s) [64,68,69].

As shown in Figure 7, when applying different PCMs for each case, the range of the PMV value was 73.0–80.3% for Case 2 and 68.6–76.0% for Case 3. Figure 10 presents the results regarding the three optimal energy cases, Case 1, 2 and 3 with 24–26 °C, 23–25 °C and 24–26 °C, respectively. For thermal comfort, Case 2 showed the highest percentage, which is approximately 80.3% followed by Case 3 with approximately 76%. Therefore, Case 2 is superior to Case 3 in terms of the portion where the PMV falls within the range of −1 to 1. The reason why Case 2 shows a higher percentage of PMV than Case 3 is that winter is the season when a PCM is more effective in the Korean climate. In winter, the PMV is measured based on the operating temperature of the PMV of approximately

20–24 °C, that is, an average of 22 °C [70]. The internal heating temperature is set to operate at 24 °C. The set temperature of the PCM is 24–26 °C, which is also higher than the PMV standard ranges. The winter temperature is often maintained above 24 °C because the internal heating system tends to maintain temperatures greater than or equal to 24 °C and in Case 3, the performance of insulation and infiltration is quite good. Therefore, the PMV is relatively higher than the reference value. In addition, the occupant has the advantage of wearing thinner clothes on the inside in winter and staying in a better thermal condition than usual [71].

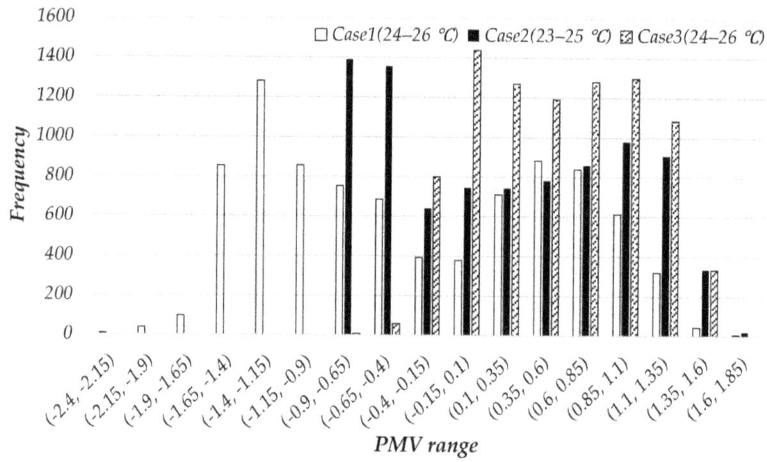

Figure 10. Frequency diagram of PMV in three cases.

When the PMV range was subdivided and its frequency was examined, as shown in Figure 10, the absolute value range of the PMV changes was smaller for Case 3 than for Case 2. In other words, the area in Case 3 is displayed at the frequency of 9 stages in total from (−0.65, −0.4) to (1.35, 1.6) compared to the area in Case 2, which is displayed at a frequency of 11 stages in total from (−0.9, −0.65) to (1.6, 1.85). This suggests that Case 3 can maintain the internal temperature of the room more effectively within the appropriate range, resulting in a "flattening effect" [55] on the internal room temperature.

4. Conclusions

This study examined how a PCM combined with a new-*Hanok* envelope, particularly the wall composite, can affect the building energy and thermal performance. The heating and cooling load reduction and PMV comfort range were measured and evaluated with the variation of infiltration, insulation and PCM properties. The optimal phase change temperature was also investigated in variable types of PCM applications with sensitivity analysis. The results for the design and material guidelines are as follows.

- Infiltration revealed a significant effect on the energy consumed. In particular, *Hanok*, which is constructed by mixing various materials, has considerably low infiltration, so it is important to strengthen it.
- Compared to traditional *Hanok* materials, strengthening insulation based on passive house regulation for nZEB can reduce the heating and cooling energy by approximately 73%.
- The PCM helps reduce the energy consumption when it approaches the passive house regulation material property. The heating and cooling load reduction ratio compared to the non-PCM

application are as follows: Case 1 (old-*Hanok*), 10.2%; Case 2 (Korean Building Act), 21.3%; and Case 3 (passive house regulation), 53.1%.

- The optimal set point of the PCM for energy saving has a different value depending on the material properties. The optimal phase change temperature of a PCM for a new-*Hanok* is as follows: Case 1, 24–26 °C; Case 2, 23–25 °C; and Case 3, 24–26 °C.
- In PMV analysis, the use of a PCM can narrow the comfort range and centralize the optimal point, which is 0 point. The results confirm that a PCM has a "flattening" influence inside the zone temperature.

This paper examined the feasibility of applying a PCM to *Hanok* through the sensitivity survey method. Nevertheless, more advanced research will be needed to address the weak points.

- PMV is affected significantly by the operation schedule and occupant behavior. Dividing the basic types of the occupant schedule can allow the PMV effect to be measured more accurately by applying a PCM.
- Studies of the life cycle cost and life cycle CO_2 analysis are the major points for economic analysis. The price of the PCM will need to be in the acceptable range and a long-term heating and cooling load calculation will be needed.

Acknowledgments: This study was supported by INHA UNIVERSITY Research Grant.

Author Contributions: The work presented in this article is the result of a collaboration of all authors. Jaewook Lee contributed to writing the manuscript and edited the document. Jiyoung Park critically reviewed the article. All authors have read and approved the final manuscript.

Conflicts of Interest: The authors declare no conflict of interest.

References

1. National Hanok Center Official Website. Available online: http://www.hanokdb.kr (accessed on 25 February 2018).
2. Image (Myeongjaegotak, Korea). Available online: http://first.designdb.com/imagedata/survey/P/60_6_3_P.jpg (accessed on 25 February 2018).
3. Image (Hwagyeongdang, Korea). Available online: http://www.dtoday.co.kr/news/articleView.html?idxno=92052 (accessed on 25 February 2018).
4. Kim, D. The Natural Environment Control System of Korean Traditional Architecture: Comparison with Korean Contemporary Architecture. *Build. Environ.* **2006**, *41*, 1905–1912. [CrossRef]
5. Ryu, Y.; Kim, S.; Lee, D. The Influence of Wind Flows on Thermal Comfort in the Daechung of a Traditional Korean House. *Build. Environ.* **2009**, *44*, 18–26. [CrossRef]
6. Sadineni, S.B.; Madala, S.; Boehm, R.F. Passive Building Energy Savings: A Review of Building Envelope Components. *Renew. Sustain. Energy Rev.* **2011**, *15*, 3617–3631. [CrossRef]
7. Cheung, C.; Fuller, R.; Luther, M. Energy-Efficient Envelope Design for High-Rise Apartments. *Energy Build.* **2005**, *37*, 37–48. [CrossRef]
8. Lee, J.; Jung, H.; Park, J.; Lee, J.; Yoon, Y. Optimization of Building Window System in Asian Regions by Analyzing Solar Heat Gain and Daylighting Elements. *Renew. Energy* **2013**, *50*, 522–531. [CrossRef]
9. Yang, K.; Hwang, H.; Kim, S.; Song, J. Development of a Cementless Mortar Using Hwangtoh Binder. *Build. Environ.* **2007**, *42*, 3717–3725. [CrossRef]
10. Johansen, O. *Thermal Conductivity of Soils*; Cold Regions Research and Engineering Lab: Hanover, NH, USA, 1977.
11. Abdou, A.A.; Budaiwi, I.M. Comparison of Thermal Conductivity Measurements of Building Insulation Materials under Various Operating Temperatures. *J. Build. Phys.* **2005**, *29*, 171–184. [CrossRef]
12. Dincer, I.; Rosen, M. *Thermal Energy Storage: Systems and Applications*, 2nd ed.; John Wiley & Sons, Inc.: Haboken, NJ, USA, 2010; ISBN 0470747064, 9780470747063.
13. Paksoy, H.Ö. *Thermal Energy Storage for Sustainable Energy Consumption: Fundamentals, Case Studies and Design*; Springer: Berlin, Germany, 2007; Volume 234, ISBN 1402052901, 9781402052903.

14. Cabeza, L.F.; Castell, A.; Barreneche, C.; De Gracia, A.; Fernández, A. Materials Used as PCM in Thermal Energy Storage in Buildings: A Review. *Renew. Sustain. Energy Rev.* **2011**, *15*, 1675–1695. [CrossRef]
15. Akeiber, H.; Nejat, P.; Majid, M.Z.A.; Wahid, M.A.; Jomehzadeh, F.; Famileh, I.Z.; Calautit, J.K.; Hughes, B.R.; Zaki, S.A. A Review on Phase Change Material (PCM) for Sustainable Passive Cooling in Building Envelopes. *Renew. Sustain. Energy Rev.* **2016**, *60*, 1470–1497. [CrossRef]
16. Hassan, A.; Shakeel Laghari, M.; Rashid, Y. Micro-Encapsulated Phase Change Materials: A Review of Encapsulation, Safety and Thermal Characteristics. *Sustainability* **2016**, *8*, 1046. [CrossRef]
17. Jeon, J.; Lee, J.; Seo, J.; Jeong, S.; Kim, S. Application of PCM Thermal Energy Storage System to Reduce Building Energy Consumption. *J. Therm. Anal. Calorim.* **2013**, *111*, 279–288. [CrossRef]
18. Memon, S.A. Phase Change Materials Integrated in Building Walls: A State of the Art Review. *Renew. Sustain. Energy Rev.* **2014**, *31*, 870–906. [CrossRef]
19. Mazzeo, D.; Oliveti, G.; De Simone, M.; Arcuri, N. Analytical Model for Solidification and Melting in a Finite PCM in Steady Periodic Regime. *Int. J. Heat Mass Transfer* **2015**, *88*, 844–861. [CrossRef]
20. Mazzeo, D.; Oliveti, G.; Arcuri, N. A Method for Thermal Dimensioning and for Energy Behavior Evaluation of a Building Envelope PCM Layer by Using the Characteristic Days. *Energies* **2017**, *10*, 659. [CrossRef]
21. Memon, S.A.; Cui, H.; Zhang, H.; Xing, F. Utilization of Macro Encapsulated Phase Change Materials for the Development of Thermal Energy Storage and Structural Lightweight Aggregate Concrete. *Appl. Energy* **2015**, *139*, 43–55. [CrossRef]
22. Ye, H.; Long, L.; Zhang, H.; Zou, R. The Performance Evaluation of Shape-Stabilized Phase Change Materials in Building Applications Using Energy Saving Index. *Appl. Energy* **2014**, *113*, 1118–1126. [CrossRef]
23. Sayyar, M.; Weerasiri, R.R.; Soroushian, P.; Lu, J. Experimental and Numerical Study of Shape-Stable Phase-Change Nanocomposite toward Energy-Efficient Building Constructions. *Energy Build.* **2014**, *75*, 249–255. [CrossRef]
24. Isa, M.H.M.; Zhao, X.; Yoshino, H. Preliminary Study of Passive Cooling Strategy Using a Combination of PCM and Copper Foam to Increase Thermal Heat Storage in Building Facade. *Sustainability* **2010**, *2*, 2365–2381. [CrossRef]
25. Hasan, A.; Al-Sallal, K.A.; Alnoman, H.; Rashid, Y.; Abdelbaqi, S. Effect of Phase Change Materials (PCMs) Integrated into a Concrete Block on Heat Gain Prevention in a Hot Climate. *Sustainability* **2016**, *8*, 1009. [CrossRef]
26. Kośny, J.; Biswas, K.; Miller, W.; Kriner, S. Field Thermal Performance of Naturally Ventilated Solar Roof with PCM Heat Sink. *Sol. Energy* **2012**, *86*, 2504–2514. [CrossRef]
27. Li, D.; Zheng, Y.; Liu, C.; Wu, G. Numerical Analysis on Thermal Performance of Roof Contained PCM of a Single Residential Building. *Energy Convers. Manag.* **2015**, *100*, 147–156. [CrossRef]
28. Grynning, S.; Goia, F.; Rognvik, E.; Time, B. Possibilities for Characterization of a PCM Window System Using Large Scale Measurements. *Int. J. Sustain. Built Environ.* **2013**, *2*, 56–64. [CrossRef]
29. Koláček, M.; Charvátová, H.; Sehnálek, S. Experimental and Numerical Research of the Thermal Properties of a PCM Window Panel. *Sustainability* **2017**, *9*, 1222. [CrossRef]
30. Xu, X.; Zhang, Y.; Lin, K.; Di, H.; Yang, R. Modeling and Simulation on the Thermal Performance of Shape-Stabilized Phase Change Material Floor Used in Passive Solar Buildings. *Energy Build.* **2005**, *37*, 1084–1091. [CrossRef]
31. Royon, L.; Karim, L.; Bontemps, A. Optimization of PCM Embedded in a Floor Panel Developed for Thermal Management of the Lightweight Envelope of Buildings. *Energy Build.* **2014**, *82*, 385–390. [CrossRef]
32. Mehling, H.; Cabeza, L.F. *Heat and Cold Storage with PCM*; Springer: Berlin, Germany, 2008; ISBN 3540685561, 9783540685562.
33. Kenisarin, M.; Mahkamov, K. Solar Energy Storage Using Phase Change Materials. *Renew. Sustain. Energy Rev.* **2007**, *11*, 1913–1965. [CrossRef]
34. Tyagi, V.; Pandey, A.; Kaushik, S.; Tyagi, S. Thermal Performance Evaluation of a Solar Air Heater with and without Thermal Energy Storage. *J. Therm. Anal. Calorim.* **2012**, *107*, 1345–1352. [CrossRef]
35. Kuznik, F.; Virgone, J.; Johannes, K. In-Situ Study of Thermal Comfort Enhancement in a Renovated Building Equipped with Phase Change Material Wallboard. *Renew. Energy* **2011**, *36*, 1458–1462. [CrossRef]
36. Evola, G.; Marletta, L.; Sicurella, F. A methodology for Investigating the Effectiveness of PCM Wallboards for Summer Thermal Comfort in Buildings. *Build. Environ.* **2013**, *59*, 517–527. [CrossRef]

37. Roberti, F.; Oberegger, U.F.; Gasparella, A. Calibrating Historic Building Energy Models to Hourly Indoor Air and Surface Temperatures: Methodology and Case study. *Energy Build.* **2015**, *108*, 236–243. [CrossRef]
38. Mazzarella, L. Energy Retrofit of Historic and Existing Buildings. The Legislative and Regulatory Point of View. *Energy Build.* **2015**, *95*, 23–31. [CrossRef]
39. Ventolà, L.; Vendrell, M.; Giraldez, P. Newly-Designed Traditional Lime Mortar with a Phase Change Material as an Additive. *Constr. Build. Mater.* **2013**, *47*, 1210–1216. [CrossRef]
40. Lee, J.; Park, J. Application of PCM-Hwangtoh Composites for Thermal Comfort in Newly Proposed Korean Traditional House. In Proceedings of the IAPS International Network Symposium, Daegu, Korea, 10–14 October 2011.
41. Kim, M.; Kim, Y.; Lee, Y.; Jung, Y. Comparative Analysis of Construction Productivity for Modernized Korean Housing (Hanok). *Korean J. Covstr. Eng. Manag.* **2013**, *14*, 107–114. [CrossRef]
42. Clarke, J.A. *Energy Simulation in Building Design*, 2nd ed.; Routledge: Oxford, UK, 2001; ISBN 0750650826, 9780750650823.
43. Strachan, P.; Kokogiannakis, G.; Macdonald, I. History and Development of Validation with the ESP-r Simulation Program. *Build. Environ.* **2008**, *43*, 601–609. [CrossRef]
44. Heim, D.; Clarke, J.A. Numerical Modelling and Thermal Simulation of PCM–Gypsum Composites with ESP-r. *Energy Build.* **2004**, *36*, 795–805. [CrossRef]
45. Kośny, J. *PCM-EnhancedBuilding Components: An Application of Phase Change Materials in Building Envelopes and Internal Structures*; Springer: Berlin, Germany, 2015; ISBN 3319142852, 9783319142852.
46. Yoon, K.; Yun, G.; Jeon, J.; Kim, K.S. Evaluation of Hourly Solar Radiation on Inclined Surfaces at Seoul by Photographical Method. *Sol. Energy* **2014**, *100*, 203–216. [CrossRef]
47. American Society of Heating, Refrigerating and Air-Conditioning Engineers, Inc (ASHRAE). ASHARE Standard 90.2-2007: Energy-Efficient Design of Low-Rise Residential Buildings. 2007. Available online: https://ashrae.iwrapper.com/ViewOnline/Standard_90.2-2007 (accessed on 25 February 2018).
48. Selkowitz, S.; Hitchcock, R.; Mitchell, R.; McClintock, M.; Settlemyre, K. *Comfen—Early Design Tool for Commercial Facades and Fenestration Systems*; Building Technologies Department, Lawrence Berkeley National Laboratory: Berkeley, CA, USA, 2014. Available online: https://windows.lbl.gov/sites/default/files/comfen-software-tool.pdf (accessed on 25 February 2018).
49. Crawley, D.B.; Lawrie, L.K.; Winkelmann, F.C.; Buhl, W.F.; Huang, Y.J.; Pedersen, C.O.; Strand, R.K.; Liesen, R.J.; Fisher, D.E.; Witte, M.J. Energyplus: Creating a New-Generation Building Energy Simulation Program. *Energy Build.* **2001**, *33*, 319–331. [CrossRef]
50. Kim, M.; Kim, J.; Park, H. A Study on the Conservative Building Improvements for Indoor Thermal Environment in a Traditional House of Genjaegolaek. *J. Korean Inst. Archit.* **2004**, *20*, 251–258.
51. American Society of Heating, Refrigerating and Air-Conditioning Engineers, Inc. (ASHRAE). *ASHARE Standard 55–2004: Thermal Environmental Conditions for Human Occupancy*; ASHRAE: Atlanta, GA, USA, 2004.
52. Sherman, M. The Use of Blower-Door Data. *Indoor Air* **1995**, *5*, 215–224. [CrossRef]
53. Feist, W.; Pfluger, R.; Kaufmann, B.; Schnieders, J.; Kah, O. *Passive House Planning Package 2007*; Passive House Institute: Darmstadt, Germany, 2007; Available online: http://www.passivhaus.org.uk/filelibrary/Contents_Manual.pdf (accessed on 25 February 2018).
54. International Standard Organisation (ISO). ISO 13790: 2008 Energy Performance of Buildings–Calculation of Energy Use for Space Heating and Cooling. 2008. Available online: https://www.iso.org/standard/41974.html (accessed on 25 February 2018).
55. Chen, C.; Guo, H.; Liu, Y.; Yue, H.; Wang, C. A New Kind of Phase Change Material (PCM) for Energy-Storing Wallboard. *Energy Build.* **2008**, *40*, 882–890. [CrossRef]
56. Ockendon, J.R.; Hodgkins, W. *Moving Boundary Problems in Heat Flow and Diffusion*, 1st ed.; Oxford Univeristy Press: Oxford, England, 1975; ISBN 0198533454, 9780198533450.
57. Zalba, B.; Marın, J.M.; Cabeza, L.F.; Mehling, H. Review on Thermal Energy Storage with Phase Change: Materials, Heat Transfer Analysis and Applications. *Appl. Therm. Eng.* **2003**, *23*, 251–283. [CrossRef]
58. Khudhair, A.M.; Farid, M.M. A Review on Energy Conservation in Building Applications with Thermal Storage by Latent Heat Using Phase Change Materials. *Energy Convers. Manag.* **2004**, *45*, 263–275. [CrossRef]
59. Tyagi, V.V.; Buddhi, D. PCM Thermal Storage in Buildings: A State of Art. *Renew. Sustain. Energy Rev.* **2007**, *11*, 1146–1166. [CrossRef]

60. Cabeza, L.F.; Mehling, H.; Hiebler, S.; Ziegler, F. Heat Transfer Enhancement in Water When Used as PCM in Thermal Energy Storage. *Appl. Therm. Eng.* **2002**, *22*, 1141–1151. [CrossRef]
61. Hawes, D.; Feldman, D.; Banu, D. Latent Heat Storage in Building Materials. *Energy Build.* **1993**, *20*, 77–86. [CrossRef]
62. Abhat, A. Low Temperature Latent Heat Thermal Energy Storage: Heat Storage Materials. *Sol. Energy* **1983**, *30*, 313–332. [CrossRef]
63. Fanger, P.O. *Thermal Comfort. Analysis and Applications in Environmental Engineering*; The McGraw-Hill Companies, Inc.: New York, NY, USA, 1972; ISBN 0070199159, 9780070199156.
64. Yang, K.; Su, C. An Approach to Building Energy Savings Using the PMV Index. *Build. Environ.* **1997**, *32*, 25–30. [CrossRef]
65. Hwang, R.-L.; Shu, S.-Y. Building Envelope Regulations on Thermal Comfort in Glass Facade Buildings and Energy-Saving Potential for PMV-Based Comfort Control. *Build. Environ.* **2011**, *46*, 824–834. [CrossRef]
66. Deng, Y.; Feng, Z.; Fang, J.; Cao, S.-J. Impact of Ventilation Rates on Indoor Thermal Comfort and Energy Efficiency of Ground-Source Heat Pump System. *Sustain. Cities Soc.* **2018**, *37*, 154–163. [CrossRef]
67. Yeo, M.; Yang, I.; Kim, K. Historical Changes and Recent Energy Saving Potential of Residential Heating in Korea. *Energy Build.* **2003**, *35*, 715–727. [CrossRef]
68. Hasan, M.H.; Alsaleem, F.; Rafaie, M. Sensitivity Study for the PMV Thermal Comfort Model and the Use of Wearable Devices Biometric Data for Metabolic Rate Estimation. *Build. Environ.* **2016**, *110*, 173–183. [CrossRef]
69. Ioannou, A.; Itard, L. In-Situ and Real Time Measurements of Thermal Comfort and Its Determinants in Thirty Residential Dwellings in the Netherlands. *Energy Build.* **2017**, *139*, 487–505. [CrossRef]
70. De Dear, R.J.; Brager, G.S. Thermal Comfort in Naturally Ventilated Buildings: Revisions to ASHRAE Standard 55. *Energy Build.* **2002**, *34*, 549–561. [CrossRef]
71. Saffari, M.; de Gracia, A.; Ushak, S.; Cabeza, L.F. Economic Impact of Integrating PCM as Passive System in Buildings Using Fanger Comfort Model. *Energy Build.* **2016**, *112*, 159–172. [CrossRef]

© 2018 by the authors. Licensee MDPI, Basel, Switzerland. This article is an open access article distributed under the terms and conditions of the Creative Commons Attribution (CC BY) license (http://creativecommons.org/licenses/by/4.0/).

sustainability

MDPI

Article

Influence of Adaptive Comfort Models on Energy Improvement for Housing in Cold Areas

Alexis Pérez-Fargallo [1], Carlos Rubio-Bellido [2,*], Jesús A. Pulido-Arcas [1], Inmaculada Gallego-Maya [2] and Fco. Javier Guevara-García [2]

1 Department of Building Science, University of Bio-Bio, Concepción 4030000, Chile; aperezf@ubiobio.cl (A.P.-F.); jpulido@ubiobio.cl (J.A.P.-A.)

2 Department of Building Construction II, University of Seville, 41012 Seville, Spain; gallego_inma@hotmail.com (I.G.-M.); guevara@us.es (F.J.G.-G.)

* Correspondence: carlosrubio@us.es; Tel.: +34-686-135-595

Received: 19 February 2018; Accepted: 13 March 2018; Published: 18 March 2018

Abstract: The evaluation of construction standards using adaptive thermal comfort models has a great impact on energy consumption. The analysis of a user's climate adaptation must be one of the first steps in the search for nearly/net Zero Energy Buildings (nZEB). The goal of this work is to analyze the standards recommended by the Chile's Construction with Sustainability Criteria for the building of housing, applying the ASHRAE 55-2017 and EN 15251:2007 adaptive comfort models in social housing. The study produces concrete recommendations associated with construction strategies, to increase the number of hours the user finds themselves with acceptable thermal comfort levels, without repercussions for energy consumption. Sixteen parametric series were evaluated with a dynamic simulation of the most common prototype of social housing in the Bio-Bio Region. The study shows that thermal comfort conditions can be increased through a combination of improvement measures compared to the ECCS standard (Construction Standards with Sustainability Criteria): 27.52% in the case of applying EN 15251:2007 and 24.04% in the case of ASHRAE 55-2017.

Keywords: climate adaptation; social housing; thermal comfort; adaptive comfort

1. Introduction

Buildings consume between 30% and 40% of the world's energy [1]. According to the International Energy Agency this could increase to 38.4 PWh by 2040 [2]. For this reason, energy consumption in buildings and energy consumption per capita are no longer indicators of economic prosperity and social well-being [3]. Currently, there is a combination of non-renewable energy resource depletion and global warming; therefore more energy efficient building development based on the use of renewable energy is being demanded [4]. The economic slowdown and climate change have forced those who are politically and scientifically responsible to rationalize the use of energy across the entire world [5,6], tending ever more towards drastically reducing energy consumption via so-called nearly/new Zero Energy Buildings (nZEB) [7–9]. The University of Chile's Studies and Energy Program (PRIEN in Spanish), estimated that the building sector will represent 18% of the total national energy efficiency potential for 2020, surpassed only by the industrial and mining sectors [10].

Energy consumption of buildings is associated with multiple factors, with the user's thermal indoor comfort demands standing out. An unsuitable internal temperature can mean either users are uncomfortable or that energy consumption increases. Thus, the building's sustainability level (i.e., minimal energy consumption) can be greatly impacted by this parameter if it is not suitable [11]. The adaptive comfort models are based on the natural tendency of individuals to adapt their thermoregulation, clothing, metabolic rate and psychological conditions to the changing conditions of the outdoor climate under natural ventilation conditions [12]. Unlike other comfort standards, this allows a greater range of temperatures related to the outside temperatures of the previous days.

In addition, the use of active air-conditioning systems in homes is subject to discussion [13] due to the increasing demand of energy resources in a scenario of expensive energy and energy poverty being more and more of a reality [14]. This is why a large number of current studies in the field of climate adaptation of the residential sector focus on the comfort study of their occupants, trying to reduce the use of active devices and intensifying passive strategies [15,16]. As an example of this, Attia has shown that the use of adaptive comfort models can generate energy savings of between 10% and 18% in warm climates [7].

The Chilean standard Construction Standards with Sustainability Criteria (ECCS by its Spanish acronym) for homes was made by the Building Research Establishment (BRE) along with the Executive Secretary of Sustainable Construction of the Chilean Housing and Urbanism Ministry (MINVU) [17]. The ECCS consists of technical guides aimed at establishing standards and best practices for the design, construction and operation of houses, both new or extant, with the goal of improving their environmental, economic and social performances, through the definition and incorporation of sustainability criteria, based on objective and verifiable parameters [17]. Technical "sustainability" standards have been developed for the housing industry, bearing in mind the different geographical and climatic situations of Chile. Its mission is to accelerate the transition of a residential building stock with low energy performance to one with improved comfort standards. The ECCS has established that the passive thermal design of homes must guarantee that between 60% and 90% of the year, indoor temperatures are between 20–26 °C, with regard to the thermal zone the building is located in, without applying the concept of adaptive comfort marked by the ASHRAE 55-2017 or by the EN 15251:2007 [17–19].

The use of adaptive comfort models, the analysis has been set out as one of the focal points for reducing energy consumption and lowering the emission levels of buildings. In this way, diverse thermal acceptability models of spaces have been generated around the world [18,19]. The objective of this article is to analyze the potential of using the adaptive comfort models of the ASHRAE 55-2017 and EN 15251:2007 standards versus those set out by the ECCS for housing. After that, the necessary improvements that need to be applied to current construction standards to maximize the time during which the building can work under free running conditions will be studied. This is justified by the fact that heating, ventilation and air conditioned (HVAC) equipment may be used only under extreme discomfort due to the users' limited resources [20]. In order to undertake an analysis of a dynamic simulation, a commonly used prototype of social housing with light construction systems was chosen as a case-study; it is located in the Bio-Bio region, lying in the Central–South fringe of Chile. This area features a Csb climate according to the Köppen–Geiger classification, where the average temperature during the hot season is below 22 °C and the average during the cold season is below 10 °C. These temperatures, though swinging between cold and hot, cannot be considered as extremes; therefore, the application of strategies related to thermal adaptation seems plausible. This is also supported by the fact that the current Chilean National Standards (ECCS) place the Bio-Bio region in thermal zone E, where the annual comfort percentage time must be 70%, with a lower limit of 20 °C and an upper of 25 °C [17]. In this way, a study focused on this region could be of interest, not only regarding the Chilean context, for evident reasons, but also for other audiences, because similar climates can be found in populated areas, such as the Mediterranean arch, the Northeast coastal area of the United States and Australia. Both the EU and the USA have already developed their own standards for adaptive comfort, and Chile is investing a great amount of resources in developing its own construction standards, such as the aforementioned ECCS, using the aforementioned countries as a reference.

By comparing and assessing the influence of a number of construction features in the applicability of such comfort standards, this research can make a contribution, targeting not only Chilean researchers but also stakeholders from other nationalities interested in the feasibility of a knowledge transfer between their countries and Chile. Selecting the best strategies will allow the provision of affordable comfort using a limited amount of energy, thus positioning Chile on the path of sustainable development.

In order to achieve this objective, this paper is divided into the following sections: First, Section 2 describes the materials and methods of the present research. The materials consist of an analysis of the different comfort standards and a description of the selected case-study, a prototype of social housing. The research methods consist of parametric simulations, performed using the software EnergyPlus®, on a combination of cases that will allow for better implementation of adaptive thermal comfort. Section 3 analyzes the results of the simulations, considering the application of the selected cases in terms of the thermal adaptive comfort models. Lastly, Section 4 delivers the main conclusions of this study.

2. Materials and Methods

2.1. Requirements and Policies for Social Housing in Chile

The Chilean State provided more than 3,671,646 subsidies to develop social housing between 1964 and 2015, investing around 19 billion euros in this policy since 1990 [21]. As such, substantial investment has been made to reduce the habitational shortfall to improve the quality of life for the most deprived sectors of Chile. At the present time, Supreme Decrees 01 and 49 are the legislative instruments that exist to provide subsidies to access to public housing [22,23]. Homes built privately or publicly under this financing can be arranged under a variety of typologies but should have a built surface area between 36.00 and 55.00 m^2. Figure 1 provides a model for the application of Supreme Decree 49, obtained from the online databank of Housing Projects. This is used by the Bio-Bio Region's Housing and Urbanization Service (SERVIU in Spanish) [24]. The geometry of these houses is quite standardized due to the surface area and budget limitations. Typically, they have two or three bedrooms, a living–dining area, kitchen and bathroom. Their construction features are associated with the local climate and availability of resources, being quite similar in the same region. In the case of the Bio-Bio region, they are built as single-family houses and they tend to feature light construction systems resting on a concrete base. The structural walls and roof structures are made from wood with different scantling; the outer side of the walls is constructed mainly with fiber cement siding, while inside, plasterboard is the common finish; the roof has corrugated zinc sheeting. The joinery is commonly done with aluminum or PVC with simple glazing, because their construction systems tend to have a high degree of airtightness and low thermal inertia [25].

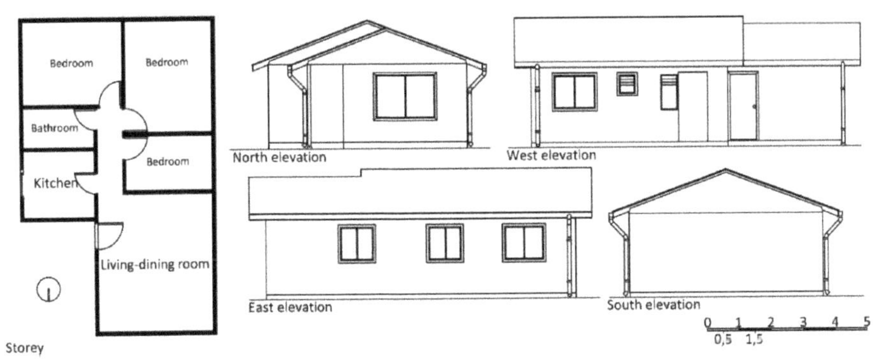

Figure 1. Housing model for the study.

Chile was the first country in Latin America to incorporate energy efficiency into their normative concepts, even though, historically, these demands had not been considered for buildings.

Currently, Art. 4.1.10. of the General Ordinance on Urbanism and Constructions (OGUC in Spanish) regulates the requirements that envelopes of homes associated with thermal transmittance must comply with. This article is mandatory, but its requirements are pretty limited [26]. The ECCS for Housing, whose requirements are more stringent, has also been produced, but, for the time being, it is optional [17]. Even so, it is necessary to point out that its standards are relatively limited when compared with other international standards (Table 1) [27].

Table 1. Limitations of transmittances, ventilation and airtightness for Concepción, OGUC (General Ordinance on Urbanism and Constructions) (title 4.1.10) and ECCS (Construction Standards with Sustainability Criteria) for Housing in the study region.

Case	U Gaps (W/m² K)			U Envelope (W/m² K)			Ventilation L/(s*person)	Airtightness (ACH50)
	<21%	21–60%	60–75%	Roof	Walls	Floor [(m²K)/W]*100		
OGUC	>3.6	2.4–3.6	<2.4	0.38	1.7	150	-	-
ECCS		3.0 [1]		0.33	0.60	45	5.2	8

[1] The maximum glazing percentage is associated with the orientation: North <50%, South <40% and East–West <30%.

Because of this, Bustamante stated that even newly built buildings will be subject to future reconditioning due to the limited associated energy efficiency demands [20]. The same authors state that despite having a participation level of 21.3% in the country's energy consumption, most homes fall below the comfort temperature in winter due to the high fuel prices and low incomes, thus showing the importance of increasing the energy efficiency and inhabitability of already existent buildings or those being built. In addition, from an economic point of view, the low thermal performance level of homes in Chile costs the state around 0.8 billion € a year [28].

2.2. *Adaptive Thermal Comfort Standards*

In the search to create systems to assess the energy behavior of buildings, states have generated indicators associated with primary energy consumption—CO_2 emissions, or net energy supplied—creating global indicators which provide data related to the construction's energy efficiency [29]. The standards generally quantify the energy efficiency of the buildings based on the energy consumption or associated CO_2 emissions. This is evident, for example, in the EN 15603:2008, EN 15217:2007 standards as well as in the ECCS [17,29,30]. A classical approach for quantifying this consumption includes simulation of a fixed set point of temperatures, operating schedule or a combination of both. The ECCS has established that comfort temperatures for the coastal Southern area (Thermal area E) where Concepción, the largest city in the Bio-Bio Region, is located, must be between 20 and 25 °C. It is necessary to indicate that this temperature model is not the most suitable for assessing social housing as it does not tend to fit the real operating conditions of social housing. These tend to work under free oscillation except during extremely cold or hot situations.

The most extended adaptive comfort models, ASHRAE 55-2017 [31] and EN 15251:2007 [11,19] were established from the RP-884 and SCATs (smart control and thermal comfort, SCATs) research projects. These models are applied in buildings without active systems, where operable windows are easily accessed, and the occupants can adapt their clothing to the thermal conditions. Both models have established that an occupant's metabolic activity must be between 1.0 and 1.3 met (1.0 met = 58 W/m^2) and that it is possible for the clothing level to be between 0.5 and 1.0 clo (1.0 clo = 0.155 m^2 K/W).

In the case of ASHRAE 55-2017, two levels of thermal comfort are established based on the occupant's acceptability percentage. The first is associated with typical demand levels of 80% (1,2) and the second when a higher level of comfort is required (90%) (3,4), where θ_{rm} is the weighted running mean temperature.

$$\text{Upper limit of the 80\% acceptability comfort zone} = 0.31 * \theta_{rm} + 17.8 + 3.5 \quad (1)$$

$$\text{Lower limit of the 80\% acceptability comfort zone} = 0.31 * \theta_{rm} + 17.8 - 3.5 \quad (2)$$

$$\text{Upper limit of the 90\% acceptability comfort zone} = 0.31 * \theta_{rm} + 17.8 + 2.5 \quad (3)$$

$$\text{Lower limit of the 90\% acceptability comfort zone} = 0.31 * \theta_{rm} + 17.8 - 2.5 \quad (4)$$

In the case of EN 15251:2007, four types of inside comfort classifications have been established with regard to the expectations the occupants have, as well as other factors which are determined by the perception of comfort and the building's age (5,6,7,8). These are as follows:

- Category I: High level of expectation, recommended for spaces used by people who are weak and sensitive with special requirements, like the disabled, sick, very young children or the elderly (acceptability range—90%).
- Category II: Normal level of expectation, should be used for new or remodeled buildings (acceptability range—80%).
- Category III: Acceptable and moderate level of expectation, can be used for existing buildings (acceptability range—65%).
- Category IV: Values outside the criteria of the previous categories. This category should only be accepted during a limited part of the year (acceptability range <65%).

The following equations have been defined for Categories I and II which establish acceptable comfort limits (90% Category I and 80% Category II) and will be used to calculate the comfort zone limits of the housing under study:

$$\text{Upper limit of Category I comfort zone} = 0.33 * \theta_{rm} + 18.8 + 2 \quad (5)$$

$$\text{Lower limit of Category I comfort zone} = 0.33 * \theta_{rm} + 18.8 - 2 \quad (6)$$

$$\text{Upper limit of Category II comfort zone} = 0.33 * \theta_{rm} + 18.8 + 3 \quad (7)$$

$$\text{Lower limit of Category II comfort zone} = 0.33 * \theta_{rm} + 18.8 - 3 \quad (8)$$

$$\theta_{rm} = (\theta_{ed-1} + 0.8 * \theta_{ed-2} + 0.6 * \theta_{ed-3} + 0.5 * \theta_{ed-4} + 0.4 * \theta_{ed-5} + 0.3 * \theta_{ed-6} + 0.2 * \theta_{ed-7})/0.8 \quad (9)$$

In both methods, for the calculation of the average operating temperature (θ) of a concrete day, the outside average temperatures of the seven previous days are used, with θ_{ed-1} being the average daily outside temperature of the previous day; θ_{ed-2} is the average outside temperature of the day before that and so on and so forth using Equation (9). These average daily temperatures throughout the 365 days of the year are compared with the operating temperatures of the 8760th h of the year and are evaluated within the 80% and 90% limits of both ASHRAE 55-2017 and Category I and II of EN 15251:2007.

The adaptive comfort models have a use limitation related to the average outside operating temperature. When the temperature falls below 15 °C for EN or 10 °C for ASHRAE, a minimum comfort temperature is established. For EN 15251, the minimum thermal comfort temperatures are 20 °C for Category II and 21 °C for Category I, and for ASHRAE 55-2017, these are 18.4 °C for the 90% thermal acceptability and 17.4 °C for 80% thermal acceptability. This means that the equations used to calculate the adaptive comfort temperatures cannot be used during the entire year in moderately cold climates like that of Concepción.

2.3. Methodology

The current research's methodology is focused on evaluating the percentage of time during which social housing falls within comfort ranges based on the different demands. For this, the results obtained through the ECCS comfort levels and the adaptive comfort models of ASHRAE 55-2017 and EN 15251:2007 were compared. After evaluating the differences between the application of the different models, a battery of strategies was used with the goal of improving the base case. The goal of this second study is to analyze which one of the implemented improvements is more effective when it comes to extending the free-running time, thus determining the potential of using adaptive comfort models versus the static models.

The battery of strategies was simulated parametrically in EnergyPlus® simulation software [32] using the prototype of social housing depicted in Figure 1, located in the city of Concepción (Chile). The case was chosen from the Project Bank of the Bio-Bio Region's Housing and Urbanization Service (SERVIU) due to being the most representative among the typologies used [24]. Priority was given to the choice from this document, due to these all being detached houses and having a higher surface area of the envelope exposed, as this assumes, a priori, more unfavorable thermal behavior in this climate [33].

Parametric simulations were carried out using the housing prototype depicted in Figure 1. Table 2 represents constructive standards for this prototype, which are grouped into transmittances, ventilation, infiltration and special solutions. At first, a base case was compiled with the following considerations: Transmittance (U values) were adjusted to the minimum values per ECCS, which is still not mandatory but much more restrictive than the already enforced Art. 4.1.10 of the OGUC. Ventilation rates were adjusted on a 24 h basis, in accordance with the minimum recommended values per the ECCS, while infiltration rates were adjusted also as per the minimum values required by the ECCS.

In this way, taking case 1 as the base case, several improvements in these four areas have been implemented and simulated. The first group of improvements consisted of elevating U values (cases 2–5); the second group consisted of adjusting ventilation rates depending on the season of the year (cases 6–8); the third group consisted of implementing an ON-OFF ventilation scheme over a 24 h period, which is described in Figure 2 (case 9); the fourth group reduced infiltration to 0.35 ACh (case 10); the fifth group implemented two special solutions for the building skin: a high absorbance material and a high thermal mass (cases 11–12). These first 12 cases are defined as "simple parametrization cases". After analyzing the results from these 12 different possibilities, those ones with the best results were selected and therefore combined into four additional parametric cases, denoted as "complex parametric cases". Case 13 combines the results from case 8 plus case 10; case 14 combines cases 5, 9 and 10; case 15 combines cases 5, 9, 10 and 11; and lastly, case 16 combines case 9 plus case 10. Finally, the results were analyzed to prioritize the implementation of enhanced construction standards in terms of the percentage of time during which the house can be in a free-running state.

Table 2. The model's parametrization data.

Case	U Openings (W/m^2 K)	U Envelope (W/m^2 K)			Ventilation L/(s*person) [1]				Infiltration (ACh)	Special Solutions
		Roof	Walls	Floor [(m^2K)/W]*100	Time	Months 5,6,7,8	Months 4,9,10	Months 1,2,3,11,12		
1	3.16	0.33	0.50	45	24 h	5.2	5.2	5.2	1	-
2	2.68	0.33	0.50	45	24 h	5.2	5.2	5.2	1	-
3	1.94	0.33	0.50	45	24 h	5.2	5.2	5.2	1	-
4	3.16	0.2	0.35	0.3	24 h	5.2	5.2	5.2	1	-
5	3.16	0.13	0.18	0.15	24 h	5.2	5.2	5.2	1	-
6	3.16	0.33	0.50	45	24 h	5.2	5.2	10	1	-
7	3.16	0.33	0.50	45	24 h	3.5	3.5	10	1	-
8	3.16	0.33	0.50	45	24 h	2.0	3.5	10	1	-
9	3.16	0.33	0.50	45	ON-OFF	2.0	3.5	10	1	-
10	3.16	0.33	0.50	45	24 h	5.2	5.2	5.2	0.35	-
11	3.16	0.33	0.50	45	24 h	5.2	5.2	5.2	1	T. absorptance 0.9
12	3.16	0.33	0.50	45	24 h	5.2	5.2	5.2	1	Δ thermal mass
13	3.16	0.33	0.50	45	24 h	2.0	3.5	10	0.35	-
14	3.16	0.13	0.18	0.15	ON-OFF	2.0	3.5	10	0.35	-
15	3.16	0.13	0.18	0.15	ON-OFF	2.0	3.5	10	0.35	Δ thermal mass
16	3.16	0.33	0.50	45	ON-OFF	2.0	3.5	10	0.35	-

[1] Months in which each ventilation flow is applied.

Regarding the ventilation rates, additional comments should be made. In mild climates such as the one found in the Bio-Bio region, natural ventilation plays a crucial role in thermal comfort [34], in accordance with adaptive thermal comfort standards, such as ASRAE 55-2017 and EN 15251. Therefore, the base case for natural ventilation has been set as 5.2 L/s per person, per ECCS, during all times of the year. After that, this scheme has been tuned step-by-step using different possibilities in order to maximize the potential of ventilation as a method for achieving comfort. Primarily, the ventilation rate has been adapted on a yearly basis, according to a classification that includes very cold (May, June, July and August), cold (April, September and October) and warmer months (January, February, March, November and December). After that, the former rate has been further adjusted on a 24 h basis. In addition, internal loads have been considered in all of the models. Occupation, lighting, equipment and ventilation loads (Table 3) as well as schedules (Figure 2) have been obtained from the ECCS (Figure 3).

Table 3. Internal heat loads for the models.

	Living-Dining Room	Kitchen	Bedroom	Bathroom	Corridor
Illumination (W/m^2)	23	13	12	13	5
Occupation (W/m^2)	8.9	8.9	8.9	8.9	8.9
Equipment (W/m^2)	12.40	12.40	12.40	-	12.40

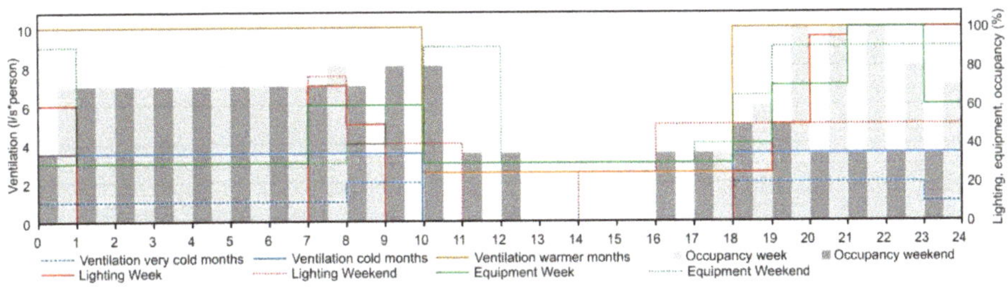

Figure 2. Occupation, lighting and equipment and ventilation schedules.

3. Results

3.1. *Analysis of Comfort in Social Housing Following the Evaluation Model Used*

The adaptive comfort models establish different temperature ranges depending on the outdoor temperature. According to the evaluation proposed by ASHRAE 55-2017 and EN 15251:2007, it is considered that the building is within comfort limits if the following criteria is met: First, in the 90% range (Category I for EN) occupants may be slightly mild or cold; this range is reduced in the 80% range (Category II for EN) [18,19]. Therefore, if criteria for category II are met, the building can be catalogued as "comfortable" without the use of active conditioning systems.

The housing model under free oscillation has been simulated to compare the comfort standards established for the ECCS with adaptive models, evaluating indoor temperatures by different thermal well-being criteria (Figures 3 and 4). It is necessary to indicate that the comfort model defined by ECCS, which we could call "static", has a less extensive temperature range than the adaptive models, being 5 °C, independent of the conditions [17]. ASHRAE's adaptive model has a daily range of 7 °C for 80% acceptability and 5 °C for 90%. The EN has a daily range of 6 °C as a maximum for Category II and 4 °C for Category I. However, their limits are not constant and adapt depending on the outdoor temperatures, as can be seen in Figures 3 and 4. This means that adaptive standards are closer to the thermal behavior of social housing under the free oscillation system [7].

The buildings located in the city of Concepción mainly have problems associated with low temperatures (Figures 3 and 4), reaching indoor temperatures close to 10 °C during the winter months. Overheating issues can be seen during the hottest months, but these are minimal compared with those of the colder temperatures. For this reason, the improvement measures have mainly been focused on the colder period, although summer measures were included (Table 2).

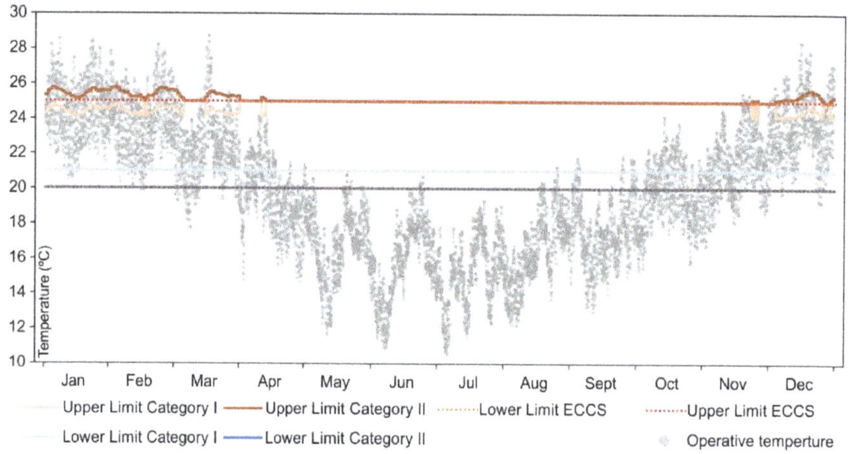

Figure 3. Operative temperature of the housing (case 1) using the limits established by ECCS and by EN 15251:2007.

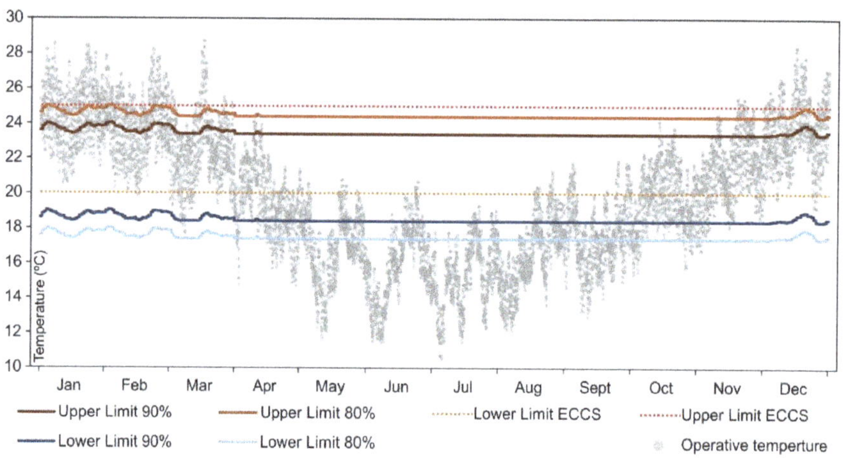

Figure 4. Operative temperature of the housing (case 1) using the limits established by ECCS and by ASHRAE 55-2017.

It can be seen in Figures 3 and 4 that the static comfort model (ECCS), with its a smaller temperature range, has worse results than the adaptive models. According to ECCS, the building is within adequate temperature ranges 39% of the time. However, when applying EN's adaptive system, this percentage is 41.23% (Figure 3), considering that the temperatures within the ranges of Types I and II are suitable for avoiding artificial condition systems. According to ASHRAE (Figure 4), thermal comfort is achieved 59.66% of the time, using the 80% limits. With regard to the thermal discomfort conditions, it is necessary to indicate that the EN 15251:2007 standard, using Category II, does not reduce the time during which cold conditions are observed, if compared with ECCS (52.31%). However, regarding ASHRAE, this time is reduced by 22.82%. All three models indicated short periods of time associated with high temperatures: 8.70% for ECCS, 6.46% for EN and 10.86% for ASHRAE.

The ECCS for housing indicates that a building can dispense with air-conditioning systems if it falls within comfort ranges at least during 70% of the time, establishing, in this way, the aforementioned temperatures [17]. On comparing the results, it is seen that to reach this percentage, it would be necessary to increase the comfort conditions by 31% in the case of ECCS, 28.77% for EN 15251:2007 and 10.34% for ASHRAE 55-2017. Because of this, the simple application of ASHRAE's adaptive thermal comfort model has relevant potential: a dwelling can work under free oscillation during an important part of the year. This would also occur if Category III of the EN norm is applied. However, it has been decided to use Category II so that the thermal acceptability percentage is similar between both adaptive standards.

3.2. *Analysis of the Improvement of the Standards for Low Energy Consumption Housing*

The ECCS establishes some basic standards for the building of homes, defining the transmittance, thermal mass, internal loads, ventilation, infiltration and use profile requirements for the climate zone of Concepción (Zone E). A first analysis of an ECCS based model (Case 1) was done to improve current standards, identifying the associated indoor comfort problems of the model and proposing 11 improvements to the base model and four combinations with the basic cases that would offer the best results.

Cases 2, 3, 4 and 5 are associated with improvements to the envelope's thermal transmittances; cases 6, 7, 8 and 9 are associated with the housing's natural ventilation flow; case 10 is associated with infiltration; and cases 11 and 12 are associated with special improvements related with to the absorptance of the roof and the increase in the housing's thermal mass. It has been considered suitable to analyze results using the static comfort model and to compare these with the adaptive models that offer a higher use potential of the housing's indoor temperatures under free oscillation and a closer match to the housing's real operation.

Improvements 2 and 3 consisted of reducing the wall opening's transmittances, including PVC joinery and glazing with an 6/12/6 air chamber, with model 3 having low emission. The results indicated that considerably improving the glazed surfaces under 3.16 W/m^2 K does not increase the home's comfort time. In fact, it even reduces it when applying the standards of the ECCS and the European Norm (EN 15251:2007) −0.03% and −0.10%, respectively; however, for the American standard (ASHRAE 55-2017), the comfort times are increased by 0.89% and 1.27%, respectively. Due to the limited impact of these improvements on the building, these have not been considered for the cases with combined measures (Table 2). Likewise, the reduction of the thermal transmittances of the opaque enclosures does not mean an important increase in the housing's comfort, increasing it by 1.87% and 1.44% for EN; 1.34% and 0.11% for ECCS, and 1.60% and −0.24% for ASHRAE. Unlike what would occur with the improvements in glazed surfaces, the low impact of the reduction of transmittances means is associated with a reduction in the percentage of time with low temperatures (5.35% and 8.45% ASHRAE; 4.47% and 7.68% ECCS; 4.47% and 7.68% EN) and a similar increase in the time that the housing would be over the thermal comfort limits (3.76% and 8.69% ASHRAE; 3.14% and 7.57% ECCS; 2.60% and 6.24% EN) (Table 4). Therefore, despite the fact that this improvement does not lead to an increase in the comfort conditions, it has been considered appropriate to include it, given that the high temperatures can be adjusted via free-cooling with natural ventilation.

Following the analysis of the results of the base case (Figures 3 and 4), it was seen to be appropriate to establish better standards for the housing's operation during summer and winter. In the base case, the time percentage where the thermal comfort limits were exceeded due to overheating was relatively small, compared to the problems associated with low temperatures, despite these representing 10.86% for ASHRAE, 8.70% for ECCS and 6.46% for EN. Because of this, the months and hours of the years in which the indoor operating temperatures are higher than the comfort temperatures have been analyzed, and it has been determined that the natural ventilation must be increased to 10 L/(s*person) during November, December, January, February and March and reduced in the coldest months (May, June, July and August).

In models 6 and 7, the natural ventilation flows are adjusted for the months of November, December, January, February and March, from 5.2 L/(s*person) to 10 L/(s*person) and in case 7 the ventilation flow of the coldest months is reduced from 5.2 L/(s*person) to 3.5 L/(s*person). As can be seen in Table 4, the adjustment of natural ventilation rates during hot periods increases the comfort time by 11.10% for ASHRAE, 3.54% for ECCS and 2.41% for EN. The flow reduction in the coldest months also considerably increases the housing's thermal comfort hours, by 3.76% for EN 4.85% for ECCS and 12.20% for the American standard. Because of this, this measure forms part of the four combined models, i.e., a ventilation rate for these models of 2.0 L/(s*person) for the very cold months, 3.5 for the colder months and 10.0 for the hot months.

Table 4. Percentage of time in different environmental situations using the comfort model.

ECCS (Zone E)							
Case	Cold (%)		Comfort (%)		Hot (%)	Total Comfort (%)	Difference
1	52.31		39.00		8.70	39.00	
2	54.30		38.96		6.74	38.96	−0.03
3	54.22		39.11		6.67	39.11	0.11
4	47.83		40.33		11.84	40.33	1.34
5	44.62		39.11		16.27	39.11	0.11
6	54.25		42.53		3.22	42.53	3.54
7	52.77		43.85		3.38	43.85	4.85
8	51.47		45.15		3.38	45.15	6.15
9	47.29		47.79		4.92	47.79	8.79
10	44.66		39.92		15.42	39.92	0.92
11	50.66		38.26		11.07	38.26	−0.73
12	55.17		42.75		2.08	42.75	3.76
13	39.04		55.24		5.72	55.24	16.24
14	10.35		70.40		19.25	70.40	31.40
15	20.10		70.23		9.67	70.23	31.23
16	35.07		56.68		8.25	56.68	17.68
EN 15251:2007							
Case	Cold (%)	Slightly Cool (%)	Comfort (%)	Slightly Warm (%)	Hot (%)	Total Comfort (%)	Difference
1	52.31	7.75	28.29	5.19	6.46	41.23	
2	54.30	7.95	28.12	5.07	4.57	41.13	−0.10
3	54.22	8.05	28.28	4.94	4.51	41.27	0.03
4	47.83	7.90	28.61	6.60	9.06	43.11	1.87
5	44.62	8.04	26.61	8.03	12.71	42.67	1.44
6	54.25	11.66	29.03	2.96	2.11	43.64	2.41
7	52.77	12.01	29.91	3.07	2.24	44.99	3.76
8	51.47	12.49	30.73	3.07	2.24	46.29	5.06
9	47.29	12.44	32.60	4.41	3.25	49.45	8.22
10	44.66	7.83	28.57	6.13	12.81	42.53	1.30
11	50.66	7.50	27.90	4.94	9.00	40.34	−0.89
12	55.17	7.00	32.69	4.81	0.33	44.50	3.26
13	39.04	14.16	38.37	4.37	4.06	56.89	15.66
14	10.35	10.35	55.42	7.73	16.14	73.50	32.27
15	20.10	15.27	49.25	10.26	5.11	74.78	33.55
16	35.07	13.90	39.81	5.05	6.18	58.76	17.52
ASHRAE 55-2017							
Case	Cold (%)	Slightly Cool (%)	Comfort (%)	Slightly Warm (%)	Hot (%)	Total Comfort (%)	Difference
1	29.49	9.83	42.69	7.13	10.86	59.66	
2	30.86	9.90	43.56	7.09	8.60	60.55	0.89
3	30.66	9.90	43.82	7.20	8.41	60.92	1.27
4	24.13	9.87	42.59	8.79	14.61	61.26	1.60
5	21.04	9.32	40.83	9.27	19.54	59.42	−0.24
6	25.02	11.42	53.24	6.10	4.22	70.75	11.10
7	23.74	11.21	54.29	6.36	4.39	71.86	12.20
8	20.53	10.54	58.18	6.36	4.39	75.08	15.42
9	16.75	10.24	59.25	7.16	6.61	76.64	16.99
10	21.47	9.00	42.71	8.70	18.13	60.40	0.74
11	28.09	10.01	41.21	7.28	13.40	58.50	−1.15
12	35.75	7.92	43.29	9.85	3.18	61.06	1.40
13	10.90	8.95	64.21	8.23	7.71	81.39	21.74
14	0.47	1.88	55.66	16.87	25.11	74.42	14.76
15	0.00	0.90	70.10	16.07	12.92	87.08	27.42
16	8.70	7.74	63.66	8.66	11.23	80.07	20.41

In case 9, three operation schedules have been incorporated, depending on the month. The occupation of the housing and its use have been considered in the schedule (See Figure 2). The main flow regulation schedule is produced between 18:00 and 10:00, as residential buildings are mainly used during this time, being also the period when heating issues occur. This improvement produces an 8.22% increase in the housing's comfort time according to EN and 16.99% for ASHRAE, indicating that this is an acceptable measure for reducing the building's indoor temperatures when being assessed with both standards. Therefore, it was decided it would form part of cases 14, 15 and 16.

With regard to air infiltration, it is necessary to indicate that the ECCS establishes an air leak class of 1 ACH, considering the improvement associated to the lower temperature which, for case 10, is reduced to 0.35 ACH. In this case, according to EN, the indoor conditions are improved 1.30% of the time, increasing the discomfort due to heat situations by 6.35% and reducing the temperatures under the comfort limits by 7.65%. In case 10, according to ASHRAE, the improvement is produced 0.74% of the time, with the heat situation increasing by 7.27% and cold decreasing by 8.01%. Just as occurred with the reduction of the thermal transmittances, it has been considered appropriate to include this in the combined measures, as the high temperatures can be adjusted using free-cooling with natural ventilation.

Models 11 and 12 consider passive measures associated with the envelope's thermal behavior, the first associated with improving the gains through the roof's radiation to improve indoor conditions associated with the cold, and the second, with the increase in the housing's thermal mass to increase its storage and thermal damping capacity. The application of improvement 11 does not assume an increase in the comfort conditions as can be seen in Table 4 (EN −0.89%, ECCS −0.73% and ASHRAE −1.15%).

Model 12 would have a negative impact of 3.26% for EN, while for the American standard, it would mean a comfort increase of 1.40%, partially reducing the problems associated with overheating. For ECCS, the thermal mass increase would mean reducing the overheating by 6.62%. It is necessary to indicate that the ECCS establishes that, for the thermal zone where the case study is located, the housing must have a light thermal mass 0–70 (kJ/m^{2}K), but the results show that it is suitable to increase the calorific capacity by between 200–400 (kJ/m^{2}K) to avoid overheating problems. Therefore, this solution has been implemented within one of the combined cases.

Model 13 has considered all the natural ventilation improvements which means an increase in comfort and reduction of infiltration; it produces a natural ventilation rate of 2.0 L/(s*person) during the coldest months, 3.5 L/(s*person) during cold months and 10.0 L/(s*person) during the warm months, including a leak rate of 0.35 ACH. For model 16, the ventilation operation schedule evaluated in case 9 has been incorporated. Model 13, according to EN and ASHRAE improves the comfort conditions by 15.66% and 21.74%, respectively, and model 16 by 17.52% (EN) and 20.41% (ASHRAE). Thus, it is seen that, according to the European Model, it is essential that the housing's natural ventilation flows is regulated in accordance with the outdoor conditions; however, for the American model, this would mean practically no benefit whatsoever.

Model 14 has been formed by the combination of the measures of models 5, 9, 10, thus regulating the ventilation and the infiltration and reducing the thermal transmittance of the envelope's opaque enclosures. The grouping of the measures offers a satisfactory result as it increases the housing's comfort time by 32.27% using EN, 31.40% using ECCS and 14.76% according to ASHRAE; therefore, the combination of these improvements means a very important comfort increase for EN and ECCS; however, for ASHRAE there is not such an important increase in the comfort conditions (Table 4).

The combination of these measures with the thermal mass increase (Model 15) means that the building's behavior increases considerably with this model 74.78% of the time, being within the comfort values established by EN 15251:2007 for Category II which would mean an increase of 33.55% versus case 1. With regard to ASHRAE, this model is within the comfort values 87.08% of the time, applying the ranges for 80% of thermal acceptability, which would mean an increase of 27.42% versus case 1. In the case of ECCS, this would mean an increase of 31.23% with a total percentage within the 70.23% limits.

On analyzing Figure 5, the housing's indoor operative temperatures versus the average outdoor operative temperature are seen, calculated based on the EN 15251:2007 and ASHRAE 55-2017 norms. It can be seen that only 2209 h are outside the Category II limits and 1132 h out of the 8760 h of the year are outside ASHRAE's 80% limits. In the American norm's case, this is always due to heat. However, on applying ECCS's temperature range, that figure would be increased to 2608 h, of which 1761 h would be due to the cold. It is necessary to indicate that if Category III of the EN 15251:2007 norm had been applied, case 15 would have almost 100% of the time in comfortable conditions, with 7 h under the comfort limits and 22 h over the limits.

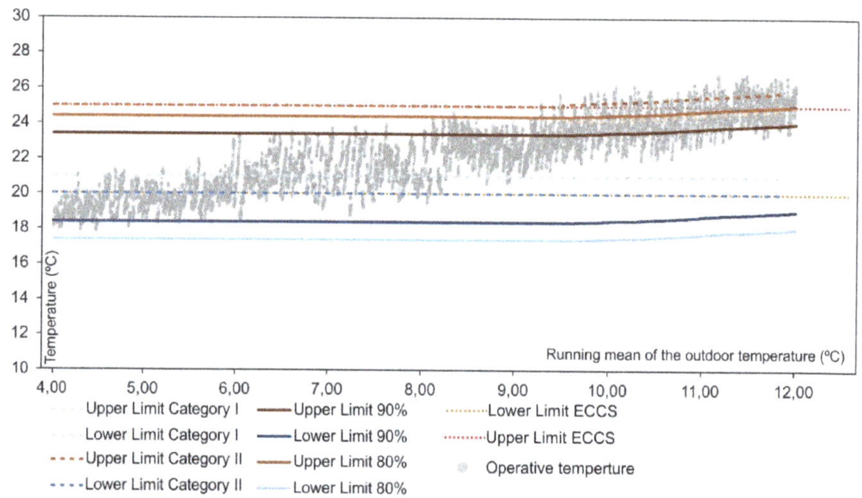

Figure 5. Spread of operative temperatures and mean outdoor temperatures of case 15 versus the limits set by ECCS, by EN 15251:2007 and by ASHRAE 55.

This indicates that the adaptive comfort models have a lot of potential to evaluate the potential of a building to operate under free oscillation without considering active energy consumption measures but, at the same time, very different results can be obtained depending on the model chosen for the evaluation.

Therefore, to have housing that can be naturally ventilated in Concepción, it is necessary to increase the requirements associated with the transmittances, ventilation, infiltration and thermal mass, with the housing being able to operate for 70% of the year without an air-conditioning system, even reaching 100% if Category III of the EN 15251:2007 norm is applied.

4. Conclusions

This research focused on clarifying which construction strategies have a greater influence on the thermal comfort of inhabitants, considering a social housing prototype located in the region of the Bio-Bio and analyzed under two international standards (EN 15251:2007 and ASHRAE 55-2017 adaptive comfort models) and one national standard (Chilean ECCS). Looking at the results, the following conclusions can be drawn.

Results for the three standards analyzed are coherent. The base case (case 1) was always amongst the worst options regarding comfort; the complex cases (13–16) offered the best results and the in-between cases (2–12) offered similar patterns of improvement. This tells that results in numerical terms can be different depending on the standard, but all of them will point in the same direction, so that the three of them can find application in this context and the improvements will also have a similar effect even when considering different criteria.

Going deeper, it can be highlighted that the ASHRAE standard is the least flexible, so to speak, when different constructive improvements are applied to increase the period of time in thermal comfort. On the other hand, the Chilean ECCS shows greater dependence on the application of such improvements; the European standard is in between both of them. It is remarkable how a standard with such fixed comfort threshold temperatures (ECCS) is in need of constructive improvements, whereas those with adaptive limits are not. In other words, adaptive thermal comfort implies a non-dependence on constant upgrading of constructive systems.

The analysis has clarified that, under the three standards, a social dwelling prototype located in the region of the Bio-Bio will reach it best operation under three complex cases (cases 14–16). Regarding construction, this would mean having low U values for the roof, walls and floors, but not necessarily for openings. The most important feature, present in the three of them, is having an adapted ventilation scheme on a yearly and a 24 h basis, and reduced airtightness; additionally, thermal mass could be an asset, but its effect would not be so notable, except for when the ASHRAE standard is used. These conclusions pose an interesting debate about the real effectiveness of a constant increase in U standards in mild climates. In this case, it has been proven that, despite high U values being beneficial, an improved ventilation scheme is much more effective when trying to reduce dependence on HVAC systems.

It is possible to reach 80%, or nearly 80%, of time in thermal comfort in this climate under certain situations. The remaining time, discomfort is associated primarily with a cold environment, rather than with hot weather; at this point, it has been proven that heating systems would play a significant role for 20–30% of the year, but always in combination with a carefully designed ventilation scheme and improved airtightness.

The results of this research are oriented towards the tight relationship between building construction and building operation, focusing on social dwellings, whose inhabitants operate in close relation to adaptive comfort. The results can be beneficial, for public policies that aim to decrease the energy consumption of buildings. Not only better constructive standards, but also intelligent and adapted patterns of use, are of capital importance when trying to reach this objective. International standards can find application in different national contexts, always having in mind the particularities, necessities and level of development of each socioeconomic reality.

Acknowledgments: This work forms part of the FONDECYT 3160806 research project, "Study of the viable energy improvement standard for social housing in a situation of energy poverty via the evaluation of post-occupational adaptive comfort and its progressive implementation", financed by the National Science and Technology Research Commission (CONICYT). The authors Alexis Pérez-Fargallo and Carlos Rubio-Bellido would like to acknowledge the VI PPIT-US for support this research and internationalization.

Author Contributions: All the authors contributed equally in the present research.

Conflicts of Interest: The authors declare no conflict of interest.

References

1. United Nations Environment Programme (UNEP). *Building Design and Construction: Forging Resource Efficiency and Sustainable Development*; UNPE: Nairobi, Kenya, 2012.
2. International Energy Agency (IEA). World Energy Outlook 2013. Available online: https://www.iea.org/newsroom/news/2013/november/world-energy-outlook-2013.html (accessed on 15 March 2018).
3. Nicol, F. Comfort and energy use in buildings—Getting them right. *Energy Build.* **2007**, *39*, 737–739. [CrossRef]
4. Santamouris, M. Innovating to zero the building sector in Europe: Minimising the energy consumption, eradication of the energy poverty and mitigating the local climate change. *Sol. Energy* **2016**, *128*, 61–94. [CrossRef]
5. European Union. European Commission Directive 2002/91/EC of the European Parliament and of the council of 16 December 2002 on the energy performance of buildings. *Off. J. Eur. Union* **2002**, 65–71. [CrossRef]
6. European Union. European Commission Directive 2010/31/EU of the European Parliament and of the Council of 19 May 2010 on the energy performance of buildings. *Off. J. Eur. Union* **2010**, 13–35. [CrossRef]
7. Attia, S.; Carlucci, S. Impact of different thermal comfort models on zero energy residential buildings in hot climate. *Energy Build.* **2015**, *102*, 117–128. [CrossRef]
8. Salata, F.; Golasi, I.; Domestico, U.; Banditelli, M.; Lo Basso, G.; Nastasi, B.; de Lieto Vollaro, A. Heading towards the nZEB through CHP+HP systems. A comparison between retrofit solutions able to

increase the energy performance for the heating and domestic hot water production in residential buildings. *Energy Convers. Manag.* **2017**, *138*, 61–76. [CrossRef]

9. Burattini, C.; Nardecchia, F.; Bisegna, F.; Cellucci, L.; Gugliermetti, F.; De Lieto Vollaro, A.; Salata, F.; Golasi, I. Methodological approach to the energy analysis of unconstrained historical buildings. *Sustainability* **2015**, *7*, 10428–10444. [CrossRef]
10. Programa de Estudios e Investigaciones en Energía (PRIEN). Estimación del Potencial de Ahorro de Energía, Mediante Mejoramientos de la Eficiencia Energética de los Distintos Sectores. Available online: http://old.acee.cl/576/article-59077.html (accessed on 15 March 2018).
11. Nicol, J.F.; Humphreys, M.A. Adaptive thermal comfort and sustainable thermal standards for buildings. *Energy Build.* **2002**, *34*, 563–572. [CrossRef]
12. Humphreys, M.A.; Rijal, H.B.; Nicol, J.F. Updating the adaptive relation between climate and comfort indoors; new insights and an extended database. *Build. Environ.* **2013**, *63*, 40–55. [CrossRef]
13. Peacock, A.D.; Jenkins, D.P.; Kane, D. Investigating the potential of overheating in UK dwellings as a consequence of extant climate change. *Energy Policy* **2010**, *38*, 3277–3288. [CrossRef]
14. Thomson, H.; Snell, C. Quantifying the prevalence of fuel poverty across the European Union. *Energy Policy* **2013**, *52*, 563–572. [CrossRef]
15. Van Hooff, T.; Blocken, B.; Hensen, J.L.M.; Timmermans, H.J.P. Reprint of: On the predicted effectiveness of climate adaptation measures for residential buildings. *Build. Environ.* **2015**, *83*, 142–158. [CrossRef]
16. Oropeza-Perez, I.; Petzold-Rodriguez, A.H.; Bonilla-Lopez, C. Adaptive thermal comfort in the main Mexican climate conditions with and without passive cooling. *Energy Build.* **2017**, *145*, 251–258. [CrossRef]
17. Ministry of Housing and Urbanism of Chile (MINVU). Código de Construcción Sustentable. Available online: http://csustentable.minvu.gob.cl/consulta-publica/ (accessed on 15 March 2018).
18. The American Society of Heating, Refrigerating and Air-Conditioning Engineers (ASHRAE). *ASHRAE Standard 55-2013 Thermal Environmental Conditions for Human Occupancy*; ASHRAE Inc., Ed.; American Society of Heating, Refrigerating and Air Conditioning Engineers: Atlanta, GA, USA, 2013; ISBN 1041-2336.
19. Comité Européen de Normalisation (CEN). *EN 15251:2007 Indoor Environmental Input Parameters for Design and Assessment of Energy Performance of Buildings Addressing Indoor Quality, Thermal Environment, Lighting and Acoustics*; European Committee for Standardization: Brussels, Belgium, 2007.
20. Bustamante, W.; Cepeda, R.; Martínez, P.; Santa María, H. Eficiencia energética en vivienda social: Un desafío posible. In *Camino al Bicentenario—Propuestas para Chile*; Pontificia Universidad Catolica de Chile: Santiago de Chile, Chile, 2009; pp. 253–282.
21. Ministry of Housing and Urbanism of Chile (MINVU). Estadisticas históricas. Available online: http://www.minvu.cl/opensite_20061113165249.aspx (accessed on 15 March 2018).
22. Ministry of Housing and Urbanism of Chile (MINVU). *Ds 01. Reglamento Del Sistema Integrado de Subsidio Habitacional*; MINVU: Santiago de Chile, Chile, 2011.
23. Ministry of Housing and Urbanism of Chile (MINVU). *Ds 49. Reglamento Del Programa Fondo Solidario de Elección de Vivienda*; MINVU: Santiago de Chile, Chile, 2011.
24. Ministry of Housing and Urbanism of Chile (MINVU). *Viviendas de Modalidad Construcción en Sitio Propio con Proyecto Tipo*; MINVU: Santiago de Chile, Chile, 2011.
25. Citec, U.B.B.; Decon, U.C. *Manual de Hermeticidad al Aire de Edificaciones*, 1st ed.; Trebilcock, M., Ed.; Fondef: Concepción, Chile, 2012; ISBN 978-956-9275-27-2.
26. Ministry of Housing and Urbanism of Chile (MINVU). Artículo 4.1.10. Exigencias de acondicionamiento térmico de la Ordenanza general de urbanismo y construcciones. In *DS 47 Ordenanza General de la ley General de Urbanismo y Construcciones*; MINVU, Ed.; MINVU: Santiago, Chile, 2007.
27. Kunkel, S.; Kontonasiou, E.; Arcipowska, A.; Mariottini, F.; Atanasiu, B. *Indoor Air Quality Thermal Comfort and Daylight. Analysis of Residential Building Regulations in Eight EU Member States*; Buildings Performance Institute Europe (BPIE): Bruselas, Belgium, 2015.
28. Figueroa, R.; Bobadilla, A.; Besser, D.; Días, M.; Arriagada, R.; Espinoza, R. Air infiltration in Chilean housing: A baseline determination. In Proceedings of the PLEA2013—29th Conference, Sustainable Architecture for a Renewable Future, Munich, Germany, 10–12 September 2013; p. 6.

29. Comité Européen de Normalisation (CEN). *EN 15217:2007 Energy Performance of Building. Methods for Expressing Energy Performance and for Energy Certification of Building*; European Committee for Standardization: Brussels, Belgium, 2007.
30. Comité Européen de Normalisation (CEN). *EN 15603:2008 Energy Performance of Building. Overaall Energy Use and Definition of Energy Ratings*; CEN: Bruselas, Belgium, 2008.
31. De Dear, R.J.; Brager, G.S. Thermal comfort in naturally ventilated buildings: Revisions to ASHRAE Standard 55. *Energy Build.* **2002**, *34*, 549–561. [CrossRef]
32. US Department of Energy. *EnergyPlus Engineering Reference: The Reference to EnergyPlus Calculations*; US Department of Energy: Washington, DC, USA, 2010; article-id:10579266.
33. Pérez Fargallo, A.; Calama Rodríguez, J.; Flores Alés, V. Comparativa de resultados de rehabilitación energética para viviendas en función del grado de mejora. *Informes de la Construcción* **2016**, *68*. [CrossRef]
34. Rubio-Bellido, C.; Pulido-Arcas, J.A.; Ureta-Gragera, M. Aplicabilidad de estrategias genéricas de diseño pasivo en edificaciones bajo la influencia del cambio climático en Concepción y Santiago, Chile. *Hábitat SustenTable* **2015**, *5*, 33–41.

© 2018 by the authors. Licensee MDPI, Basel, Switzerland. This article is an open access article distributed under the terms and conditions of the Creative Commons Attribution (CC BY) license (http://creativecommons.org/licenses/by/4.0/).

sustainability

MDPI

Article

Solar Heat Gain Reduction of Ventilated Double Skin Windows without a Shading Device

Bokyoung Koo [1,*], Keonho Lee [1], Youngsub An [2] and Kyudong Lee [2]

[1] Korea Institute of Civil Engineering and Building Technology, 10223 Goyang-si, Korea; lee1ncdh@kict.re.kr
[2] Kolon Global Corp., 21984 Incheon-si, Korea; ysan@kolon.com (Y.A.), kudeng@kolon.com (K.L.)
* Correspondence: koo.bok@kict.re.kr; Tel.: +82-31-910-0143

Received: 13 November 2017; Accepted: 27 December 2017; Published: 28 December 2017

Abstract: With global efforts to strengthen various energy-saving policies for buildings to reduce greenhouse gas emissions, in South Korea, new laws and regulations have been in force since May 2015 to install shading devices in public buildings and to include the solar heat gain coefficient (SHGC) reduction performance of shading devices in the evaluation of building performance. By making a ventilated air layer outer glass and inner glass to lower the temperatures of the air layer and glass surface, it is possible to reduce the amount of heat flowing into the building while maintaining the same level of light transmission as plain window systems. This study proposes a double-skin façade window with a 20 mm ventilated air cavity, and assumes that insolation inflow indoors would be reduced through ventilation in the air cavity. The artificial solar lab test results show that the SHGC can be lowered through ventilation by 28% to 52.9%. Additionally, in an outdoor test cell experiment, the results show that the mean temperature was 0.6 K and the peak temperature was 0.9 K lower with ventilation in the air cavity than that without ventilation in the air cavity.

Keywords: building envelope; double façade window; SHGC; artificial solar labs; outdoor chamber test

1. Introduction

Demand for curtain wall systems has increased rapidly due to the activation of high-rise buildings. However, curtain wall buildings made of metal, such as aluminum, have limitations in meeting the national insulation standards, and indoor overheating due to solar thermal heat gain in summer is another problem. The double-skin façade has arisen as an alternative to overcome the disadvantages of such a curtain wall system, and various studies have been conducted since the 2000s [1].

However, due to the nature of the double-skin façade, which adds a single envelope to the existing envelope, the increase in existing construction costs is inevitable and, thus, efforts to reduce the material cost by reducing the thickness of the double envelope system have been conducted.

Meanwhile, as the importance of building energy performance increases, the standards on building envelopes have been raised and windows, which are the most vulnerable part in the envelope, have double burdens: one being improvements to the energy efficiency rating, and the other is the installation of shades.

The Korean Government has strengthened the energy reduction policies in building sectors continuously with the aim of passive houses by 2017 and zero-energy buildings by 2025 while preparing for energy reduction measures in building sectors for greenhouse gas reduction.

In particular, windows with first-grade energy efficiency ratings should be purchased in priority during new building construction to play a leading role in energy saving for public buildings, and new buildings or building extensions whose total floor area is greater than 3000 m^2 should acquire the first-grade building energy efficiency ratings. Furthermore, when public buildings whose total floor area is more than 3000 m^2 install windows at outer walls, or use glass materials at outer walls during new building or remodeling, solar control devices, such as shades, should be installed or building owners should take other action to reduce solar heat gain and increase energy efficiency.

The Korean government is encouraging the installation of exterior shades, but this is limited in application with high initial cost. For this reason, it seems that the application of a slim form of double-skin window that has proper thermal resistance and reduces solar heat gain can be applied as an alternative.

Therefore, this study searched for a solution to the high cost problem of the existing double-skin façade system by means of reducing the width of the intermediate air cavity. To maintain a thickness of the window at the existing level of window thickness, a double-skin façade window whose intermediate cavity was 20 mm was proposed, and the reduction performance of indoor solar heat acquisition through ventilation through the intermediate cavity was evaluated.

2. Working Principle

2.1. Literature Reviews on Existing Double-Skin Façade Systems

A double-skin façade consists of a normal façade, an air cavity, and an additional external skin usually made of glass. It is common to apply a shading system within the cavity between the two layers of the façade [2].

Commonly, the most efficient running scenario for a double-skin façade is to close the cavity vent in winter and to open the cavity vent in summer. Sanchez et al. [3] approved the scenario with computer simulations. In the case of glazing properties and combination, Guardo et al. [4] determined that a further decrease in total solar energy transmission would be possible by increasing the outside solar reflectance of the shading screen and replacing the internal glazing for a low-emissivity glass can duplicate the reduction of the solar load gain. Perez-Grande et al. [5] insisted that a reduction of external glazing transmissivity by 55% can lead to a 40% enhancement in the reduction in the solar load gain.

However, the installation of the shading device can increase the size of the window, which can increase the construction cost. It is also ironic to abandon the transmittance of the glazing in order to reduce the heat gain.

Studies on cavity depth [6–9] mainly focused on a double skin with a cavity layer of at least 30 cm to 150 cm.

In this study, we proposed a double skin window that can maintain the level of transmittance at the level of the general window while reducing the amount of indoor solar energy heat gain, and a performance experiment was conducted to evaluate the effect of cooling load reduction on a double skin window with a minimum 2 cm cavity layer.

2.2. Principle of Solar Heat Gain through Ventilation via an Intermediate Air Cavity

The ventilated double skin window proposed in this study consists of single outer glass, double inner glass, and a 20 mm air layer between the two glasses. This air layer is ventilated at all times because the upper and lower parts of single glass are exposed to the outside air. It can be concluded that, with this configuration, the heat flowing into the room can be reduced by convection, and this principle can be explained as follows:

The solar heat gain coefficient (SHGC) refers to the ratio of solar heat penetrating through windows compared to the solar energy reaching the window per unit area in a given incident angle and environmental conditions (indoor temperature, outdoor air temperature, wind speed, and insolation). The SHGC can be expressed via the following equation:

$$\mathrm{SHGC} = \tau_S + N_i \alpha_S \quad (1)$$

τ_S: Insolation that transmits the window;

N_i: Ratio of absorbed calorie radiated to the indoor; and

α_S: Solar heat absorption in the single-pane fenestration system.

Existing windows have heat transfer through conduction and radiation thereby constantly having heat transfer to the indoor side (Figure 1a). However, the proposed window can reduce heat transfer by naturally ventilating the inside of the air cavity whose temperature increases due to conduction (Figure 1b). That is, N_i and α_S values can be reduced through ventilation of the air cavity. Thus, the SHGC in the proposed window can be maintained at a lower value than that of the exiting window.

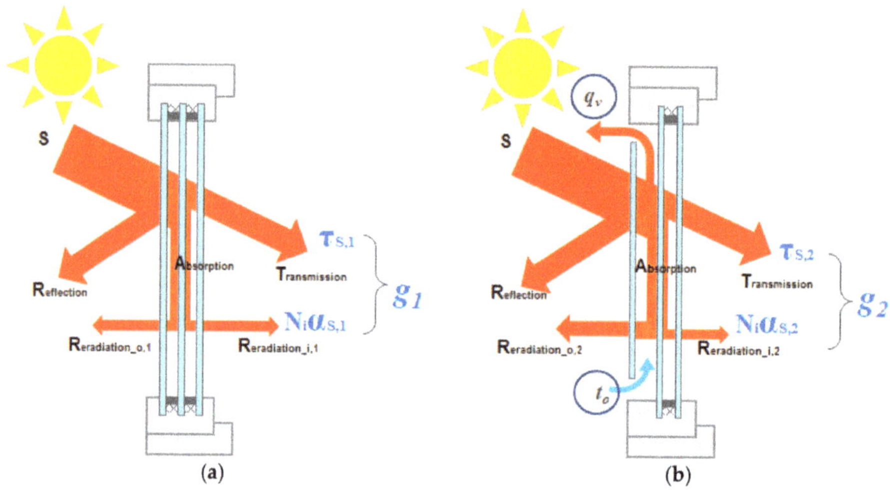

Figure 1. Concept of solar radiation rate reduction through cavity ventilation: (**a**) general single skin; and (**b**) ventilated double skin.

3. Artificial Solar Lab Test

3.1. Overview

3.1.1. Experimental Equipment and Measurement Plan

The artificial solar lab in which the SHGC can be calculated through samples using artificial sun is divided into an indoor chamber (heating box) and an outdoor chamber.(Figures 2 and 3) Figure 2 shows the configuration. The wall, bottom, and ceiling of the chamber are made with sandwich panels filled with 75 mm-thick urethane foam. Each of the chambers controls the temperature and humidity via a thermo-hygrostat and the temperature outside the chamber can also be controlled through a separate thermo-hygrostat. The air conditioned at the thermo-hygrostat is supplied to the indoor chamber through 12 diffusers in the ceiling, and discharged through the outlet in the wall. The conditioned air is supplied and exhausted through the upper and lower grills in the outdoor chamber and an artificial solar device consisting of 250 W metal halide lamps is installed, directed towards the sample placed between the indoor and outdoor chambers. In this artificial solar lab, it is possible to control factors, such as the indoor and outdoor temperature, and solar radiation, which can affect SHGC, so that a steady state can be maintained. It is also easy to measure consistent SHGC because it is not affected by altitude, longitude, and seasonal solar spectrum changes. Summary of the facilities installed in the artificial solar laboratory chamber is shown in Table 1.

The maximum sample size that can be installed is 2 m × 2 m. Thermocouples are installed at 21 positions on each surface of the inner and outer sides of the sample, 15 positions in the indoor chamber, three positions in the outdoor chamber, three positions in the thermostatic chamber, and at each of the main duct and outlet of the heating, ventilation, and air conditioning (HVAC) to measure the temperature. In addition, a pyranometer is installed on the sample side facing the outdoor chamber to measure insolation reaching the sample.

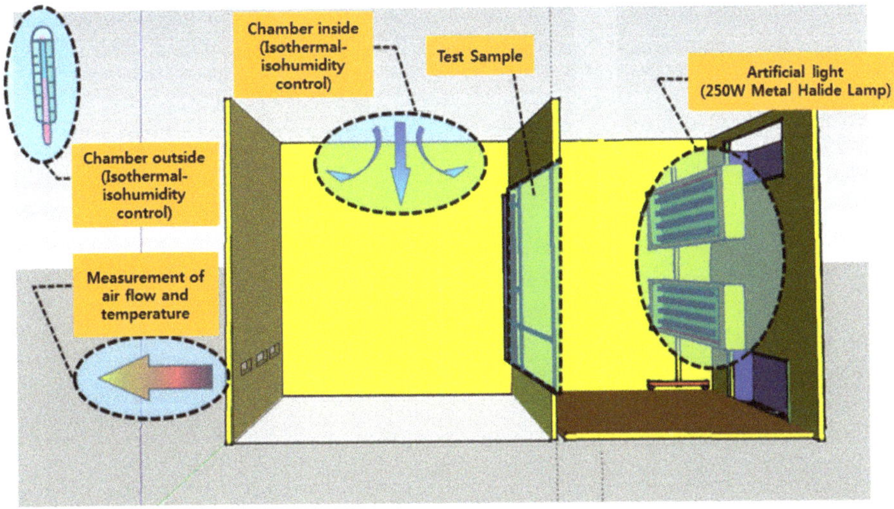

Figure 2. Schematic diagram of the artificial solar lab.

(a) (b) (c)

Figure 3. Photo of the artificial solar lab: (**a**) indoor chamber; (**b**) outdoor chamber; and (**c**) artificial lights.

Table 1. Summary of the facilities installed in the artificial solar laboratory test chamber.

	Size	**(W) 4725 × (D) 3450 × (H) 2650**
Indoor chamber	Materials of the inside and outside	SS400, 0.54 T, Color Steel.
	Insulation material	Urethane foam, 75 mm
	Floor	Access floor + 50 mm expanded polystyrene
	Temperature range	15 °C to 30 °C
	Temperature: cooling	26 °C (±1 °C)
	Temperature: heating	22 °C (±1 °C)
Outdoor chamber	**Size**	**(W) 2600 × (D) 3450 × (H) 3770**
	Materials of the inside and outside	SS400, 0.54 T, Color Steel.
	Insulation material	Urethane foam, 75 mm
	Floor	Urethane panel + plywood + linoleum
	Air circulation fan	Sirocco Fan, 12″ × 6″, four each
	Temperature range	−10 °C to 40 °C
	Temperature: cooling	32 °C (±1 °C)
	Temperature: heating	−5 °C (±1 °C)
	Wind speed	0–4.7 m/s (max. 80.0 rpm)
	Artificial sun light	Max. 1000 W/m^2 (adjustable by the number of lights on/off control)

The equation that calculates the SHGC is presented in Equation (1), which is derived via the insolation reaching the surface of the window system, insolation introduced to the indoor chamber, and the calories introduced due to difference between the indoor and outdoor temperatures:

$$\mathrm{SHGC} = \frac{q_s - q_{U-\mathrm{factor}}}{A_S E_S} \tag{2}$$

where:

q_s: heat flux through the test specimen (W);
$q_{U-\mathrm{factor}}$: heat flux through the window specimen (W);
A_S: Projected area of the window specimen (m^2);
E_S: Solar radiation energy incident to the window specimen (W/m^2);
q_S is equivalent to Equation (3):

$$q_S = q_{\mathrm{walls}} + q_{SP} + q_{\mathrm{fl}} + q_{\mathrm{fluid}} + q_{AUX} \tag{3}$$

where:

q_{walls}: heat flux through solar calorimeter walls (W);
q_{SP}: heat flux through surround panel (W);
q_{fl}: heat flux by flanking loss (W);
q_{fluid}: heat removed by fluid heat extraction system (W);
q_{AUX}: heat input into solar calorimeter by pumps and fans (W).

Temperatures of the test sample were measured at nine places on both the outside and the inside of the outer glass, air temperature was measured at one position inside the 20 mm air cavity, and nine positions each in the outside and inside of the double-skin façade. A mean temperature at the measured positions was calculated. However, the temperature of the 20 mm air cavity was measured at only one point in the center of the air cavity.

In order to reduce the error of the experimental results, the experiment was conducted more than three times for the same case. Each experiment was conducted for about eight hours for each case, and terminated when the SHGC value was found to be convergent.

During testing, insolation, temperature, power, flow rate, wind velocity, wind direction, and power of the HVAC were measured according to NFRC201 7.3 [10], and the final results were derived when a steady state was reached. The items below describe the criteria to determine the steady state. The steady state in the system was checked after measuring each factor five times to calculate a mean value and their variation was investigated to determine the steady state of the system.

- Maintaining a temperature of outdoor air chamber: within ±2 °C;
- Checking a thermal gradient in the inside and outside chamber: less than 2 °C; and
- Variation of air volume: within ±5%.
- Supply air temperature: within ±1 °C.

Since the indoor chamber temperature is always maintained at the set temperature, it is determined by the supply air temperature instead of the indoor chamber temperature. Due to the characteristics of the HVAC system, the supply air temperature shows a periodic rise and fall pattern (approximately in every 25 min). If this pattern is observed more than three times, it is regarded as the steady state. One period refers to a time taken from the minimum to the peak temperature and then back to the minimum temperature again.

3.1.2. Experimental Alternatives

Window configurations for the target windows with performance below 1.0 W/m^2·K (the first grade window in window energy efficiency rating) were selected in accordance with Korean energy performance regulations for windows.

The target windows to be assessed are a window (Case A) consisting of outer glass (clear single glass) + 20 mm air cavity + inner glass (6 mm low-e glass + 12 mm argon gas + 6 mm low-e glass), and a window (Case B) consisting of outer glass (clear single glass) + 20 mm air cavity + inner glass (6 mm reflective glass + 12 mm air + 6 mm low-e glass). The target windows to be assessed and an overview of the cases are presented in Table 2. The thermal transmittance and SHGC of the entire glasses in Case A are 0.936 W/m^2·K and 0.233, assuming that ventilation is not conducted, and those in Case B are 0.947 W/m^2·K and 0.200. The thermal transmittance and SHGC of the inner glass in Case A are 1.683 W/m^2·K and 0.203, and those of Case B are 1.695 W/m^2·K and 0.141.

The emissivity of low-e coating glass, which is applied to Cases A and B, is 0.837 on the outside and 0.013 on the inside, and the reflectivity is 0.364 on the outside, and 0.592 on the inside. The transmittance is 0.239. The emissivity of reflective glass, which is applied to Case B, is 0.837 on the outside and 0.536 on the inside, and the reflectivity is 0.132 on the outside, and 0.259 on the inside. The transmittance is 0.136.

Table 2. Test chamber cases.

Cases	Outer Glass	Inner Glass	Cavity Ventilation
A-1	Clear single glass	Low-e glass + Argon + Low-e glass	X
A-2			O
B-1	Clear single glass	Reflective glass + Air + Low-e glass	X
B-2			O

Note: X: No ventilation, O: ventilation.

3.2. Experiment Results of Case A

For each case, the surface temperature at the window and air temperature in the air cavity were measured according to whether ventilation at the air cavity is available or not. Each case had three or more measurements, and the cases that converged to the steady state were analyzed. The measurement positions are shown in Figures 4 and 5, and the temperature measurement results are presented in Table 3 and Figure 6.

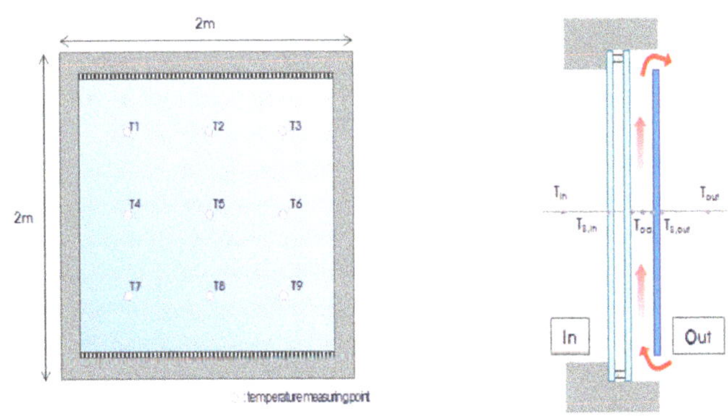

(a) (b)

Figure 4. Temperature measurement points of the specimen installed in the artificial solar lab. (**a**) Elevation of the specimen; (**b**) section of the specimen.

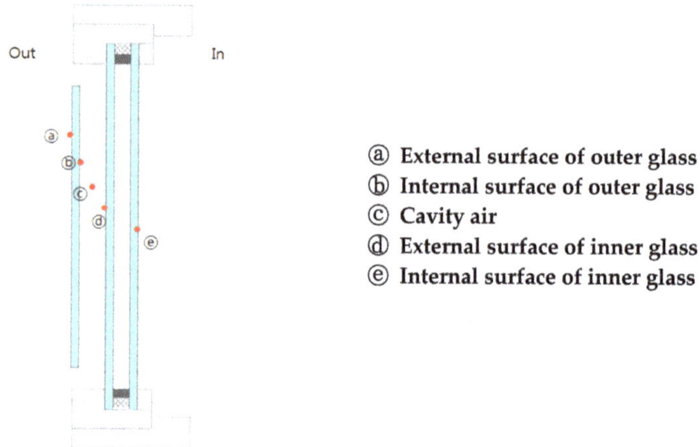

Figure 5. Temperature measurement points.

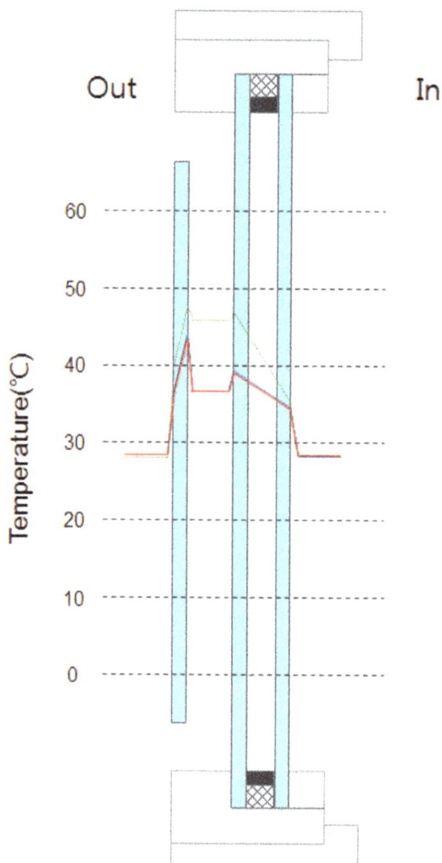

Figure 6. Temperature of measuring point of Case A: red and blue solid lines indicate the first and second test of Case A-1. The green and purple dotted lines indicate the first and second test of Case A-2.

Table 3. Temperature results of Case A (°C).

Cases	Cavity Ventilation	Test No.	Outdoor Chamber	ⓐ Point	ⓑ Point	ⓒ Point	ⓓ Point	ⓔ Point	Indoor Chamber
A-1	X	1st	28.6	36.1	44	36.8	39.4	34.5	28.2
		2nd	28.5	35.7	43.6	36.6	39.2	34.4	28.2
		Avg.	28.6	35.9	43.8	36.7	39.3	34.5	28.2
A-2	O	1st	28.1	39.3	47.6	45.9	46.1	35.2	28.3
		2nd	28.1	39.4	47.7	45.9	46.2	35.5	28.3
		Avg.	28.1	39.4	47.7	45.9	46.25	35.4	28.3

The results of SHGC values for each case and a graph of values applied to the calculation are shown in Figures 7 and 8 and Table 4. The SHGC values are calculated by measuring the elements measured for the SHGC value at one-minute intervals, averaging the values for 10 min every one minute, and then averaging the values again at 50-min intervals. The SHGC value when it is judged that the steady state has been reached is calculated according to the steady state determination standard in Section 3.1. The *X*-axis in Figures 7 and 8 are the order of the measurement values, and one number means the average value for about one hour.

In Case A, when the air cavity was ventilated, a temperature inside the cavity was higher than that without ventilation. A temperature difference inside the air cavity was around 10 K with and without ventilation. The air velocity in the 20 mm air cavity was about 0.3 m/s.

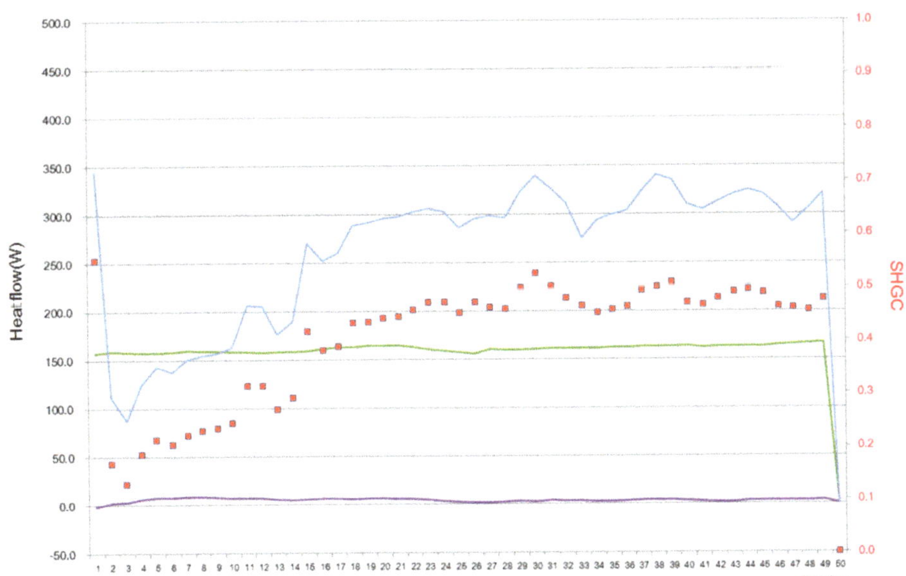

(a)

Figure 7. *Cont.*

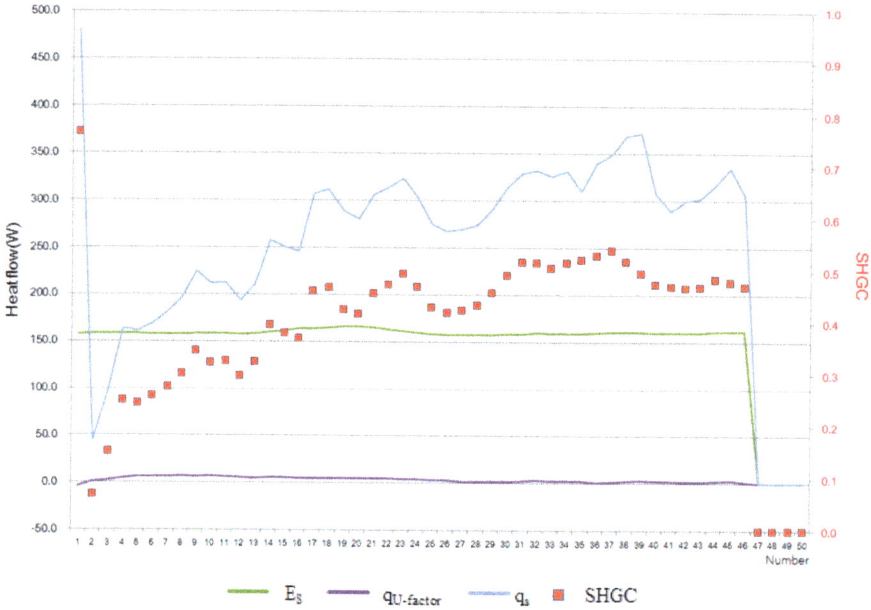

(**b**)

Figure 7. SHGC output graph of A-1: (**a**) first test of A-1 (SHGC = 0.46); and (**b**) the second test of A-1 (SHGC = 0.46).

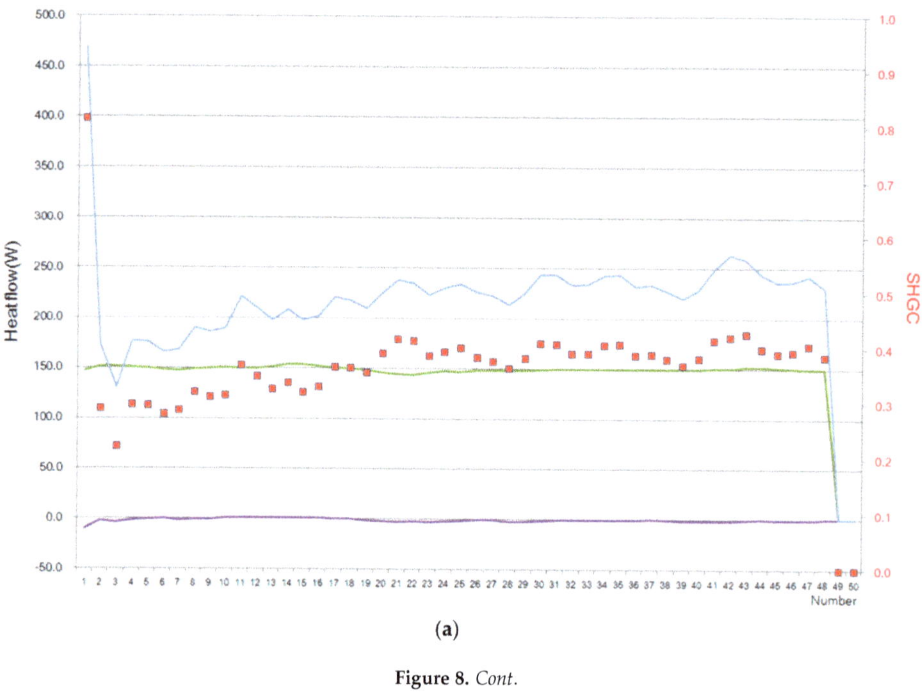

(**a**)

Figure 8. *Cont.*

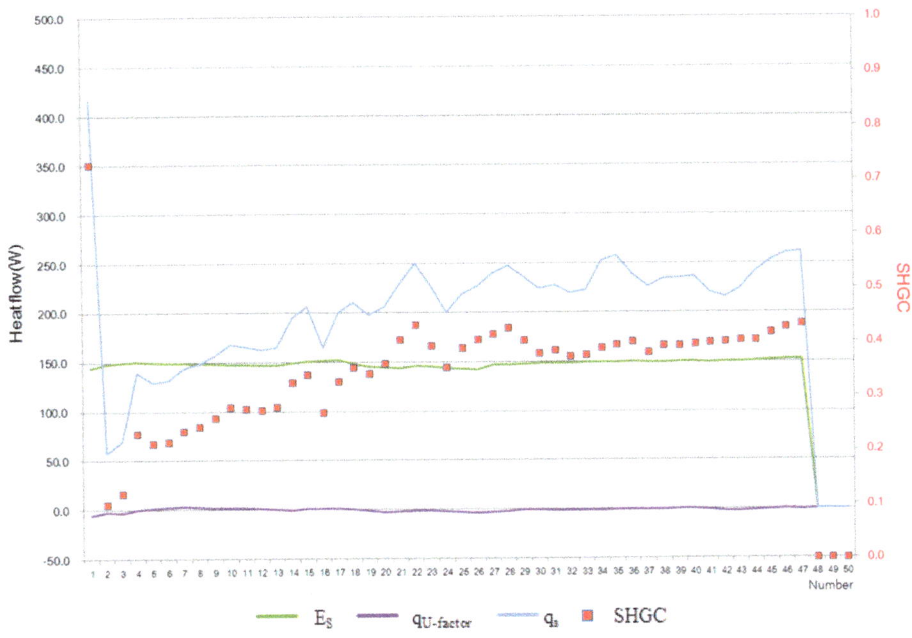

(b)

Figure 8. SHGC output graph of A-2: (**a**) first test of A-2 (SHGC = 0.38); and (**b**) the second test of A-2 (SHGC = 0.37).

The SHGC in Case A was around 0.46 without ventilation in the air cavity, and around 0.38 with ventilation, indicating the reduction effect on SHGC by about 28% through ventilation. As a result, the selectivity (ratio of visible light transmittance to SHGC) that was calculated in consideration of visible light transmittance (the verified value using the Window 7.4 simulation tool (Lawrence Berkeley National Laboratory, Berkeley, CA, USA) instead of measured value) was increased by about 20% from 0.63 to 0.76.

Table 4. Results of SHGC and selectivity of Case A.

Cases	Cavity Ventilation	SHGC			Visual Transmittance (T_{vis})	Selectivity ($T_{vis}/SHGC_{avg}$)
		1st Test	2nd Test	Average		
A-1	X	0.46	0.46	0.46	0.29	0.63
A-2	O	0.38	0.37	0.38		0.76

3.3. Experiment Results of Case B

In Case B, for each case, surface temperature at the window and air temperature in the air cavity were also measured according to whether ventilation at the air cavity is available or not. Each case had three or more measurements, and the cases that converged to the steady state were analyzed. The measurement results of temperature are presented in Table 5 and Figure 9.

In Case B, the inner glass consisted of reflective glass and low-e glass. When the air cavity was ventilated, the temperature inside the cavity was lower than that without ventilation. A temperature difference inside the air cavity was around 4 K with and without ventilation.

Due to the characteristic of reflective glass that absorbs and emits more solar heat than reflects and transmits the solar heat, a temperature inside the air cavity was increased more when the air cavity was not ventilated.

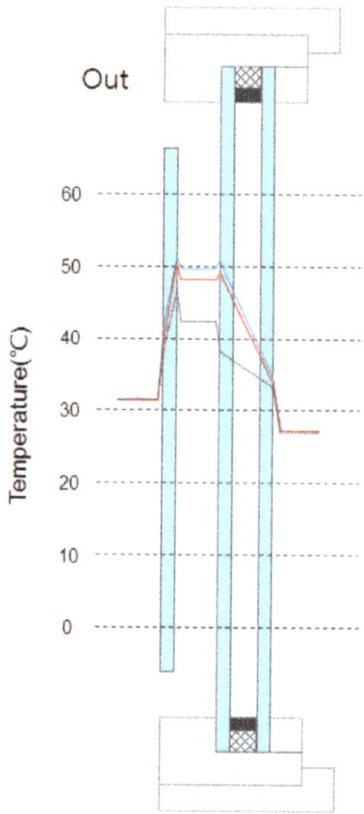

Figure 9. Temperature of measuring point of Case B: red and blue solid lines indicate the first and second tests of case B-1. Green and purple dotted lines indicate the first and second test of case B-2.

Table 5. Temperature results of Case B (°C).

Cases	Cavity Ventilation	Test No.	Outdoor Chamber	ⓐ Point	ⓑ Point	ⓒ Point	ⓓ Point	ⓔ Point	Indoor Chamber
B-1	X	1st	31.4	41.3	51.4	49.8	50.9	34.4	27.2
		2nd	31.3	38.9	50.4	48.2	49.6	34.1	27.2
		Avg.	31.4	40.1	50.9	49.0	50.3	34.3	27.2
B-2	O	1st	31.4	37.9	46.7	42.5	38.3	33.4	27.2
		2nd	31.5	37.9	46.7	42.5	38.4	33.2	27
		Avg.	31.5	37.9	46.7	42.5	38.4	33.3	27.1

The results of SHGC values for each case and a graph of values applied to the calculation are shown in Figures 10 and 11 and Table 6. The SHGC values are calculated by measuring the elements measured for the SHGC value at one-minute intervals, averaging the values for 10 min every one minute, and then averaging the values again at 50-min intervals. The SHGC value when it is judged that the steady state has been reached is calculated according to the steady state determination standard in Section 3.1. The *X*-axis in Figures 10 and 11 is the order of the measurement values, and one number represents the average value for about one hour.

The SHGC in Case B was around 0.17 without ventilation in the air cavity, and around 0.08 with ventilation, indicating the reduction effect on SHGC by about 52.9% through ventilation. At this time, the air velocity in the 20 mm air cavity was about 0.3 m/s.

Furthermore, when selectivity (the ratio of visible light transmittance to SHGC) was calculated in consideration of visible light transmittance (the verified value using the Window 7.4 simulation tool (Lawrence Berkeley National Laboratory, Berkeley, CA, USA) instead of the measured value), it was increased by about 114% through ventilation, which maintained the comfortable indoor light environment while reducing cooling energy significantly.

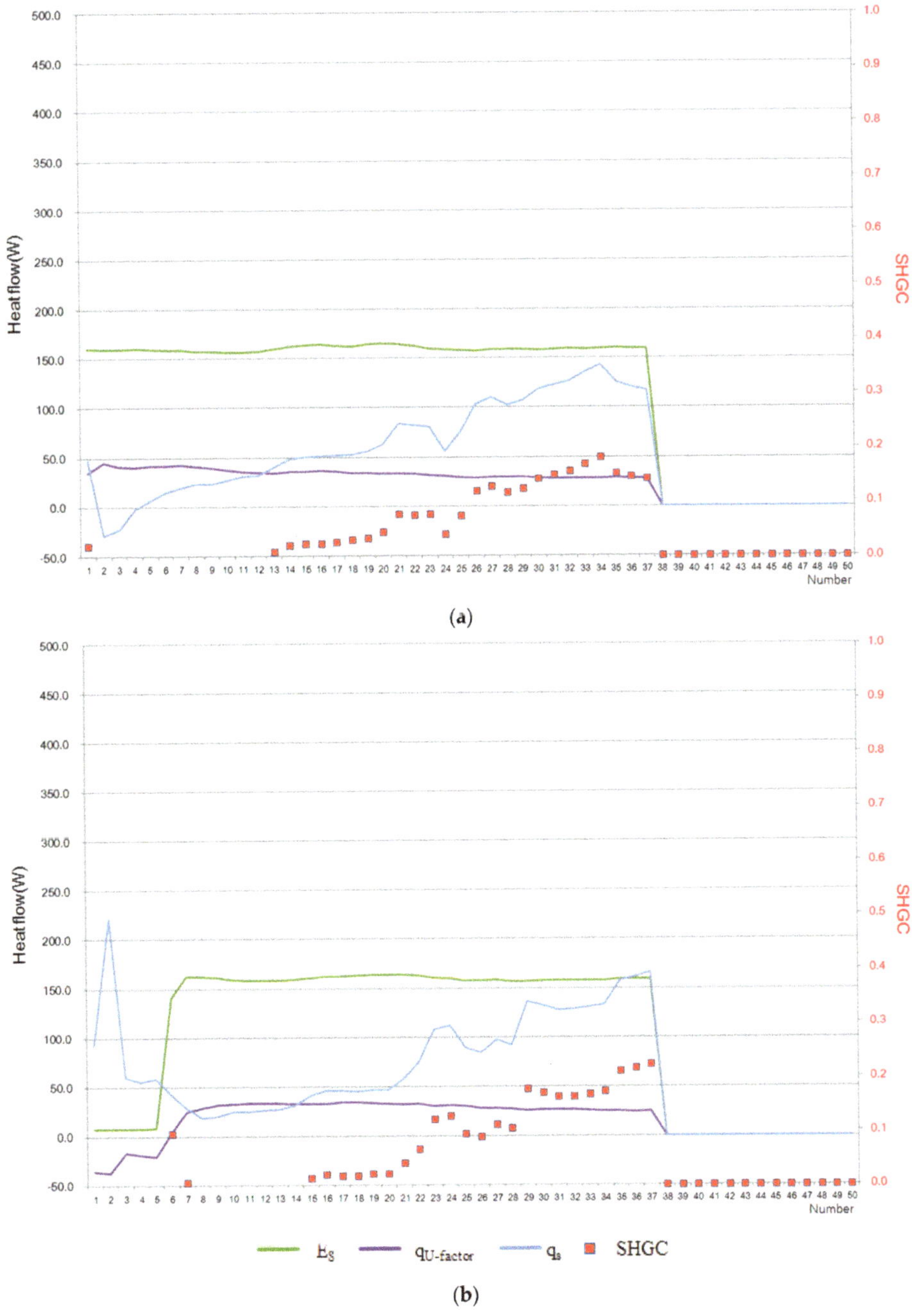

Figure 10. SHGC output graph of B-1: (**a**) first test of B-1 (SHGC = 0.16); and the (**b**) second test of B-1 (SHGC = 0.18).

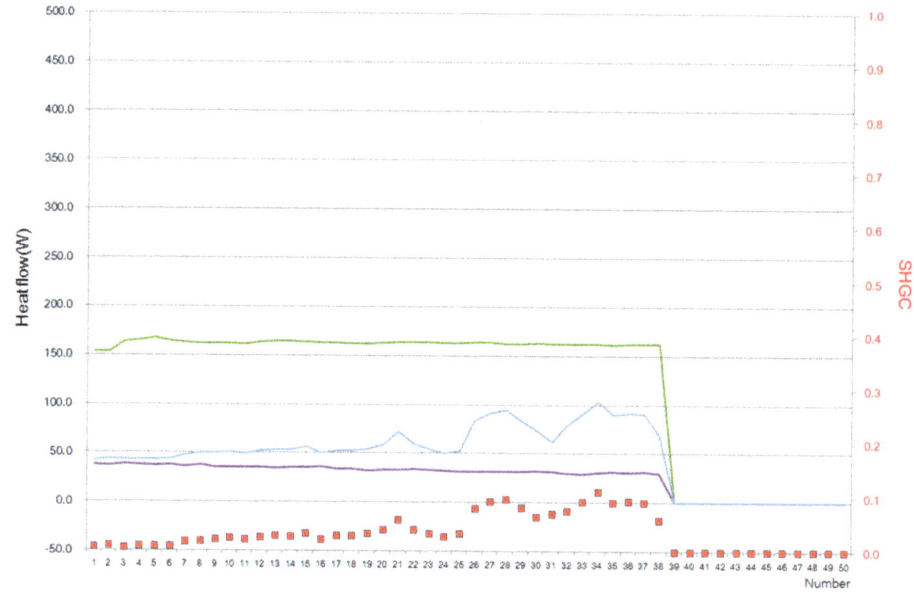

(**a**)

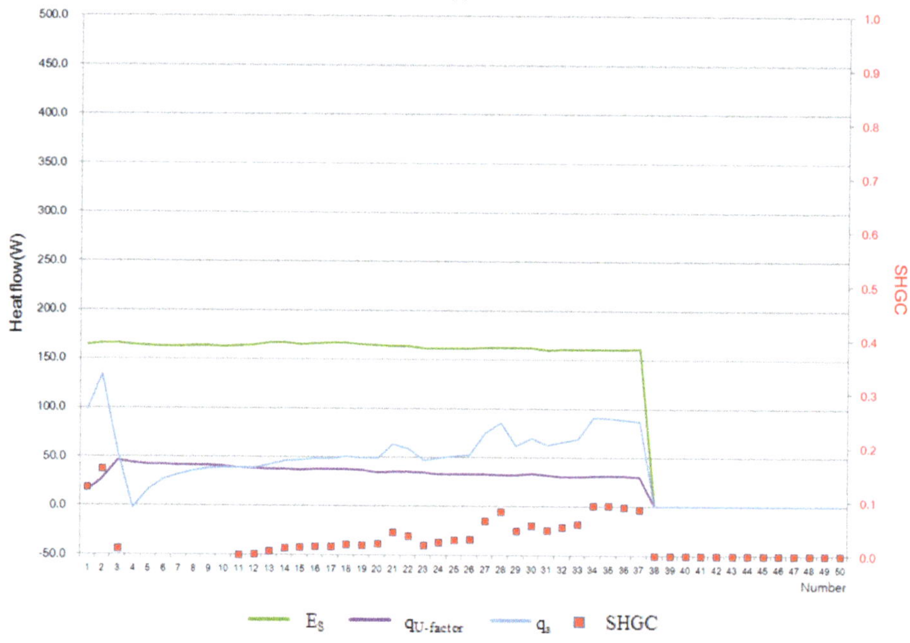

(**b**)

Figure 11. SHGC output graph of B-2; (**a**) first test of B-2 (SHGC = 0.09); and the (**b**) second test of B-2 (SHGC = 0.07).

Table 6. Results of SHGC and selectivity of Case B.

Cases	Cavity Ventilation	SHGC			Visual Transmittance (T_{vis})	Selectivity ($T_{vis}/SHGC_{avg}$)
		1st Test	2nd Test	Average		
B-1	X	0.16	0.18	0.17	0.13	0.76
B-2	O	0.09	0.07	0.08		1.63

4. Outdoor Test Cell Experiment

4.1. Overview

The test cell experiment exposed to the outdoor was conducted in the performance lab in the technical research center of the K Construction Company.

The performance lab was three-story building and the test cell was located in the third floor (Figure 12). It had two rooms for experimental and control groups and its dimension were: width = 2870 mm, depth = 4600 mm, and height = 2300 mm. The size of the window used in the experiment for the two groups was width = 2870 mm and 2300 mm, respectively.

The windows to be assessed used triple low-e glass (thermal transmittance 1.22 W/m^2·K (6 mm clear glass + 12 mm air + 6 mm clear glass + 12 mm air + 6 mm low-e glass)) and a ventilated double-skin window (thermal transmittance 1.38 W/m^2·K, with 20 mm air cavity ventilated (6 mm clear single glass + 20 mm air cavity + 6 mm clear glass + 12 mm air + 6 mm low-e glass)) in the control group.

The outdoor temperature was measured around the outer wall of the test cell, and temperatures were measured at two positions on the outer surface of the window, two positions inside the air cavity, and two positions inside of the window. Furthermore, air temperatures inside were measured at five positions and a mean value was calculated, thereby determining the effect of insolation introduced through the window on the indoor temperature. Moreover, cooling requirements were analyzed through power consumption, thereby evaluating the performance of the cooling energy reduction in double-skin façade window with 20 mm air cavity proposed in this study. Finally, indoor comfort PMV was measured to analyze the effect on indoor comfort. PMV (Predicted Mean Vote) predicts the mean comfort response of a larger group of people on the following 7-point thermal sensation scale. +3: hot, +2: warm, +1: slightly warm, 0: neutral, −1: slightly cool, −2: cool, −3: cold.

The overview of the window used in the experiment is presented in Table 7, and the experimental cases divided by operation mode for each window type are presented in Table 8.

The measurement was conducted at mid-May when summer just started and from mid-June to early July during summer for 25 days.

Table 9 shows the measurement equipment used in the experiment.

Table 7. Cases of outdoor test cell experiment.

Cases	Glass	Cavity Ventilation	Chamber Air Conditioning
I-1	Triple low-e glass	X	Off
I-2	Ventilated double skin window	O	Off
II-1	Triple low-e glass	X	On
II-2	Ventilated double skin window	O	On

Table 8. Measurement period and experiment case overview.

Cases	Measurement Period	Glass	Cavity Ventilation	Chamber Air Conditioning	Indoor Temperature
I-1	15~23 May 2016	Triple low-e glass	X	Off	-
I-2		Ventilated double skin window	O	Off	-
II-1	21 June~4 July 2016	Triple low-e glass	X	On	28.0 °C
II-2		Ventilated double skin window	O	On	28.0 °C

Table 9. Measurement equipment.

Equipment	Specification	
Air conditioner	Product type Cooling capacity Cooling power consumption	Inverter heating/cooling equipment 3000 W/W 950 W
	Cooling period energy consumption efficiency	4.4 W/W
PMV meter	Product type	testo480
	Temperature: Pt100	Measuring range: −100 to +400 °C Resolution: 0.01 °C
	Temperature: TC Type K (NiCr-Ni)	Measuring range: −200~+1370 °C Accuracy: ±(0.3 °C to 0.1% of mv) Resolution: 0.1 °C
	Humidity: Capacitive	Measuring range: 0–100% RH(Relative Humidity) Resolution: 0.1% RH
	Velocity: Vane anemometer	Measuring range: 0.6 to +50 m/s (vane anemometer Ø 16 mm) 0.1 to +15 m/s (vane anemometer Ø 100 mm) Resolution: 0.1 m/s (vane anemometer Ø 16 mm) 0.01 m/s (vane anemometer Ø 100 mm)
	Velocity: Hot wire anemometer	Measuring range: 0 to +20 m/s Resolution: 0.01 m/s
	Comfort level measurement	Measuring range: 0 to +5 m/s Resolution: 0.01 m/s

Figure 12. Location of the window test cell.

The measurement positions including surface temperature in the window are shown in Figure 13, and the measurement positions of air temperature inside the test cell are shown in Figure 14.

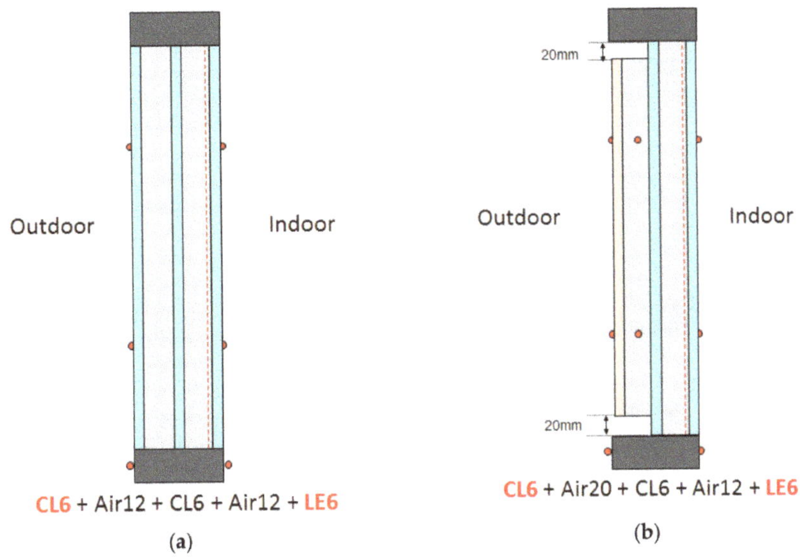

Figure 13. Measurement point of window temperature: (**a**) triple low-e glass; and (**b**) ventilated double-skin window.

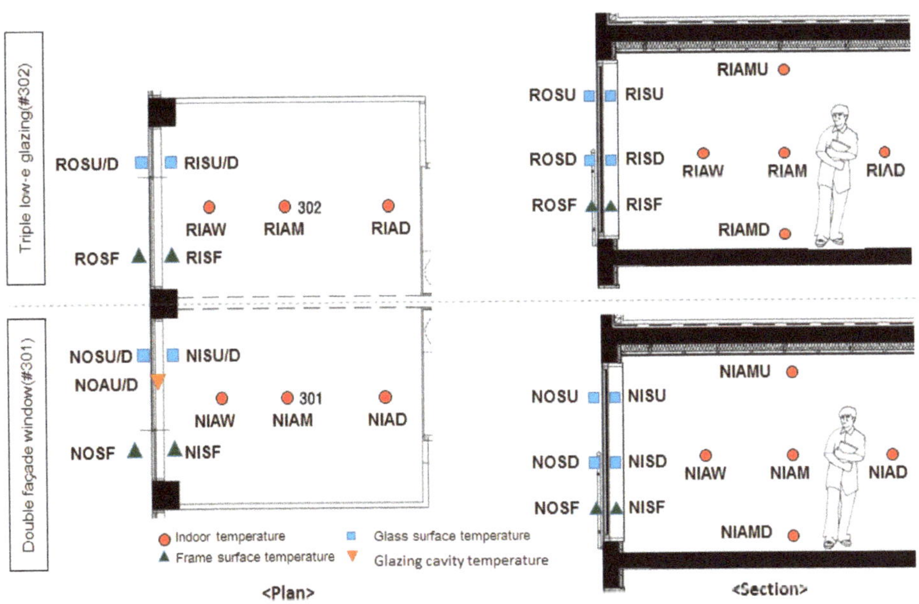

Figure 14. Room interior temperature sensor location.

4.2. Temperature Evaluation

4.2.1. Indoor Temperature and Power Consumption

The mean indoor temperature and cumulative power consumption for each case are presented in Table 10.

The measurement results of Case I measured during May showed that the mean temperature was 0.6 K and the peak temperature was 0.9 K lower with ventilation in the air cavity, even though the heat transmittance was relatively high, than that without ventilation in the air cavity. The measurement results of Case II during June and July showed that the mean temperature was 0.3 K and the peak temperature was 0.2 K lower with ventilation in the air cavity than that without ventilation in the air cavity.

Considering the error of the temperature measurement equipment, it can be said that there is almost no difference. However, the power consumption difference of 3% between II-1 and II-2 can be considered to mean that air cavity ventilation has the effect of reducing solar heat gain.

Table 10. Results of average room temperature and cumulative power consumption measurement.

Cases	Outdoor Average Temperature (24 H)	Indoor Temperature			Cumulative Power Consumption
		24 H Average	9 H Avg. (09:00~18:00)	Max. Temperature (24 H)	
I-1	23.5 °C	28.3 °C	29.7 °C	36.0 °C	-
I-2	23.5 °C	27.9 °C	29.1 °C	35.1 °C	-
II-1	24.8 °C	26.6 °C	26.7 °C	28.0 °C	19,939 kWh
II-2	24.8 °C	26.2 °C	26.4 °C	27.8 °C	19,350 kWh

4.2.2. PMV Values

The measurement results of indoor thermal comfort for each experimental case are presented in Table 11.

The measurement results of Case I showed that a mean and peak PMVs were lower by 0.10 and 0.24, respectively, with ventilation in the air cavity than that without ventilation. The measurement results of Case II showed that a mean and peak PMVs were lower by 0.01 and 0.00, respectively, with ventilation in the air cavity than that without ventilation. The reason for the lower change in PMV in Case II was due to the control of the indoor temperature.

Table 11. Results of the indoor comfort measurement.

Cases	PMV		
	24 H Average	9 h Avg. (09:00~18:00)	Max. PMV (24 H)
I-1	1.44	1.88	3.60
I-2	1.32	1.78	3.36
II-1	0.85	1.04	1.50
II-2	0.85	1.03	1.50

5. Conclusions

This study proposed a double-skin façade window with a 20 mm air cavity, and assumed that insolation inflow indoors would be reduced through ventilation in the air cavity. To achieve this, windows were installed in chambers in the artificial solar lab and real building test cells and their performances were evaluated.

(1) The artificial solar lab test result showed that the SHGC in Case A was around 0.46 without ventilation in the air cavity, and around 0.38 with ventilation, indicating the reduction effect

on SHGC by about 28% through ventilation. The SHGC in Case B was around 0.17 without ventilation in the air cavity, and around 0.08 with ventilation, indicating the reduction effect on SHGC by about 52.9% through ventilation.

(2) In the test cell experiment exposed to the outdoors, the measurement results of Case I showed that a mean temperature was 0.6 K and the peak temperature was 0.9 K lower with ventilation in the air cavity than that without ventilation in the air cavity. The measurement results of Case II in the experiment showed that a mean temperature was 0.3 K and the peak temperature was 0.2 K lower with ventilation in the air cavity than that without ventilation in the air cavity. Furthermore, this study verified that cumulative power consumption was reduced by around 3% through ventilation in the air cavity and the effect on cooling energy reduction in actual buildings can be obtained.

Acknowledgments: This research was supported by the Research Grant from KICT through the Korea Agency for Infrastructure Technology Advancement funded by the Ministry of Land, Infrastructure and Transport of the Korean government (project No. 17TBIP-C111666-02).

Author Contributions: Bokyoung Koo and Keonho Lee designed the artificial solar lab experiments and analyzed the data; Youngsub An and Kyudong Lee conceived and designed the outdoor test cell experiments and analyzed the data; Bokyoung Koo wrote the paper.

Conflicts of Interest: The authors declare no conflict of interest.

References

1. Cho, K.; Cho, D.; Yoo, J. Analysis of solar heat gain coefficients of slim-type double skin window systems using a solar simulator and a simulation. *J. Archit. Inst. Korea Struct. Constr.* **2017**, *33*, 53–60.
2. Barbosa, S.; Ip, K. Perspectives of double skin facades for naturally ventilated buildings: A review. *Renew. Sustain. Energy Rev.* **2014**, *40*, 1019–1029. [CrossRef]
3. Sanchez, E.; Rolando, A.; Sant, R.; Ayuso, L. Influence of natural ventilation due to buoyancy and heat transfer in the energy efficiency of a double skin façade building. *Energy Sustain. Dev.* **2016**, *33*, 139–148. [CrossRef]
4. Guardo, A.; Coussirat, M.; Egusquize, E.; Alavedra, P.; Castilla, R.A. CFD approach to evaluate the influence of construction and operation parameters on the performance of Active Transparent Facades in Mediterranean climates. *Energy Build.* **2009**, *41*, 534–542. [CrossRef]
5. Perez-Grande, I.; Meseguer, J.; Alonso, G. Influence of glass properties on the performance of double-glazed facades. *Appl. Therm. Eng.* **2005**, *25*, 3163–3175. [CrossRef]
6. Rahmani, B.; Kandar, M.Z.; Rahmani, P. How double skin façade's air-gap sizes effect on lowering solar heat gain in tropical climate? *World Appl. Sci. J.* **2012**, *18*, 774–778.
7. Radhi, H.; Sharples, S.; Fikiry, F. Will multi-façade systems reduce cooling energy in fully glazed buildings? A Scoping study of UAE buildings. *Energy Build.* **2013**, *56*, 179–188. [CrossRef]
8. Torres, M.; Alavedra, P.; Guzman, A.; Cuerva, E.; Planas, C.; Clemente, R.; Escalona, V. Double skin facades-cavity and exterior openings dimensions for saving energy on Mediterranean climate. In Proceedings of the Building Simulation, Beijing, China, 3–6 September 2007.
9. Gratia, E.; De Herde, A. Greenhouse effect in double-skin façade. *Energy Build.* **2007**, *39*, 199–211. [CrossRef]
10. NFRC (National Fenestration Rating Council). *NFRC 201 Interim Standard Test Method for Measuring the Solar Heat Gain Coefficient of Fenestration Systems Using Calorimetry Hot Box Methods*; National Fenestration Rating Council: Greenbelt, MD, USA, 2003.

© 2017 by the authors. Licensee MDPI, Basel, Switzerland. This article is an open access article distributed under the terms and conditions of the Creative Commons Attribution (CC BY) license (http://creativecommons.org/licenses/by/4.0/).

MDPI AG
Grosspeteranlage 5
4052 Basel
Switzerland
Tel.: +41 61 683 77 34

Sustainability Editorial Office
E-mail: sustainability@mdpi.com
www.mdpi.com/journal/sustainability

Disclaimer/Publisher's Note: The title and front matter of this reprint are at the discretion of the Guest Editors. The publisher is not responsible for their content or any associated concerns. The statements, opinions and data contained in all individual articles are solely those of the individual Editors and contributors and not of MDPI. MDPI disclaims responsibility for any injury to people or property resulting from any ideas, methods, instructions or products referred to in the content.

www.ingramcontent.com/pod-product-compliance
Lightning Source LLC
Chambersburg PA
CBHW061840210926
51850CB00006B/937
9783725858750